Mobile Technologies and Augmented Reality in Open Education

Gulsun Kurubacak
Anadolu University, Turkey

Hakan Altinpulluk
Anadolu University, Turkey

A volume in the Advances in Educational Technologies and Instructional Design (AETID) Book Series

IGI GLOBAL
DISSEMINATOR OF KNOWLEDGE

www.igi-global.com

Published in the United States of America by
IGI Global
Information Science Reference (an imprint of IGI Global)
701 E. Chocolate Avenue
Hershey PA, USA 17033
Tel: 717-533-8845
Fax: 717-533-8661
E-mail: cust@igi-global.com
Web site: http://www.igi-global.com

Library of Congress Cataloging-in-Publication Data

Names: Kurubacak, Gulsun, 1964- editor. | Altinpulluk, Hakan, 1986- editor.
Title: Mobile technologies and augmented reality in open education / Gulsun Kurubacak and Hakan Altinpulluk, editors.
Description: Hershey, PA : Information Science Reference, 2017. | Includes bibliographical references.
Identifiers: LCCN 2016054992| ISBN 9781522521105 (hardcover) | ISBN 9781522521112 (ebook)
Subjects: LCSH: Mobile communication systems in education. | Virtual reality in education. | Augmented reality. | Open education.
Classification: LCC LB1044.84 .M627 2017 | DDC 371.33--dc23 LC record available at https://lccn.loc.gov/2016054992

This book is published in the IGI Global book series Advances in Educational Technologies and Instructional Design (AE-TID) (ISSN: 2326-8905; eISSN: 2326-8913)

British Cataloguing in Publication Data
A Cataloguing in Publication record for this book is available from the British Library.

All work contributed to this book is new, previously-unpublished material. The views expressed in this book are those of the authors, but not necessarily of the publisher.

For electronic access to this publication, please contact: eresources@igi-global.com.

Advances in Educational Technologies and Instructional Design (AETID) Book Series

Lawrence A. Tomei
Robert Morris University, USA

ISSN:2326-8905
EISSN:2326-8913

Mission

Education has undergone, and continues to undergo, immense changes in the way it is enacted and distributed to both child and adult learners. From distance education, Massive-Open-Online-Courses (MOOCs), and electronic tablets in the classroom, technology is now an integral part of the educational experience and is also affecting the way educators communicate information to students.

The **Advances in Educational Technologies & Instructional Design (AETID) Book Series** is a resource where researchers, students, administrators, and educators alike can find the most updated research and theories regarding technology's integration within education and its effect on teaching as a practice.

Coverage

- Classroom Response Systems
- E-Learning
- Digital Divide in Education
- Online Media in Classrooms
- Web 2.0 and Education
- Virtual School Environments
- Social Media Effects on Education
- Bring-Your-Own-Device
- Educational Telecommunications
- Adaptive Learning

IGI Global is currently accepting manuscripts for publication within this series. To submit a proposal for a volume in this series, please contact our Acquisition Editors at Acquisitions@igi-global.com or visit: http://www.igi-global.com/publish/.

The Advances in Educational Technologies and Instructional Design (AETID) Book Series (ISSN 2326-8905) is published by IGI Global, 701 E. Chocolate Avenue, Hershey, PA 17033-1240, USA, www.igi-global.com. This series is composed of titles available for purchase individually; each title is edited to be contextually exclusive from any other title within the series. For pricing and ordering information please visit http://www.igi-global.com/book-series/advances-educational-technologies-instructional-design/73678. Postmaster: Send all address changes to above address.

Titles in this Series

For a list of additional titles in this series, please visit: www.igi-global.com

Integrating an Awareness of Selfhood and Society into Virtual Learning
Andrew Stricker (The Air University, USA) Cynthia Calongne (Colorado Technical University, USA) Barbara Truman (University of Central Florida, USA) and Fil Arenas (The Air University, USA)
Information Science Reference • copyright 2017 • 383pp • H/C (ISBN: 9781522521822) • US $200.00 (our price)

Intercultural Responsiveness in the Second Language Learning Classroom
Kathryn Jones (Lamar University, USA) and Jason R. Mixon (Lamar University, USA)
Information Science Reference • copyright 2017 • 327pp • H/C (ISBN: 9781522520696) • US $175.00 (our price)

Handbook of Research on Transformative Digital Content and Learning Technologies
Jared Keengwe (University of North Dakota, USA) and Prince Hycy Bull (North Carolina Central University, USA)
Information Science Reference • copyright 2017 • 453pp • H/C (ISBN: 9781522520009) • US $230.00 (our price)

Digital Tools for Seamless Learning
Süleyman Nihat Şad (Inonu University, Turkey) and Martin Ebner (Graz University of Technology, Austria)
Information Science Reference • copyright 2017 • 398pp • H/C (ISBN: 9781522516927) • US $200.00 (our price)

Multiculturalism and Technology-Enhanced Language Learning
Dara Tafazoli (University of Córdoba, Spain) and Margarida Romero (Université Laval, Canada)
Information Science Reference • copyright 2017 • 339pp • H/C (ISBN: 9781522518822) • US $190.00 (our price)

Handbook of Research on Writing and Composing in the Age of MOOCs
Elizabeth A. Monske (Northern Michigan University, USA) and Kristine L. Blair (Youngstown State University, USA)
Information Science Reference • copyright 2017 • 457pp • H/C (ISBN: 9781522517184) • US $270.00 (our price)

Integration of Cloud Technologies in Digitally Networked Classrooms and Learning Communities
Binod Gurung (New Mexico State University, USA) and Marohang Limbu (Michigan State University, USA)
Information Science Reference • copyright 2017 • 329pp • H/C (ISBN: 9781522516507) • US $200.00 (our price)

Exploring the New Era of Technology-Infused Education
Lawrence Tomei (Robert Morris University, USA)
Information Science Reference • copyright 2017 • 388pp • H/C (ISBN: 9781522517092) • US $200.00 (our price)

IGI GLOBAL
DISSEMINATOR OF KNOWLEDGE
www.igi-global.com

701 E. Chocolate Ave., Hershey, PA 17033
Order online at www.igi-global.com or call 717-533-8845 x100
To place a standing order for titles released in this series, contact: cust@igi-global.com
Mon-Fri 8:00 am - 5:00 pm (est) or fax 24 hours a day 717-533-8661

Editorial Advisory Board

List of Reviewers

Table of Contents

Detailed Table of Contents

This chapter intends to advance a sociocultural perspective to frame the integration of augmented reality technology for the learning of mathematics. It is discussed the design of an application already produced with this technology. The purpose of the application is to promote the development of spatial ability while also reviewing the visualization of some calculus topics at college. The author contends the benefits of focusing in the graphical representation to approach mathematics knowledge, especially because digital devices are part of the modern culture. Students today are familiar to explore in digital media, and this should be an opportunity to inquire the power of digital design to engage in mathematical thinking.

Mobile augmented reality offers important opportunities for learning. Moreover, it may represent new challenges for teachers and researchers. Implementing an augmented reality (AR) or a virtual reality (VR) learning experience involves the exploration of unusual pedagogical and technological boundaries. According to recent studies, it would be more productive to consider the augmented reality as a concept rather than an educational technology. This chapter is devoted to analyze a high school class of 23 students invited to use AR and VR tools to create their own study material. They are about 16-year-old attending Istituto Tecnico Tecnologico "Eustachio Divini" in San Severino Marche, Italy. The basic idea of the trial is to create a short printed document augmented with the technologies of AR and VR. The experience is evaluated using tests and direct observation. The aim is to observe the impact of augmented mobile learning and to demonstrate that AR and VR study material may represent a new communication object adequate to teach future students.

Use of technology in education has been widespread in the last decade, thanks to developments and improvements in information and communication technologies, especially in mobile devices. Among the fields in which mobile devices play important roles, education is one of the leading ones. Mobile devices help teachers and learners access educational resources when needed. To increase the reality of virtual learning environments on mobile devices, Augmented Reality (AR) technologies were introduced for mobile platforms, and the term Mobile Augmented Reality (MAR) arose. MAR opens a new door for educators and trainers to experience new methods of teaching for mobile learners. In this chapter, educational use of AR on mobile devices will be explained. Throughout the content of the chapter, readers will be informed about how AR applications changed people's teaching and learning styles.

Augmented reality (AR) brings flexibility and openness to learning. It also strengthens connections between learners in multiple contexts by adding a new dimension to mobile learning. As well, AR facilitates the learning process of the learners by bringing different disciplines together with education. This chapter focuses on technologies, tools, programs, operating systems, a number of standards as well as some parts of web infrastructure standards for use in mobile AR applications and then discusses the current state, issues, and direction of the development and the use of these apps. The purpose of this chapter is to provide information on that can be used by developers of mobile AR applications for learning and to assist educators and instructional designers in developing mobile augmented reality apps for learning by using AR development tools that provide them to create custom mobile AR applications without programming skills.

Ecosystems are particular areas in nature where all living and nonliving components interact with each other and their environment. The term has also been used as a metaphor in scientific and social contexts. Learning ecosystem is one of these which studies the components and interactions of learning processes. Augmented reality is among the components of a (mobile) learning ecosystem. Potentials of integrating augmented reality in mobile learning are not denied, yet there are concerns that these might turn into short living fashion items if their long term consequences are not considered. Defining a mobile learning ecosystem, clarifying the position of augmented reality component within, describing its relations with other components and searching for a balance in these interactions would be an answer to these concerns. This chapter gives an answer by presenting mobile learning ecosystem and augmented reality definitions derived from a Delphi study carried out in 2016 in Turkey. The results and discussions present a "good fit" framework for a viable mobile learning ecosystem.

Objective of this chapter is to introduce an Augmented Reality based intelligent mobile application (M-Learning application) to support courses of Computer Education. In the study, it was aimed to provide an alternative way of improving M-Learning experiences by employing both Augmented Reality and Artificial Intelligence based approaches in a common environment. Briefly, the application is able to use an intelligent mechanism evaluating students' several dynamic learning parameters to match them with the most appropriate course materials provided over the system. So, each student can encounter with appropriate course materials matching with their states over the application system. The related course materials include both AR based ones and standard ones as uploaded by teachers. An evaluation based flow has been run in the study by using the developed application through courses of Computer Education and the obtained results have shown that the application is effective and successful enough at improving students' learning experiences and achieving a good Open Computer Education.

With the introduction of Google Cardboard, a combination of mobile devices, Virtual Reality (VR) and making was created. This "marriage" opened a wide range of possible, cheap Virtual Reality applications, which can be created and used by everyone. In this chapter, the potential of combining making, gaming and education is demonstrated by evaluating an implemented math-game prototype in a school by pupils aged 12-13. The aim of the virtual reality game is to solve math exercises with increasing difficulty. The pupils were motivated and excited by immerging into the virtual world of the game to solve exercises and advance in the game. The results of the evaluation were very positive and showed the high motivational potential of combining making and game-based learning and its usage in schools as educational instrument.

The evolution and broad ownership of mobile devices has led to an increased interest to integrate the benefits of mobile learning and augmented reality applications. New possibilities for teaching and learning provided by augmented reality have been increasingly recognized by educational researchers. Mobile augmented reality can provide rich contextual learning for individuals. Currently virtual reality and augmented reality applications are used for training in fields as diverse as trades, military, entertainment, education and health. This chapter will explore different dimensions of mobile augmented reality and exemplify their potential for education. Therefore, how mobile augmented realty applies to education and training domain, and the potential impact on the future of education will be explained. Current status, opportunities and challenges of mobile augmented reality in education, mobile augmented reality applications will be included.

Introducing an intelligent augmented reality based M-learning application designed and developed for improving engineering students' English language skills, this chapter reports a work-in-progress that focuses on system design procedure. The application consists of Artificial Intelligence (AI) based functions to ensure an effective learning flow while taking advantages of game-based learning by providing a story board structure with a content tree. Four design principles "fair use, flexible use, fault tolerance, educational climate" by Elias in addition to Stockwell and Hubbard's principles for mobile supported language learning have been taken into account. Furthermore, the proposed system here employs an effective approach combining both real and virtual environments to achieve an Augmented Reality based learning experiences for students. After the introduction of the application, the chapter outlines how it will be processed in the future.

In an educational perspective, unlike other disciplines, hands-on practice is difficult to come by in Civil Engineering. By providing a student with a realistic 3D simulation, we propose a concept that improves the understanding of the individual and eliminate guess-work entirely. Our platform that makes education fun and interactive by eliminating the constraints of a conventional teaching environment by incorporating Virtual Reality (VR) or Augmented Reality (AR) as a tool. AR can help the target audience visualize a model (to scale) in all three dimensions in the palm of their hand. This chapter explores the use of interactive 3D game environments in design visualization in Building Information Modeling (BIM) by adopting various available software packages and APIs. VR will allow the prospective customer to enter and explore a structure before it is constructed. This can be achieved by making use of a powerful game engine, in this case, Unity3D. In this chapter, we will describe ways to pivot Unity's functions towards the benefit of civil engineering.

In the year 2016, flipped learning became an innovative learning approach. The students, in flipped learning, prepare to their courses watching the videos, listening to podcasts, and reading the articles prepared for themselves, and have a prior knowledge. In flipped learning approach, efficiency of learning materials is related to providing permanent learning, facilitating the learning and being interactive. For that reason, the students should be supported with stronger learning materials rather than the aforementioned materials. The most current one among these materials includes Augmented Reality implementations that have been prominent in recent years. In this chapter, flipped learning approach and augmented reality concept will be emphasized, augmented reality implementations possible to be used in flipped learning will be exemplified, and limitations will also be discussed.

Today the world revolves around technology in daily tasks. Society uses technology to communicate, trade, for business, politics, and to education. Through education, Augmented Reality (AR) takes on a new purpose. AR enables students to reach a higher level of education, and teachers to efficiently engage students to see real-world situations. The use of technology in classrooms is currently on the rise; but how are teachers using AR within their classrooms to engage and education children? This chapter discusses literature and research supporting the AR affordances in K-12 Education.

Foreword

It is an immense honor to share some preliminary remarks in this Foreword to this exciting collection of chapters in this publication titled *Mobile Technologies and Augmented Reality in Open Education.*

Before I share some thoughts on the focus of this publication, I would like to share with you how I approached writing this Foreword – what it is, and more importantly, what this Foreword is *not*. We have all opened a book and possibly either skipped the Foreword because it often serves a function to introduce the reader in the scope of the collection of chapters, possibly the rationale for their selection, and a summary of what the chapters include, who wrote them representing which institutions or stakeholders. Personally, I find such a foreword often as a brief overview as an expanded table of contents. I would then use such a foreword as guide on which chapters to read.

Possibly also in this category of forewords as introductory overviews of the chapters to follow, the person who writes the foreword congratulates the editors and authors and laud their insights and contributions to the discourses surrounding the focus of the publication.

Then there is another type of foreword that attempts to provide a map of the broader discourses surrounding the focus of the publication. Often such a foreword functions as a first chapter and is written by someone who would add gravitas to the publication. Such a foreword is then often about the voice of the author who have been asked to write the foreword as laying claim to being an expert in the scope of the publication.

I am afraid that I am neither famous and the fact that I have been asked to contribute this foreword may not add gravitas or stature to this selection of chapters. I also want to shy away of writing an unofficial first chapter mapping the need for and state of the field pertaining to the potential, challenges and unresolved issues of using mobile technologies and augmented reality in open education. I had no insight or say in the selection of chapters, or in drafting the call for papers. What I do find interesting is not the reference to the two technologies namely mobile technologies and augmented reality, but the fact that the authors of the chapters in this volume were provided with a specific focus namely how these technologies focus in the context of *open education.*

So, with the above disclaimers out of the way, allow me to share some brief thoughts that may provide a lens through which the chapters in this volume can be read and re-read. I have no doubt that the editors of this volume selected the chapters in this volume for their contribution to the theme. My hope is that this foreword opens up a space for critical and reflective conversations and that we do not accept the chapters in this volume as the final word, but as an invitation for engagement.

No technology is neutral and in a volume dedicated to exploring mobile technologies and augmented reality, we have to ask ourselves who has vested interests in having these two technologies on the agenda, who benefits, who determines the rules that will determine and shape their use, who are excluded from these technologies and why, and finally, how will these technologies serve *learning*. No technology is neutral and their introduction and use in education often serve commercial interests without due consideration of who is excluded, who determines the agenda for there use, and the ethical considerations in using or not using them.

On a personal level, I am increasingly concerned that we often see and embrace technological advances without due consideration of the above questions. Often the adoption of technologies in education is driven by stakeholders who do not, necessarily, have the interests of learning and openness in mind, but who sell us snake oil. Educators, students and more importantly those in management and procurement functions in our educational institutions often see education as in need of 'fixing' with the technologies in question proposed as the solution to all of the complexities education faces.

Finally, allow me to congratulate the publisher, the editors and the authors of the chapters in this volume for this publication. May this volume of chapters allow us to ask critical and difficult questions for which there may not be (easy) answers. May this collection start a conversation about the potential, usefulness, impact, appropriateness and ethical issues pertaining to the use of mobile technologies and augmented reality in open education.

Paul Prinsloo
University of South Africa (UNISA), South Africa

Paul Prinsloo *is a Research Professor in Open and Distance Learning (ODL) in the College of Economic and Management Sciences, University of South Africa (Unisa). His academic background includes fields as diverse as theology, art history, business management, online learning, and religious studies. Paul is an established researcher and has published numerous articles in the fields of teaching and learning, student success in distance education contexts, learning analytics, and curriculum development. His current research focuses on the collection, analysis and use of student data in learning analytics, graduate supervision and digital identity.*

Preface

Augmented reality is an interactive technology which makes it possible to combine virtual and real, instead of creating a completely virtual world, and to view such combination in the same medium. Although the ideas on augmented reality has an historical background, its technological development is relatively new. We see that augmented reality applications have been developed in the recent years for wearable technologies and especially for mobile technologies rather than for desktop/laptop computers. Strengthening of mobile technologies with various sensors and hardware and ease of producing mobile application software have led to rapid widespread use of mobile devices in all segments of society. With such spread of mobile technologies, augmented reality applications are downloaded and used by everyone and such applications are even developed without any need to advanced software knowledge.

Augmented reality is one of the popular subjects frequently studied in the field of educational technologies in the recent years. Augmented reality applications have been used intensively in educational environments with their interactive, interesting and entertaining elements. Augmented reality applications enrich the physical world through various three-dimensional interactions and a number of studies show that these applications have positive effects on learners in different aspects. However, we see that only a limited number of studies examine the use of augmented reality in open education. Augmented reality may be used in open and distance learning where learners and instructors are distant to one another. However, there are not many studies, especially books and projects, dealing with this issue. Therefore, this book has been prepared to fill that gap in this field: it aims to provide suggestions to learners, academics, software developers and institutions' managers with respect to use of augmented reality (especially mobile augmented reality applications) in educational environments (especially in the scope of open education). The book consists of 16 chapters and authors analyze the possibility to use many augmented reality components from smart glasses to mobile applications. A brief summary of the subjects handled in each chapter and the significance of these chapters are given below.

Chapter 1 discusses the positive and negative aspects of possible outcomes from the public use of smart glasses which are frequently used in augmented reality applications. It analyzes the social aspect of smart glasses which is one of the most well-known and widely used wearable technologies. It aims to determine how the consumers interpret and evaluate the potential opportunities and threats associated with public use of smart glasses. As there are fewer studies in this field, the authors aim to fill that gap in the literature.

Chapter 2 addresses to developers of augmented reality software, introducing educational applications of augmented reality and provides information on development processes of such applications. In Chapter 2, it was presented the various limitations and advantages of augmented reality revealed by some empirical studies in the literature. In addition, it was given information about augmented reality development tools/programs, add-on packages and presented development stages for an exemplary augmented reality book page. The use of the Unity and Vuforia was explained as the development tools.

Chapter 3 intends to advance a sociocultural perspective to frame the integration of augmented reality technology for the learning of mathematics. It is discussed the design of an application already produced with this technology. The purpose of the application is to promote the development of spatial ability while also reviewing the visualization of some calculus topics at college. Patricia Salinas Martinez contends the benefits of focusing in the graphical representation to approach mathematics knowledge, especially because digital devices are part of the modern culture.

Chapter 4 is devoted to analyze a high school class of 23 students invited to use Augmented Reality and Virtual Reality tools to create their own study material. They are about 16-year-old attending Instituto Tecnico Tecnologico "Eustachio Divini" in San Severino Marche, Italy. The basic idea of the trial is to create a short printed document augmented with the technologies of Augmented Reality and Virtual Reality. The experience is evaluated using tests and direct observation. The aim is to observe the impact of augmented mobile learning and to demonstrate that Augmented Reality and Virtual Reality study material may represent a new communication object adequate to teach future students.

In Chapter 5, educational use of augmented reality on mobile devices is explained. Throughout the content of the chapter, readers are informed about how augmented reality applications changed people's teaching and learning styles. Besides, it presents the history, application platforms, educational use, advantages and disadvantages of augmented reality. Additionally, detailed information on mobile augmented reality is analyzed through in-depth literature search.

Chapter 6 focuses on technologies, tools, programs, operating systems, a number of standards as well as some parts of web infrastructure standards for use in mobile augmented reality applications and then discusses the current state, issues, and direction of the development and the use of these apps. The aim of Chapter 6 is to provide information on that can be used by developers of mobile augmented reality applications for learning and to assist educators and instructional designers in developing mobile augmented reality apps for learning by using augmented reality development tools that provide them to create custom mobile augmented reality applications without programming skills.

Chapter 7 gives an answer by presenting mobile learning ecosystem and mobile augmented reality definitions derived from a Delphi study carried out in 2016 in Turkey. The results and discussions present a "good fit" framework for a viable mobile learning ecosystem.

Chapter 8 introduces an augmented reality based intelligent mobile application to support courses of Computer Education. In the study, it was aimed to provide an alternative way of improving M-Learning experiences by employing both augmented reality and artificial intelligence based approaches in a common environment.

In Chapter 9, the potential of combining making, gaming and education is demonstrated by evaluating an implemented math-game prototype in a school by pupils aged 12-13. The aim of the virtual reality game is to solve math exercises with increasing difficulty. The pupils were motivated and excited by immerging into the virtual world of the game to solve exercises and advance in the game. The results of the evaluation were very positive and showed the high motivational potential of combining making and game-based learning and its usage in schools as educational instrument.

Chapter 10 deals with various aspects of mobile augmented reality and its potential in education. This chapter covers a comprehensive literature search and provides detailed information on mobile augmented reality application examples.

Chapter 11 deals with design and development of artificial intelligence based "intelligent Augmented Reality based M-Learning application" and its effects on the English language skills of engineering students.

Chapter 12 explores the use of interactive 3D game environments in design visualization in Building Information Modeling (BIM) by adopting various available software packages and APIs. Virtual reality will allow the prospective customer to enter and explore a structure before it is constructed. This can be achieved by making use of a powerful game engine, in this case, Unity3D. In Chapter 12, the authors describe ways to pivot Unity's functions towards the benefit of civil engineering.

Chapter 13 integrates "flipped learning" and augmented reality approaches and brings a new dimension to use of augmented reality in education and distance education. In this context, it provides some information especially on the augmented reality hardware and software.

In Chapter 14, authors discuss literature and research supporting the augmented reality, affordances in K-12 Education. The purpose of this chapter is to articulate the affordances of using augmented reality in K-12 classrooms. This is thematic review of the literature to understand what themes are developing from the publish literature. This chapter begins with a definition of augmented reality. This is followed by the unpacking of four trends in how augmented reality can support teaching and learning. These affordances are that augmented reality supports; authentic learning, contextualized learning, student-centered learning and enables students to better visualize subject content. Finally, the projected future path of augmented reality is revealed to show where augmented reality may benefit future students.

Chapter 15 emphasizes the importance of mobile augmented reality in "online nursing education" and follows an interesting and quite different approach. It presents proposals to use mobile augmented reality applications in "online nursing education" to increase the quality of nurses and to provide a better education service.

Chapter 16 provides first general information about video games and video gaming sector and then explains the details of design, development and marketing of such games. This chapter categorizes the concept of video game systematically and explains each category and type of games in details. Also, it aims to create awareness among researchers on video game development process.

This book consists 16 chapters and has been prepared with contributions by many authors from various countries. 28 authors, mostly from Turkey, and from the USA, Germany, Mexico, Italy, Austria and India put a great effort and time to create this book. The book not only handles the augmented reality mobile applications used in education and open education but also provides, in connection with the augmented reality, a detailed analysis of the virtual reality applications, video games, wearable technologies, smart glasses and artificial intelligence based applications. This book serves as a reference book to academics, educators, managers, software application developers and students who study on augmented reality and make research on this field. It aims to provide available academic information on augmented reality and contribute to this field by filling the gap in education.

Gulsun Kurubacak
Anadolu University, Turkey

Hakan Altinpulluk
Anadolu University, Turkey

Acknowledgment

We have many individuals to thank for their impressive efforts, help, and encouragement on our book. First, we would like to praise the people at the IGI Global. Joshua Herring, Lindsay Johnston and Jan Travers, our editors, provided a constructive framework for the revision. Their invaluable suggestions are very important for us to give the final shape of our book. Not only did they help us to make critical decisions about carefully structuring our book, but also they provided useful and informative feedback to increase our commitment and effort to complete our book.

We would like to thank "The Scientific and Technological Research Council of Turkey" for their support in preparing this book. This book was supported under project number "115K627".

We would like to thank each our authors for their excellent contributions. We would also to express our appreciations to the Editorial Advisory Board (EAB) members and reviewers separately.

Gulsun Kurubacak
Anadolu University, Turkey

Hakan Altinpulluk
Anadolu University, Turkey

Chapter 1
Are Wearables Good or Bad for Society?
An Exploration of Societal Benefits, Risks, and Consequences of Augmented Reality Smart Glasses

Daniel W. E. Hein
University of Bamberg, Germany

Philipp A. Rauschnabel
University of Michigan – Dearborn, USA

Jennah L. Jodoin
University of Michigan – Dearborn, USA

Björn S. Ivens
University of Bamberg, Germany

ABSTRACT

Recent market research forecasts predict that a new form of wearable devices will soon influence the media landscape: Augmented Reality Smart Glasses. While prior research highlights numerous potentials in personal and professional settings of smart glasses, this technology has also triggered several controversies in public discussions, for example, the risk of violating privacy and copyright laws. Yet, little research addresses the questions of whether smart glasses are good or bad for societies, and if yes, why. This study conducts exploratory research to contribute to narrowing this gap. Based on a survey among consumers, the authors identify several societal benefits and risks that determine consumers' evaluation of the anticipated and desired success of smart glasses. These findings lead to numerous important implications for consumers, scholars, managers, and policy makers.

INTRODUCTION

A recent study by Goldman Sachs (2016) asserts that augmented reality (AR) smart glasses are the latest step in an evolution of digitization of reality and a large developing future market. What began with stationary computers that were temporarily online to receive and display information to their mostly business users in the last century turned into a network dominated by user-generated content (UGC), with

DOI: 10.4018/978-1-5225-2110-5.ch001

users being mainly consumers that are permanently online and access the internet through a multitude of devices (Hennig-Thurau et al., 2010). Smart glasses mark the current latest step in this evolution as they hold the potential to merge online and offline – not just through ubiquitous and permanent online accessibility, but rather by merging online senses directly with consumers' vision (Rauschnabel, Brem, & Ro, 2015). This results in entirely new ways for consumers to interact with the Internet and its content.

Smart glasses represent a radically new invention with an important number of potential use cases for both civil and business life (Berque & Newman, 2015; Hein & Rauschnabel, 2016; Moshtaghi et al., 2015). However, because they are so new and their uses so unexplored, part of their success will depend on the hopes and fears consumers hold with regard to their usage and their proliferation (Lee, Bojanova, & Suder, 2015). Consumers do not just care about their mere self-interest, but rather include interests of the environment and society in the decision making (Doane, 2001). This trend of ethical consumerism can be defined as "spending that makes a positive difference in the world" (Witkowski & Reddy, 2010). It may become problematic with regard to smart glasses. When trying to predict smart glasses adoption, the technology acceptance literature would typically be of high relevance. However, early research suggests that smart glasses possess the potential to radically change society and the rules and norms it operates on for better and for worse (Wassom & Bishop, 2015). Anecdotal evidence for this derives from various newspaper articles, press commentaries and articles that discuss (predominantly negative) consequences of this technology for societies. However, the academic literature does not provide findings on what these potential consequences are and how these factors influence consumers' evaluations from a societal perspective. Absent such knowledge, managers may find it difficult to promote the use of smart glasses while policy makers may have difficulties in developing appropriate legislation. Finally, for theory, the lack of pre-market knowledge on societal consequences might hinder the understanding of the diffusion of smart glasses, as prior research has shown that consumers tend to include various ethical (Jacobsen & Dulsrud, 2007) and social (Venkatesh, Thong, & Xu, 2012; Davis, Bagozzi, & Warshaw, 1989) variables in their decision making.

In this chapter, we aim at increasing the understanding of societal factors associated with smart glasses. We are particularly interested in providing answers to the following research questions (RQs):

RQ1: How do consumers evaluate the potential opportunities and threats of smart glasses for society?
RQ2: How are these potential opportunities and threats related to consumers' desired and anticipated success of smart glasses?

The remainder of this chapter is organized as follows: First, we provide a definition of smart glasses as well as findings and use cases from prior research. Then, we briefly review technology acceptance theories and their applications on smart-glasses. This is followed by a review of ethical consumerism and consumer citizenship studies. For the empirical part of this chapter we present an exploratory survey study. Survey data was gathered and factor analyzed in order to extract factors that consumers use to evaluate smart glasses. In subsequent analyses, we empirically assessed how these factors relate to consumers' anticipated and desired success of smart glasses, particularly, the degree to which they anticipate and hope that smart glasses will become a successful technology. The chapter ends by discussing the results and providing managerial implications on how to react to the study's outcome.

RESEARCH BACKGROUND

Smart Glasses: Definition and Use Cases

Smart glasses are wearable computing devices with integrated AR features. AR is a "medium in which digital information is overlaid on the physical world, that is, in both spatial and temporal registration with the physical world and that is interactive in time" (Craig, 2013, p.20). AR is not a very novel concept. Applications have been developed for stationary and mobile devices for years. An example of a smartphone app is Wikitude. With this app, users can look at a famous building. Wikitude then automatically includes relevant Wikipedia information in the view field on the smartphone's screen.

Wearables are a sub category of mobile devices that is attached to a user's body. Often, wearables share several physical and design-related similarities with fashion items (e.g., a smartwatch looks similar to a traditional watch). This is why some authors term them as 'fashnology' (e.g., Rauschnabel et al., 2016).

Smart glasses are wearable devices that are worn like regular glasses and possess the ability to merge the physical environment with virtual information within the view field of the AR technology user. Typically, they are equipped with various sensors that gather information about the user's situational context, a WiFi-antenna to receive and send online information, a small memory, a processing unit and a small screen located in front of one eye or integrated into one or both of the translucent lenses. The processing unit allows the smart glasses to operate various recognition technologies to give the user context-relevant information on his/her social and spatial surrounding (Rauschnabel, Brem, & Ro, 2015). These AR devices need to be clearly distinguished from their virtual reality (VR) counterparts, that possess an opaque screen and do not allow for an overlay of virtual and physical reality, but that conceal the user's vision within the device shutting him off from all exterior visual stimuli. Typical models of VR devices include Oculus Rift and Samsung Gear VR. This chapter solely focuses on AR smart glasses.

Smart glasses represent a topic of growing interest in multiple disciplines. Researchers from various fields have made efforts to understand and describe use potentials for their respective domains, for instance, for maintenance processes (Quint & Loch, 2015; Yang & Choi, 2015), in customer service (Depari et al., 2015) or for medical settings, in which either doctors can share information live and remotely in surgery (Albrecht et al., 2014; Muensterer, Lacher, Zoeller, Bronstein, & Kübler, 2014), implicitly for administering a signal system for surgeons (Geißler et al., 2013) or as a means of augmenting patients who suffer from diseases that call for permanent behavioral monitoring, as is the case for instance for diabetes (Wall, Ray, Pathak, & Lin, 2014). In terms of educational purposes, the technology offers great potential by providing instructional approaches to solving complex tasks or visualizing information, making it interactive and more intuitive to grasp and thus usable in both basic and advanced education (Wu, Lee, Chang, & Liang, 2013; Yuen, Yaoyuneyong, & Johnson, 2011). Open education may benefit as AR smart glasses allow actual work on virtual components. Fine motor skills, for example, that depend on material resources to be practiced on, to a degree can become independent, as smart glasses can simply replace material while still allowing authentic interaction and manipulation of virtual representations (Spitzer & Ebner, 2016).

In brief, the new technology possesses the potential to improve open education, R&D activities, overall business process efficiency, or even allow for the creation of entirely new business models (Rauschnabel, Brem, & Ro, 2015). An example of a concrete activity that smart glasses can augment business with are specific apps which ultimately may fuel an entirely new business model: these apps provide contextualized advertisements shown to consumers as a new form of push-marketing and can be used for a new way of offering market research, as described by Hernandez, Jimenez, and José Martin (2010). A theoretical example for the improvement of knowledge management processes by means of smart glasses can be found with Hein and Rauschnabel (2016). They introduce a generic multi-level framework for the introduction of smart glasses as building blocks in a corporate knowledge management system that remains open to the adaptation of smart glasses for other business-related contexts.

The available devices that can be used for implementation are about to become as numerous as their potential use cases. Google Glass 2.0 as seen on Table 1, for example, is an enterprise edition of smart glasses. Google's perceived intention is to benefit companies in the manufacturing and other sectors. Other smart glasses are for entertainment or niche market purposes. For example, Everysight is for mountain bikers, who could use the smart glasses for calculating distance or managing rough terrain. Other glasses, like the Microsoft Hololens, also provide hedonic benefits like gaming. Table 1 provides an overview of some exemplified smart glasses.

Table 1. Overview of smart glasses

Company	Glasses	Release Date	Price points	Features
Carl Zeiss Optics	Zeiss Smart Lens	Estimated: 2017	unknown	Better optics Improved look
Epson	Moverio BT-300	Late 2016	To be announced BT-200 (previous model): $699.99	Front facing camera Lightest on the market OLED display (HD) Fits over normal glasses 5m pixel camera
Google	Glass 2.0 Enterprise Edition	unknown	unknown	Front light with camera Waterproof Wireless connectivity
Microsoft	Microsoft HoloLens	3/30/2016	$3,000.00 (beta version)	3D content enabled HoloStudio – video-editing Skype, Gaming etc. Actiongram-3D visual effects
ODG	ODG R-7	6/9/2015 (announced)	$2,750.00	Bluetooth Autofocus camera Magnetic lens Swappable lenses
Sony	SmartEyeGlass	3/27/2015	$899.00	Gyrosope Accelerometer Light sensor Camera
Everysight	Raptor	2016	To be announced	Advertised for use while biking. Overlays information on, for examples, distance, speed, elevation, navigation, time etc. Built-in camera

Acceptance Theories

Scholars attempt to understand new technologies by resorting to models such as the technology acceptance model (TAM) (Rauschnabel, Brem, & Ivens, 2015; Spagnolli, Guardigli, Orso, Varotto, & Gamberini, 2014). The TAM sees the adoption of any technology as the result of perception and evaluation processes of a technology's usefulness and its perceived ease of use. More recent research on AR technologies acknowledges their multi-faceted character and turns to more advanced models of technology adoption (e.g., the "unified theory of acceptance and use of technology", UTAUT) that further elaborate on the drivers of adoption (e.g. East & Havard, 2015; Yiwen, Li, & Luo, 2015). These approaches differ from the original TAM by considering factors beyond usefulness and ease of use as relevant adoption intention drivers. Furthermore, these models are based on the assumption of enterprise instead of personal use, which brings the aspect of cost tolerance and other differences into play. The latest model by Venkatesh, Thong, and Xu (2012) provides researchers with the adoption drivers in in personal settings, such as hedonic motivations. While extant technology acceptance theories provide a solid and robust framework to study a user's adoption intention, they do not aim at understanding the impact a technology has societies, which has been shown to be an influential factor in consumption decisions (Bijker & Law, 1992; Doane, 2001). Supported by prior research (Shaw, Newholm, & Dickinson, 2006), we propose these awareness of social consequences (i.e. societal benefits and risks) to affect consumers' individual evaluation of smart glasses.

LITERATURE REVIEW

Although research on smart glasses is still relatively new, few studies have looked at consumers' reactions to them. We organize these findings around the UTAUT2 model (Venkatesh et al., 2012) and discuss the degree to which they can be suitable to explain societal acceptance. Drivers that are connected to societal questions of smart glass adoption will be discussed in more detail than those without relevance to our research question.

Performance Expectancy and Hedonic Motivation

Performance expectancy and hedonic motivation both directly refer to the intentions that drive people to use a technology. Performance expectancy refers to motivations whose cause lies outside of the user's personality – an extrinsic motivation. In contrast, hedonic motivation refers to the potential enjoyment one might perceive without benefitting from any increase in efficiency or benefits other than well-being (Ratneshwar & Mick, 2005; Venkatesh, Morris, Davis, & Davis, 2003; Venkatesh et al., 2012). Often, these performance expectancies are termed 'perceived usefulness' (Davis, 1989), 'functional benefits' (Rauschnabel, Brem & Ivens, 2015) or 'utilitarian benefits' (Chitturi, Raghunathan, & Mahajan, 2008). Performance expectancies are functional and task-oriented expectations about a technology that are associated to performance improvements. In contrast, hedonic motivations cover the enjoyment and entertainment factor of technologies (Venkatesh et al., 2012).

For various sorts of wearable devices, like smart glasses, studies have shown that performance expectancies positively influence adoption and use in personal settings (Gu, Wei, & Xu, 2015; Rauschnabel & Ro, 2016) and suggested them in the context of enterprises (Hein & Rauschnabel, 2016). To the best of our knowledge, only one study has looked at the hedonic benefits smart glasses can offer and shown its empirical relevance (Rauschnabel et al., 2016). Although hedonic motivations play a role in the adoption process, we do not include these drivers into our model as they bear no relevance with regard to the ethical consumer. However, utilitarian benefits could impact evaluations from a societal perspective.

Effort Expectancy

User friendliness of the technology in question has been introduced in various forms into the technology acceptance literature. Originally, the focus lay on the *'perceived ease of use'* of a respective technology, as office software was typically the subject of research and its use often less than intuitive with Graphical User Interfaces not yet invented (Davis, 1989). Subsequent models changed this notion when describing the degree of complexity associated with technology use (Thompson, Higgins, & Howell, 1991) and unifying theories further abstracted this acceptance aspect to the concept of effort expectancy (Venkatesh et al., 2003), which was tailored to more user-friendly systems that would call for a limited amount of effort to learn effective operation. However, we assume that the effort expectancy does not pose a matter of interest to ethically motivated users.

Social Influence

Social influence (syn: social norms or normative believes) are indicative of social influence and represent personal beliefs a person holds regarding whether a behavior is desired or frowned upon by those who are closest to him/her (Fishbein & Ajzen, 1975; Weiz, Anand, & Ernst, 2016). Smart glasses can evoke a negative effect of social influence, as they can interfere with the normal way people interact with each other in conversations in an undesired way (Due, 2015). For instance, smart glasses can be used to record people without their knowledge, which would be an infringement to their personal rights, or users may dedicate their attention to online content while being in conversation. Given normal circumstances, both examples would be considered as socially unacceptable behavior. In fact, the term "glasshole" emerged for smart glasses users who – willingly or unwillingly – violate these norms of social behavior (Lawler, 2013). This presence of social norms within society antagonizing smart glass use may pose a problem to their large-scale adoption (Weiz, Anand, & Ernst, 2016). Two other studies provide support for this. First, Rauschnabel, Brem and Ivens (2015) study descriptive norms and show that people who think that using smart glasses will be common among their peers are more likely to adopt them. Rauschnabel and Ro (2016) complement this by showing that social influence, as in TAM studies, is positively related to smart glasses adoption intention.

Social influence, as it is considered in UTAUT2, only refers to the desires of thirds regarding whether a person should show a specific adoption behavior and how this wish influences the adopter (Fishbein & Ajzen, 1975; Venkatesh et al., 2003). The mechanisms could work differently when studying the societal perspective. Smart glasses can fundamentally change how people communicate, because they introduce a potential information asymmetry into every conversation where one is permanently online, connected to various databases and equipped with such a device, while the other is not (Due, 2015).

We argue that whereas, in the past, scientists and technologists discussed amongst each other who to blame for potentially unwanted consequences of their inventions (Koepsell, 2010), responsibility has shifted to the consumer who now decides how to use a new technology. Ethically motivated consumers recognize their responsibility reaching further than the mere control for an ecologic supply chain design or use and deposit of products. They integrate the awareness of potential influences on society into their reasoning process and their evaluation of a technology, making the question whether to adopt a question of consumer ethics (Vitell, 2003). With the ethically responsible consumer active in today's markets, we research those scenarios that consumers think of when pondering how smart glasses may affect society at large and how these projections alter adoption intentions and expectations.

Facilitating Conditions

Facilitating conditions refer to a user's ability to access resources that are relevant to the adoption of a technology, such as financial and knowledge resources (Venkatesh et al., 2012). With regard to smart glasses, these facilitating conditions should play a role because, for the near future, they will remain a high-price high-tech product. However, manufacturers will be eager to gather use information about use cases and how to improve early models through their products and respective support, presumably granting early adopters generous access to support and other knowledge resources needed in order to efficiently operate them. Still, neither the availability of financial resources nor the knowledge resources do relate to questions that supersede the individual and thus could be of interest to ethically oriented consumers. Thus, this factor shall not be discussed further as it does not contribute to the understanding of society adoption with regard to ethical consumerism.

Price Value

Price value refers to the commitment of users who compare the monetary sacrifice for the purchase of the technology with the benefits they may expect from the purchase in a cognitive trade-off (Venkatesh et al., 2012). According to UTAUT2, perceived price value determines the behavioral intention to use the purchased technology. Again, this factor is well-established, yet presumably does not constitute a driver relevant to ethically motivated consumers and has not yet been studied in the context of smart glasses.

Experience and Habit

Experience and habit also contribute to the adoption intention of any user which has been acknowledged only in UTAUT2 (Venkatesh et al., 2012). The difference between these two lies in their character; experience engulfs a temporal component, covering the amount of time someone already spent using a particular technology, with operationalizations ranging from the time passed since the initial use to various levels of experience at different points in time (Venkatesh et al., 2003; Venkatesh et al., 2012). Habit, in contrast, is the extent to which people show behavior automatically as a consequence of having learned it. From a logical point of view, this requires some prior experience making experience a necessary condition (Venkatesh et al., 2012), which is, in the current stage of smart glasses penetration, scarce. What could the role of habit and experience from a societal perspective be in the future? People might get used to them and adjust their behavior. For example, during the turn of the millennium, many people were skeptical about the use of mobile phones for societies and might have used this reason to not adopt a mobile phone. Current cellphone penetration rates of close to, or even above, 100%, indicate that this fear seemed to have decreased or lost in importance.

Privacy

Just like the wearer's surrounding, a user's level of privacy may exert an influence on the adoption intention, as privacy can be seen as a "psychological as well as a social and a political requirement" (Mann, Nolan, & Wellman, 2003, S. 334). The question whether a user's own privacy is safe may ultimately be a technological question, as informational safety in corporate contexts is a precondition to any adoption (Hein & Rauschnabel, 2016). For private contexts, different cultures have shown to have different levels of tolerance when it comes to informational uncertainty and will to disclosure (Petronio, 2002). However, a given level of mistrust against technology is endemic in many populations. It manifests itself in behaviors such as laptop users blocking the integrated webcam with Post-its. For a camera, microphone and a whole range of other sensors worn on the head, this mistrust may become even more severe. Prior studies do not report a statistically significant effect of privacy risks on adoption intention or attitude measures (Rauschnabel & Ro, 2016). Rauschnabel et al. (2016) studied the role of two different conceptualizations of privacy risks of smart glasses. What they labeled 'personal privacy risks' describes the extent to which smart glasses threaten a user's own privacy (for instance, if a hacker get access to the camera). The authors then show that personal privacy concerns do not affect adoption intention. However, public privacy concerns (i.e. the extent to which using smart glasses can threaten other people's privacy) do. They conclude that when it comes to smart glasses, people tend to care more about other people's privacy than about their own. In this study, we look at privacy concerns from a third perspective: Is smart glasses penetration related to perceived 'societal privacy'? That is, does overall public privacy decrease and does this matter to consumers?

Fashnology-Related Factors

Adding to the drivers hypothesized in UTAUT2, we stress that smart glasses are not just a technology to be 'used', they are also products to be 'worn' like fashion. That is, factors that determine fashion decisions are important predictors of smart glasses' adoption. Particularly, "Smart glasses [...], as any wearable devices, [are] also a new form of fashion accessory for users" (Rauschnabel, Brem, & Ro, 2015). Research suggests using the term 'Fashnology' for smart glasses – a conceptual combination of fashion and technology. Fashion is characterized by high visibility and serves as means of constructing a social identity through the symbolic value it conveys (Davis, 1992) and therefore also needs to meet a specific level of comfort.

Prior research provides first evidence. For example, the influence of design on usage intention has been confirmed for wearable headphones (Reinelt, Hadish, & Ernst, 2016). Hein and Rauschnabel (2016) discuss that smart glasses need to meet certain expectations of physical appearance (i.e., design) and comfort (i.e., weight, well-fitting). For wearable devices, the aspect of wearing comfort is even more true as part of its functional value is a direct result of it being attached to the human body in an ergonomic way. If users perceive the wear of a wearable device to be physically or emotionally uncomfortable or its use overly tiring, this will negatively affect willingness-to-use (Hein & Rauschnabel, 2016; Bodine & Gemperle 2003). Chun et al. (2016) and Kim, Shin, and Park (2015) also show that these self-presentation related factors also hold for other wearables, such as smart watches. However, wearing fashion-like smart glasses in everyday life that integrate virtual elements in one's view field could distract people. This will be addressed later.

RESEARCH QUESTIONS

Figure 1 shows the framework for this research. We propose that consumers make judgments about both positive and negative societal factors (societal benefits and risks) which they then use to anticipate the success of smart glasses and make judgments about their desire of the success (see Figure 1). Thus, in contrast to the traditional technology acceptance and adoption model, our framework does not use a consumer's personal evaluation of smart glasses, but his/her societal evaluation – that is, if he/she thinks that smart glasses are something good or bad for society.

Anticipated success is the degree to which people expect that a new technology (like smart glasses) will be an established technology in the near future. A related, but distinct construct is desired success, defined as consumers' aspiration that a technology (like smart glasses) becomes widely adopted. Societal benefits describe expected positive consequences of an established technology, whereas risks cover negative consequences. Thus, in contrast to traditional acceptance models (Venkatesh & Bala, 2008; Venkatesh & Davis, 2000), our proposed model does not focus on a user's personal benefits or risks, but on the societal evaluations, which is a novelty of this research.

Because of the exploratory nature of our research, we formulate two research questions:

RQ1: How do consumers perceive smart glasses in terms of societal benefits and risks?
RQ2: Which societal benefits and risks influence the anticipated and desired success of smart glasses?

Methodology and Research Design

Because of the novelty of the technology we use a scenario-based survey approach. We conducted computer-assisted personal interviews in which 364 students from a mid-sized German University were

Figure 1. Framework

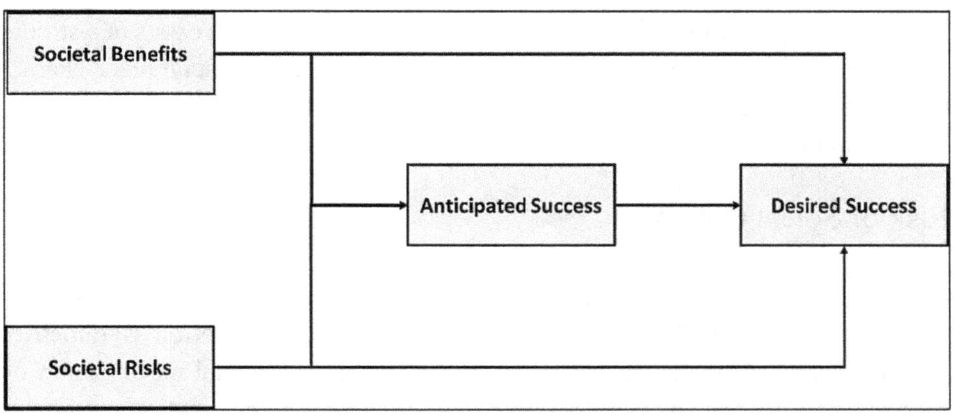

asked to take part in a survey on laptop computers in summer 2014. Candies and snacks were offered as incentives for participation. The sample consisted of 43.1% male and 56.9% females, with an average age of 22.6 years (SD=2.4). 61.7% were undergraduate students. A high educational level characterizes university students, including a high ability of critical thinking and an appropriate level of interest in new technologies, making them an appropriate sample for this context. As Google Glass was the most well-known smart glasses devices, we focused the study around Google Glass.

As prior research does not provide appropriate scales we used a very exploratory approach. More specifically, we extracted items by adopting existing scales and items to a society context (e.g. perceived usefulness, Davis, 1989), discussing items in the research team and with an AR expert who is not involved in this study, and by extracting statements from newspaper articles and discussions in online forums. We believe that the exploratory nature of this research, the lack of theory and prior research, as well as the pre-market stage of the technology justify this ad hoc approach compared to established complex scale development procedures. All items were measured on 7-point-Likert scales, anchored from 1 ("I do not agree") to 7 ("I fully agree").

Identification of Societal Benefits and Risks

26 items were subject to a series of exploratory factor analyses. After the application of various rotation methods and dropping of two items (due to low factor loadings and/or high cross-loadings across different rotation methods), we received a stable six-factor solution, consisting of three risk and three benefit factors. We assessed the reliability of the factors calculating Cronbach's alphas. All of them were close to or exceeded the recommended threshold of.70 (Nunally, 1978). Based on the corresponding items, we generated names for each factor. In the following sections, we will introduce, define and discuss each factor in detail.

Antecedents to Anticipated and Desired Market Success

Societal Risk of Loss of Awareness is the risk of society members losing their sense for processes and events happening in their immediate environment because they become distracted by the visual overlay of information provided by the smart glasses. Similar fears emerged with regard to the Sony Walkman when it became introduced to the end consumer market in 1980. Public criticism raised that users would ultimately lose touch with their surrounding and degenerate to self-enclosed and incommunicative individuals, all part of what David Riesman referred to as the "lonely crowd" (Hosokawa, 1984; Riesman, Glazer, Denney, & Gitlin, 2001). For smart glasses, this concern exists in a similar manner as described. Therefore, smart glasses hold the potential to make society a social place characterized by a loss of mutual considerateness and thoughtfulness amongst members. We name this the potential for a societal loss of awareness.

Societal Risk of Social Cohesion

The extent to which consumers expect smart glasses to diminish the amount of social behavior in a society in general. This concern is the result of consumers expecting information asymmetries to be used by users of smart glasses in opportunistic ways, thereby eroding the social norm of non-opportunistic behavior. For instance, users of smart glasses may look up personal information on their conversation partners without them knowing, and use that informational advantage to their own benefit (Rauschnabel, Brem, & Ivens, 2015). Furthermore, seeking of self-interest with guile, which is the definition of opportunism according Williamson (1985), can be greatly facilitated. With these unwanted side-effects of widespread adoption of smart glasses, we propose smart glasses may pose a threat to social cohesion.

Societal Risk of Public Privacy

This construct refers to consumers' fear that widespread use of smart glasses may make infringements to their right to informational self-determination. This is exacerbated, as most built-in cameras possess no outward-facing indicator of activity. People who are filmed or photographed cannot tell they are being recorded. Thus, this risk covers the fear of people that privacy and anonymity in daily life and public may be threatened, as everyone can become subject of recording anywhere anytime by someone linked to the Internet. These fears are not new. First, a recent study has shown that when it comes to smart glasses, people tend to care more about other people's privacy than their own (Rauschnabel et al., 2016). Second, similar concerns regarding the loss of privacy in public were raised when CCTV (closed circuit television) was introduced for public supervision in Great Britain, which consisted of large numbers of TV cameras in public spaces surveyed by public authorities (Armitage, 2002). Wearables like smart glasses pose a form of 'sousveillance' - which is surveillance directed at the public authorities surveilling its citizens to permanently exert power (Mann et al., 2003). Adding to the inventors' conceptualization, we note that sousveillance based on endemic use of wearables does not stop at focusing and exerting counter-power to authorities, but that it also focuses other citizens.

Similar to these concerns, smart glasses pose a risk to privacy in public, which we term societal risk of public privacy.

Societal Benefit of Public Safety Improvement

Smart glasses, when adopted at a large scale, can also provide some of the benefits of a CCTV. Benefits concern crime prevention as well as disaster management (Armitage, 2002). For crime prevention, CCTV works by addressing both potential offenders as well as potential victims. Potential victims will be reminded of eventual hazards, altering their behavior to a more precautious one. Potential offenders can be deterred, as the anonymity in public spaces that allows them to commit crimes is replaced by the possibility of having their identity uncovered and to allow prosecution (Armitage, 2002). For crisis management, CCTV can allow emergency managers to evaluate the situation at different locations, allowing for an efficient deployment of forces (Alvear, Abreu, Cuesta, & Alonso, 2013; Sime, 1999). Similar to CCTV, smart glasses possess the properties of being able to record felonies and, in case of emergencies, provide real-time footage which allows for a more efficient deployment of rescue forces. Thus, smart glasses hold what we call a public safety improvement potential.

Societal Benefit of Progress

Smart glasses, like any other new technology, give rise to hopes regarding how its proliferation will enable societal progress to happen. We suggest that this societal progress has a social and an economic side to it, which both are interlinked. Regarding the social side, hopes articulated include the abstract ideal of increasing social cohesion by making the world a smaller place, as people can communicate even more using smart glasses. Specifically, people who previously interacted in the context of various small-groups (e.g. families, colleagues, sports clubs) jump to a situation where they communicate with other dispersed individuals, surpassing the boundaries of any group (Mann et al., 2003). Regarding the economic side, smart glasses and their widespread use affect both business and private contexts, creating an entirely new business model. These new business models call for new technologies, new understandings and ultimately result in entirely new jobs and, potentially, job descriptions, increasing a society's welfare.

Societal Perceived Usefulness

Drawing on classical Technology Adoption Literature, we found a general assessment of perceived usefulness at a society dimension without further specifications, thus measuring a utilitarian value. This factor captures consumers' perception of how useful smart glasses may become to society.

In the appendix, Table 4 provides an overview of items and reliability measures. Appendix 2, Table 5 contains descriptive statistics of our sample.

RESULTS

RQ1: Societal Perception of Smart Glasses

To answer RQ1, we computed means across variables. Table 2 provides an overview of the means of the main sample, as well as sub-group comparisons.

Among the overall sample, respondents seem to rate the societal risk factors higher than the societal benefits. When comparing desired and anticipated success, consumers seem to rate the anticipated success higher than the desired success. We now discuss several variables along which consumers' perceptions differ:

The first variable is gender. More females demonstrate agreement in the societal loss of social cohesion (4.84 female to 4.26 male), societal loss of awareness (5.77 female to 5.24 male), and Google brand attitude (4.91 female to 4.25 male). Males, on the other hand, demonstrate agreement with their desired success for Google Glass (2.95 male to 2.12 female), anticipated success for Google Glass (4.11 male to 3.19 female), and familiarity with Google Glass (5.30 male to 4.42 female).

Second, consumers' level of familiarity with smart glasses (median split) matters. While there are no significant differences concerning benefits, high familiarity consumers tend to rate the risks higher. Surprisingly, high familiarity consumers also tend to have higher levels of anticipated and desired market success.

Third, respondents' attitude towards the manufacturer brand of the smart glasses matters. For respondents with low Google brand attitude, there is larger agreement for societal loss of awareness (5.65 low to 5.41 high) and societal loss of privacy in public (5.71 low to 5.15 high). For respondents with a more positive brand attitude, there is larger agreement for societal perceived usefulness (3.60 high to 3.00 low) and desired success (2.79 high to 2.21 low).

RQ2: Drivers of Anticipated and Desired Market Success

In order to answer RQ2, a covariance-based structural equation model was calculated to predict the desired success of Google Glass based on both negative and positive consequences for society and the anticipated success as a mediating variable using SPSS AMOS 23. Negative consequences as mentioned afore contained damage of social cohesion, the loss of awareness of individuals and a loss of society level privacy. Positive consequences, in contrast, engulfed safety improvement potential, social progress potential and a societal usefulness potential. Inspired by prior research and because of the novelty of the technology (Rauschnabel et al., 2015; Rauschnabel & Ro, 2016), we controlled for the Attitude towards the brand Google, familiarity with Google Glass, gender and age. The overall model fit was acceptable, as indicated by a significant χ^2-value (χ^2=305.53; df=212; p<.001; CFI=.98; TLI=.98, NFI=.95). Furthermore, an RMSEA value of.035 indicated the absence of substantial approximation errors. By choice of our independent variables, we were able to explain 28.9% of variance with regard to anticipated success, and, including it into the model, 63.2% of variance (R^2_{Δ}=34,4%) of desired success. The results are visualized in Figure 2 and fully documented together with their respective p-values, in Table 3.

For potential benefits to increase respondents' expected success, we identified beliefs in general usefulness (ß$_3$=.234, p=.02) and a tendency to believe in potential safety improvements enabled by smart glasses (ß$_1$=.130, p=.097). Desire for Google Glass to succeed was significantly caused by societal progress potential (ß$_8$=.283, p=.005) and the general perceived usefulness (ß$_9$=.230, p<.001).

On the downside of potentials, none of the theoretical concerns reached significance as respondents did not expect them to stop smart glasses from finding wide adoption. Still, respondents showed significant animosity towards use as they feared the loss of social cohesion caused by smart glasses (ß$_{10}$=-.170, p<.001) and a tendency towards losing of their anonymity in public (ß$_{12}$=-.137, p=.081). Expectedly, anticipated success also showed a significant impact on the desired success (ß$_{13}$=.211, p<.001).

Figure 2. SEM results; note: bold letters: p<.05; italics: p<.10; only standardized coefficients reported

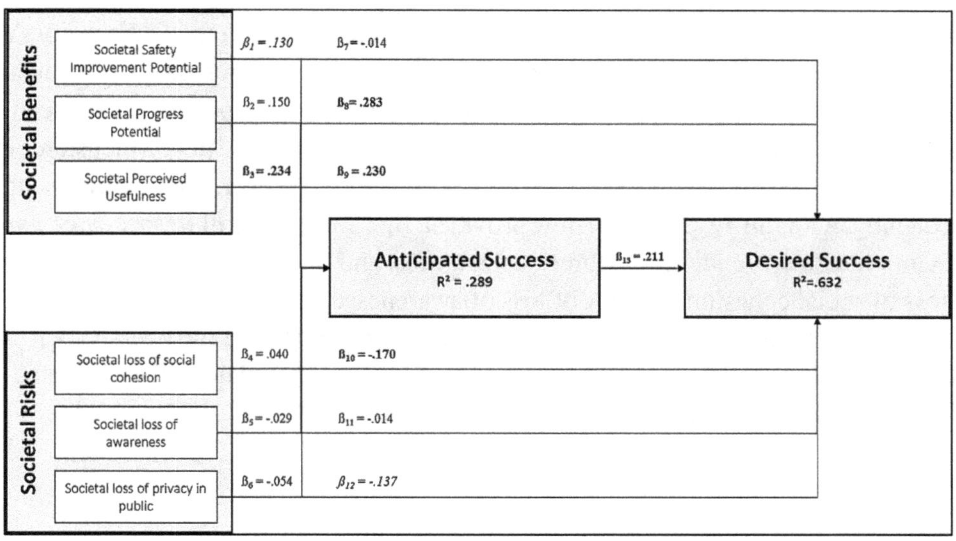

Table 2. Mean values and sub-group comparison of means

	Overall Sample		Group Comparisons								
			Gender			Familiarity			Google Brand Attitude		
	Mean	SD	M	F	p	Low	High	p	Low	High	p
Societal Benefits											
Societal Safety Improvement Potential	3.77	1.28	3.79	3.75	0.82	3.74	3.80	0.63	3.61	3.94	*0.01*
Societal Progress Potential	3.53	1.22	3.67	3.42	*0.06*	3.45	3.62	0.20	3.32	3.77	0.43
Societal Perceived Usefulness	3.28	1.42	3.38	3.21	0.27	3.18	3.40	0.15	3.00	3.60	*<0.01*
Societal Risks											
Societal loss of social cohesion	4.59	1.54	4.26	4.84	*<0.01*	4.77	4.39	*0.02*	4.83	4.32	*0.01*
Societal loss of awareness	5.54	1.32	5.24	5.77	*<0.01*	5.70	5.35	*0.01*	5.65	5.41	*<0.01*
Societal loss of privacy in public	5.45	1.35	5.37	5.51	0.35	5.44	5.46	0.87	5.71	5.15	*<0.01*
Dependent Variables											
Desired Success for Google Glass	2.48	1.42	2.95	2.12	*<0.01*	2.32	2.66	*0.02*	2.21	2.79	*<0.01*
Anticipated Success for Google Glass	3.59	1.62	4.11	3.19	*<0.01*	3.30	3.91	*<0.01*	3.50	3.68	0.28
Control Variables											
Google Brand Attitude	4.63	1.62	4.25	4.91	*<0.01*	4.75	4.49	0.13	N/A	N/A	N/A
Familiarity with Google Glass	4.80	2.41	5.30	4.42	*<0.01*	N/A	N/A	N/A	4.87	4.72	0.47
Age	22.62	2.41	22.87	22.43	*0.09*	22.36	22.92	*0.03*	22.71	22.51	0.43
Gender	0.57	0.50	N/A	N/A	N/A	0.46	0.66	*<0.01*	0.50	0.65	*0.01*

DISCUSSION

Recent forecasts predict that smart glasses have the potential to transform media behavior. Yet, little research has been conducted to understand this new technology. Published research has exclusively looked at adoption factors on a company level (Hein & Rauschnabel, 2016) or user level (Rauschnabel, Brem, & Ivens, 2015), or applications in various contexts (e.g. Albrecht et al., 2014; Muensterer et al., 2014; Tomiuc, 2014). Potential consequences for societies as a whole, however, remains an under researched area. Consequently, with this research, we now provide a first attempt to fill this research gap. Using the example of Google Glass, we identified three societal risks and three societal benefits of smart glasses: the risk of loss of social cohesion, the risk of loss of awareness, the risk of loss of privacy in public, the societal safety improvement potential, the societal progress potential and the overall societal perceived usefulness. We then integrated these factors in a theoretical model to explain both anticipated and desired success of smart glasses given the example of Google Glass.

Table 3. Estimates, significances and controls

Independent Variables	ß	p-values
DV: Anticipated Success		
Societal Safety Improvement Potential	.130	*.097*
Societal Progress Potential	.150	.210
Societal Perceived Usefulness	.234	**.002**
...Societal loss of social cohesion	.040	.642
...Societal loss of awareness	-.029	.712
Societal loss of privacy in public	-.054	.574
DV: Desired Success		
Societal Safety Improvement Potential	-.014	.830
Societal Progress Potential	.283	**.005**
Societal Perceived Usefulness	.230	***
...Societal loss of social cohesion	-.170	*.017*
...Societal loss of awareness	-.014	.828
Societal loss of privacy in public	-.137	*.081*
Anticipated Success	.211	***
Control Variables	**ß**	**p-values**
DV: Anticipated Success of Google Glass		
Google Brand Attitude	.002	.973
Familiarity	.088	*.073*
Gender	-.229	***
Age	.041	.434
DV: Desired Success of Google Glass		
Google Brand Attitude	.040	.366
Familiarity	-.002	.952
Gender	-.149	***
Age	.096	**.021**
notes: *** = p<.001 ; n = 364; standardized coefficients reported only; coding gender: o=male, 1=female.		

Implications for Scholars

Technology adoption research has come a long way since its beginnings towards the end of the 1980s. Much adjustment has been made in order to cover for changes of technologies that original models were applied to (e.g. Venkatesh & Bala, 2008; Venkatesh & Davis, 2000; Venkatesh et al., 2012). However, it is not just the technologies that change, it is the consumers that have changed with them, as technology has dissolved into the daily lives of their users (Brenner et al., 2014), with researchers today discussing even the "Internet of Things" (Porter & Heppelmann, 2015). With UTAUT2 covering aspects discussed above, research has been provided with a tool accounting for properties of many new technologies. However, it does not cover changes in the consumerscape technology finds itself in – the far-thinking, ethical consumer, who projects and extrapolates his consumption behavior into the far future and onto entire societies, often even the global society. This consumer asks himself: am I pushing society (respectively humanity) into a desirable direction? We suggest this as consumer landscapes in many industrialized countries have developed distinct cultures of ethicality (Newholm & Shaw, 2007), in which the call for short-term economic efficiency and utilitarian effectiveness has been replaced by long-term perspectives and a turn towards social values (Freestone & MacGoldrick, 2008; Shaw, Grehan, Shiu, Hassan, & Thomson, 2005). The growing importance of social norms in UTAUT2, especially among smart glasses (Rauschnabel et al., 2015; Rauschnabel & Ro, 2016) already accounts for this tendency to take the preferences and needs of others into consideration. Yet, we argue that consumer citizenship supersedes the construct of social norms. Rather, we propose the moral norm of ethically motivated consumers to extend the concept of social norms, as social and subjective norms are relative to the particular society they are exercised in. Ethically motivated users may consider people involved into their decision making as being part of a larger group than just their direct social environment, giving their decisions in the most extreme case an element of universality. Further research could focus on the question of where ethically motivated consumers draw the line, and how different groups of these consumers operationalize the "in-group" they feel responsible for with their behavior. Regardless of whether consumers feel a universal or a conservative obligation toward their own or a global society, the rise of wearables and other consumer electronics that have the potential to alter entire societies must be addressed accounting for the existence of consumer citizenship. Incorporating a factor into applied acceptance models that covers the aspect of this hypothesized norm users turn to when pondering technology adoption can serve this aim. Questions to include may cover if "people will lose sympathy for each other if this technology becomes adopted widely," "people will stop paying attention to each other if this technology becomes adopted widely or "if this technology is adopted widely, society will change for the better/worse". Some of these items are included into our exploratory study. Apparently, these items have to be adjusted to the technological context, yet we propose that they cover a new construct at the edge of technology adoption and consumer research. This construct needs further research with regard to its dimensionality, its scope, formalization and orientation. Scale development efforts could begin here, with the ultimate aim of updating the current technology acceptance literature, as the reality in which technology adoption takes place has massively shifted.

Implications for Managers

The expectations of consumers regarding how smart glasses are about to alter society through their widespread use has several implications for managers. If managers, for instance, would like to promote smart glasses and have them promoted via word-of-mouth, this is only going to happen if consumers observe the technology in a benevolent way rather than fearing it. Therefore, managers need to tackle issues related especially to the loss of social cohesion, which showed the largest impact on the desired success. This fear especially possesses a positive counterpart: the societal progress enabled by smart glasses. Contents of communication strategies thus should focus on applications that use smart glasses to deepen existing relationships and create new ones.

However, concerns regarding a loss of privacy in public and having citizens' rights infringed, respectively infringing those of others is a concern to be tackled differently. Still, these issues should be addressed in corporate communication. Managers from other industries should observe carefully the development of smart glasses. Given the right regulation, market research based on eye-movement - nowadays a costly, yet highly effective service to order (Lohse, 1997) - could experience a similar democratization like survey-based research (Pieters, Rosbergen, & Wedel, 1999; Waclawski, 2012). Presumably, the entire field of customer management will undergo a massive change, as new possibilities of approaching customers and reaching their attention may pop into existence. Some of the troubles connected to push marketing based on location-based services that aim at smartphone-usage could become obsolete through the introduction of smart glasses, achieving higher acceptance by users and opening doors to new revenue potentials (Bruner & Kumar, 2007).

For educational purposes, AR smart glasses can be adopted into suitable contexts some of which we have already mentioned. Next to educational advantages, this may stimulate the society's propensity to make beneficial use of technology in the long run, as individual's beliefs about positive use of technology will be confirmed and susceptibility to innovation will be fostered. This will help societies to remain open to yielding benefits enabled by technological innovation as abstract fears implied by technology can be countered with actual and positive results.

Implications for Policy Makers

With regard to risks, it is the legislature's turn to regulate smart glasses like any new emergent other technology in order to protect citizens' rights and safety. Smart glasses possess the advantage of having some properties and causing the same concerns like cameras built into cell-phones that are linked to social networks. These, too, offer the possibility to snap photographs of people without them knowing and posting their images online. Thereby, people's privacy rights are potentially at risk whenever a person comes into the device's line of sight. Thus, regulation can rely on already achieved and implemented regulation, like rules that apply to the case of public CCTV surveillance. However, policy makers should also account for situations in which sensitivity is needed when weighing different interests against each other: should there be exemptions to strict regulations about information rights in emergency cases? How are these cases defined? Which property or potential of smart glasses or emergency cases may justify such an exemption? Likewise, rules for daily application need to be freedom-oriented enough that innovation does not become unattractive. Empirically, regulation was shown to have ambiguous effects on innovation – sometimes hindering it, like in the case of internet-neutrality (Ehrlich, Eisenach, & Leighton, 2010), sometimes fostering it, like in the case of eco-innovations (Ghisetti & Pontoni, 2015; Jaffe & Palmer, 1997) and sometimes leading to innovation that is set to meet regulators' expectations while keeping productivity stable (Rubashkina, Galeotti, & Verdolini, 2015). These examples illustrate that legislature needs a slow hand when designing the rules to be applied onto users of smart glasses if economic and societal progress potentials shall develop. Regulating for educational usage in this vain will pose a challenging task, given the heterogeneity of users in terms of age, psychological maturity and sense of responsibility. However and in this context, it will be administrators' jobs to be aware of potential harm AR smart glasses users can exert and that can be exerted upon users and to restrict use of functions to ensure full compliance. We stress that the risks of AR smart glasses in open education are largely congruent with those present in other contexts (e.g. Hein & Rauschnabel, 2016), such as violation of informational self-determination of individuals and privacy. Yet, responsibility needs to be distributed transpa-rently and diligently, especially when AR smart glasses are used by minors. IT-admins may carry more responsibility than teachers, as the teacher's job rather consists of choosing the right method of teaching for the right topic than actually controlling IT.

Finally, like any other technology that, in the past, has proven useful to make work more efficient, smart glasses stand to adoption by officials and authorities. Advantages in disaster management can be a justifying motivation (Alvear et al., 2013; Sime, 1999).

LIMITATIONS AND FUTURE RESEARCH DIRECTION

As with any study, this research has limitations. For example, using cross-sectional, self-reported student data and focusing on Google Glass (rather than various other devices) might limit the external validity and generalizability of the findings. Moreover, benefits and risks included in this study may not reflect all potentially relevant societal benefits and risks.

Future studies should address these gaps. For example, qualitative in-depth research with consumers and experts could help identify a larger pool of societal benefits and risks. These factors could then be analyzed on various consumer and expert samples. Case studies and longitudinal research designs could also provide more details in the dynamic nature of these factors. For example, people might adjust their behavior in order to avoid these consequences. This could be grounded in balance theory and its extensions. For example, if an innovator of early adopter perceives that using smart glasses might make the world more unsocial but on the other hand could personally benefit from utilitarian benefits, he/she might adjust this negative societal evaluation. Likewise, equity theory may help explain if and how acceptance of smart glasses is dependent upon whether the judging person himself is wearing such a device, potentially cancelling out any disadvantageous information asymmetry. Evidence that can be interpreted as implying this has been found by Rauschnabel, Brem and Ivens (2015) who showed that people that expect smart glasses to become endemic in their social environment are more likely to adopt them.

Finally, scholars could be inspired by our research to develop a more holistic adoption theory. This theory could include both personal and societal evaluations of a technology and conceptualize them as antecedents to adoption.

CONCLUSION

In this study, we have shown that consumers use certain society related variable to evaluate new technologies – smart glasses. We hope that these findings inspire policy makers and other scholars to include these findings in their work.

REFERENCES

Albrecht, U.-V., Jan, U., von, , Kuebler, J., Zoeller, C., Lacher, M., & Muensterer, O. J. (2014). Google Glass for documentation of medical findings: Evaluation in forensic medicine. *Journal of Medical Internet Research*, *16*(2), 53. doi:10.2196/jmir.3225 PMID:24521935

Alvear, D., Abreu, O., Cuesta, A., & Alonso, V. (2013). Decision support system for emergency management: Road tunnels. *Tunnelling and Underground Space Technology*, *34*, 13–21. doi:10.1016/j.tust.2012.10.005

Armitage, R. (2002). *To CCTV or not to CCTV? A review of current research into the effectiveness of CCTV systems in reducing crime*. Retrieved from https://epic.org/privacy/surveillance/spotlight/0505/nacro02.pdf

Berque, D. A., & Newman, J. T. (2015). GlassClass: Exploring the Design, Implementation, and Acceptance of Google Glass in the Classroom. In R. Shumaker & S. Lackey (Eds.), Lecture Notes in Computer Science: Vol. 9179. *Virtual, Augmented and Mixed Reality* (pp. 243–250). Cham: Springer International Publishing. doi:10.1007/978-3-319-21067-4_25

Bijker, W. E., & Law, J. (1992). *Shaping technology/building society: Studies in sociotechnical change. Inside technology*. Cambridge, MA: MIT Press.

Bodine, K., & Gemperle, F. (2003). Effects of functionality on perceived comfort of wearables.*Seventh IEEE International Symposium on Wearable Computers*, 57–60. doi:10.1109/ISWC.2003.1241394

Brenner, W., Karagiannis, D., Kolbe, L., Krüger, J., Leifer, L., Lamberti, H.-J., & Zarnekow, R. (2014). User, Use & Utility Research. *Business & Information Systems Engineering*, *6*(1), 55–61. doi:10.1007/s12599-013-0302-4

Bruner, G. C. II, & Kumar, A. (2007). Attitude toward Location-based Advertising. *Journal of Interactive Advertising*, *7*(2), 3–15. doi:10.1080/15252019.2007.10722127

Chitturi, R., Raghunathan, R., & Mahajan, V. (2008). Delight by Design: The Role of Hedonic Versus Utilitarian Benefits. *Journal of Marketing*, *72*(3), 48–63. doi:10.1509/jmkg.72.3.48

Craig, A. B. (2013). *Understanding augmented reality: Concepts and applications*. Amsterdam: Morgan Kaufmann.

Davis, F. (1992). Fashion, culture, and identity. Chicago: University of Chicago Press.

Davis, F. D. (1989). Perceived Usefulness, Perceived Ease of Use, and User Acceptance of Information Technology. *Management Information Systems Quarterly*, *13*(3), 319–340. doi:10.2307/249008

Depari, A., de Dominicis, C. M., Flammini, A., Sisinni, E., Fasanotti, L., & Gritti, P. (2015). Using smartglasses for utility-meter reading.*2015 IEEE Sensors Applications Symposium (SAS)*, 1–6. doi:10.1109/SAS.2015.7133649

Doane, D. (2001). *Taking flight: The rapid growth of ethical consumerism*. London: New Economics Foundation.

Due, B. L. (2015). The social construction of a Glasshole: Google Glass and multiactivity in social interaction. *PsychNology*, *13*(2-3), 149–178.

East, M. L., & Havard, B. C. (2015). Mental Health Mobile Apps: From Infusion to Diffusion in the Mental Health Social System. *JMIR Mental Health*, *2*(1), e10. doi:10.2196/mental.3954 PMID:26543907

Ehrlich, E. M., Eisenach, J. A., & Leighton, W. A. (2010). The Impact of Regulation on Innovation and Choice in Wireless Communications. *Review of Network Economics*, *9*(1). doi:10.2202/1446-9022.1194

Fishbein, M., & Ajzen, I. (1975). *Belief, attitude, intention, and behavior: An introduction to theory and research. Addison-Wesley series in social psychology*. Reading, MA: Addison-Wesley Pub. Co.

Freestone, O. M., & MacGoldrick, P. J. (2008). Motivations of the ethical consumer. *Journal of business ethics. JOBE, 79*(4), 445–467.

Geißler, N., Schönwald, L., Strauß, G., Meixensberger, J., Korb, W., & Bartz, D. (2013). Bewertung visueller Signalgestaltung in der Chirurgie mittels Eye-Tracking / Assessment of Visual Signals in Surgery. *i-com, 12*(2).

Ghisetti, C., & Pontoni, F. (2015). Investigating policy and R&D effects on environmental innovation: A meta-analysis. *Ecological Economics, 118*, 57–66. doi:10.1016/j.ecolecon.2015.07.009

Gu, Z., Wei, J., & Xu, F. (2015). An Empirical Study on Factors Influencing Consumers Initial Trust in Wearable Commerce. *Journal of Computer Information Systems, 56*(1), 79–85. doi:10.1080/08874417.2015.11645804

Hein, D. W. E., & Rauschnabel, P. A. (2016). Augmented Reality Smart Glasses and Knowledge Management: A Conceptual Framework for Enterprise Social Networks. In A. Roßmann, M. Besch, & G. Stei (Eds.), *Enterprise Social Networks*. Wiesbaden: Springer. doi:10.1007/978-3-658-12652-0_5

Hennig-Thurau, T., Malthouse, E. C., Friege, C., Gensler, S., Lobschat, L., Rangaswamy, A., & Skiera, B. (2010). The Impact of New Media on Customer Relationships. *Journal of Service Research, 13*(3), 311–330. doi:10.1177/1094670510375460

Hernandez, B., Jimenez, J., & José Martin, M. (2010). Business management software in high-tech firms: The case of the IT services sector. *Journal of Business and Industrial Marketing, 25*(2), 132–146. doi:10.1108/08858621011017750

Hosokawa, S. (1984). The walkman effect. *Popular Music, 4*, 165. doi:10.1017/S0261143000006218

Jacobsen, E., & Dulsrud, A. (2007). Will Consumers Save The World?: The Framing of Political Consumerism. *Journal of Agricultural & Environmental Ethics, 20*(5), 469–482. doi:10.1007/s10806-007-9043-z

Jaffe, A. B., & Palmer, K. (1997). Environmental Regulation and Innovation: A Panel Data Study. *The Review of Economics and Statistics, 79*(4), 610–619. doi:10.1162/003465397557196

Kim, K. J., Shin, D.-H., & Park, E. (2015). Can Coolness Predict Technology Adoption?: Effects of Perceived Coolness on User Acceptance of Smartphones with Curved Screens. *Cyberpsychology, Behavior, and Social Networking, 18*(9), 528–533. doi:10.1089/cyber.2014.0675 PMID:26348813

Koepsell, D. (2010). On Genies and Bottles: Scientists Moral Responsibility and Dangerous Technology R&D. *Science and Engineering Ethics, 16*(1), 119–133. doi:10.1007/s11948-009-9158-x PMID:19644770

Lawler, R. (2013). *Get Ready For Even More Google Glasshole Sightings*. Retrieved from http://social.techcrunch.com/2013/01/28/glassholes/

Lee, M. R., Bojanova, I., & Suder, T. (2015). The New Wearable Computing Frontier. *IT Professional, 5*(September-October), 16–19. doi:10.1109/MITP.2015.84

Lohse, G. L. (1997). Consumer Eye Movement Patterns on Yellow Pages Advertising. *Journal of Advertising, 26*(1), 61–73. doi:10.1080/00913367.1997.10673518

Mann, S., Nolan, J., & Wellman, B. (2003). Sousveillance: Inventing and Using Wearable Computing Devices for Data Collection in Surveillance Environments. *Surveillance & Society*, *1*(3), 331–355.

Moshtaghi, O., Kelley, K. S., Armstrong, W. B., Ghavami, Y., Gu, J., & Djalilian, H. R. (2015). Using google glass to solve communication and surgical education challenges in the operating room. *The Laryngoscope*, *125*(10), 2295–2297. doi:10.1002/lary.25249 PMID:25775933

Muensterer, O. J., Lacher, M., Zoeller, C., Bronstein, M., & Kübler, J. (2014). Google Glass in pediatric surgery: An exploratory study. *International journal of surgery (London, England)*, *12*(4), 281–289. doi:10.1016/j.ijsu.2014.02.003 PMID:24534776

Newholm, T., & Shaw, D. (2007). Studying the ethical consumer: A review of research. *Journal of Consumer Behaviour*, *6*(5), 253–270. doi:10.1002/cb.225

Park, C. W., MacInnis, D. J., Priester, J., Eisingerich, A. B., & Iacobucci, D. (2010). Brand attachment and brand attitude strength: Conceptual and empirical differentiation of two critical brand equity drivers. *Journal of Marketing*, *74*(6), 1–17. doi:10.1509/jmkg.74.6.1

Petronio, S. S. (2002). *Boundaries of privacy: Dialectics of disclosure. SUNY series in communication studies*. Albany, NY: State University of New York Press.

Pieters, R., Rosbergen, E., & Wedel, M. (1999). Visual Attention to Repeated Print Advertising: A Test of Scanpath Theory. *JMR, Journal of Marketing Research*, *36*(4), 424. doi:10.2307/3151998

Porter, M., & Heppelmann, J. E. (2015, October). How smart, connected products are transforming companies. *Harvard Business Review*.

PriceWaterhouseCoopers. (2015). *The wearable future*. Author.

Quint, F., & Loch, F. (2015). Using Smart Glasses to Document Maintenance Processes. In A. Weisbecker, M. Burmester, & A. Schmidt (Eds.), *Mensch und Computer 2015. Workshopband* (pp. 203–208). Stuttgart: Oldenbourg Wissenschaftsverlag. doi:10.1515/9783110443905-030

Ratneshwar, S., & Mick, D. G. (2005). *Inside consumption: Consumer motives, goals, and desires*. London: Routledge.

Rauschnabel, P. A., Brem, A., & Ivens, B. S. (2015). Who will buy smart glasses? *Computers in Human Behavior*, *49*, 635–647. doi:10.1016/j.chb.2015.03.003

Rauschnabel, P. A., Brem, A., & Ro, Y. K. (2015). *Augmented Reality Smart Glasses. Definition, Conceptual Insights, and Managerial Importance*. unpublished working paper.

Rauschnabel, P. A., & Ro, Y. K. (2016). Augmented reality smart glasses: An investigation of technology acceptance drivers. *International Journal of Technology Marketing*, *11*(2), 123–148. doi:10.1504/IJTMKT.2016.075690

Reinelt, P., Hadish, S., & Ernst, C.-P. H. (2016). How Design Influences Headphone Usage. In C.-P. H. Ernst (Ed.), *Progress in IS. The Drivers of Wearable Device Usage* (pp. 59–68). Cham: Springer International Publishing. doi:10.1007/978-3-319-30376-5_6

Riesman, D., Glazer, N., Denney, R., & Gitlin, T. (2001). The lonely crowd: A study of the changing American character (Abridged and rev. ed.). Yale Nota bene. New Haven, CT: Yale University Press.

Rubashkina, Y., Galeotti, M., & Verdolini, E. (2015). Environmental regulation and competitiveness: Empirical evidence on the Porter Hypothesis from European manufacturing sectors. *Energy Policy*, *83*, 288–300. doi:10.1016/j.enpol.2015.02.014

Shaw, D., Grehan, E., Shiu, E., Hassan, L., & Thomson, J. (2005). An exploration of values in ethical consumer decision making. *Journal of Consumer Behaviour*, *4*(3), 185–200. doi:10.1002/cb.3

Shaw, D., Newholm, T., & Dickinson, R. (2006). Consumption as voting: An exploration of consumer empowerment. *European Journal of Marketing*, *40*(9/10), 1049–1067. doi:10.1108/03090560610681005

Sime, J. D. (1999). Crowd facilities, management and communications in disasters. *Facilities*, *17*(9/10), 313–324. doi:10.1108/02632779910278728

Spagnolli, A., Guardigli, E., Orso, V., Varotto, A., & Gamberini, L. (2014). Measuring User Acceptance of Wearable Symbiotic Devices: Validation Study Across Application Scenarios. In G. Jacucci, L. Gamberini, J. Freeman, & A. Spagnolli (Eds.), Lecture Notes in Computer Science: Vol. 8820. *Symbiotic Interaction* (pp. 87–98). Cham: Springer International Publishing. doi:10.1007/978-3-319-13500-7_7

Spitzer, M., & Ebner, M. (2016). Use cases and architecture of an information system to integrate smart glasses in educational environments. In *EdMedia:World Conference on Educational Media and Technology,* (1), 57-64.

Tomiuc, A. (2014). Navigating Culture. Enhancing Visitor Museum Experience through Mobile Technologies. From Smartphone to Google Glass. *Journal of Media Research*, *20*(3), 33–46.

Venkatesh, V., & Bala, H. (2008). Technology Acceptance Model 3 and a Research Agenda on Interventions. *Decision Sciences*, *39*(2), 273–315. doi:10.1111/j.1540-5915.2008.00192.x

Venkatesh, V., & Davis, F. D. (2000). A Theoretical Extension of the Technology Acceptance Model: Four Longitudinal Field Studies. *Management Science*, *46*(2), 186–204. doi:10.1287/mnsc.46.2.186.11926

Venkatesh, V., Morris, M. G., Davis, G. B., & Davis, F. D. (2003). User Acceptance Technology. Toward a unified view. *Management Information Systems Quarterly*, *27*(3), 425–478.

Venkatesh, V., Thong, J. Y. L., & Xu, X. (2012). Consumer Acceptance and use of information technology: Extending the unified theory of acceptance and use of technology. *Management Information Systems Quarterly*, *36*(1), 157–178.

Vitell, S. J. (2003). Consumer Ethics Research: Review, Synthesis and Suggestions for the Future. *Journal of Business Ethics*, *43*(1/2), 33–47. doi:10.1023/A:1022907014295

Waclawski, E. (2012). How I Use It: Survey Monkey. *Occupational Medicine*, *62*(6), 477–477. doi:10.1093/occmed/kqs075 PMID:22915570

Wall, D., Ray, W., Pathak, R. D., & Lin, S. M. (2014). A google glass application to support shoppers with dietary management of diabetes. *Journal of Diabetes Science and Technology*, *8*(6), 1245–1246. doi:10.1177/1932296814543288 PMID:25015954

Wassom, B., & Bishop, A. (2015). *Augmented reality law, privacy, and ethics: Law, society, and emerging AR technologies*. Amsterdam: Syngress.

Weiz, D., Anand, G., & Ernst, C.-P. H. (2016). The Influence of Subjective Norm on the Usage of Smartglasses. In C.-P. H. Ernst (Ed.), *Progress in IS. The Drivers of Wearable Device Usage* (pp. 1–11). Cham: Springer International Publishing. doi:10.1007/978-3-319-30376-5_1

Williamson, O. E. (1985). The economic institutions of capitalism: Firms, markets, relational contracting. New York: Free Press.

Witkowski, T. H., & Reddy, S. (2010). Antecedents of ethical consumption activities in Germany and the United States. *Australasian Marketing Journal*, *18*(1), 8–14. doi:10.1016/j.ausmj.2009.10.011

Wu, H. K., Lee, S. W. Y., Chang, H. Y., & Liang, J. C. (2013). Current status, opportunities and challenges of augmented reality in education. *Computers & Education*, *62*, 41–49. doi:10.1016/j.compedu.2012.10.024

Yang, T., & Choi, Y. M. (2015). Study on the Design Characteristics of Head Mounted Displays (HMD) for Use in Guided Repair and Maintenance. In R. Shumaker & S. Lackey (Eds.), *Lecture Notes in Computer Science. Virtual, augmented and mixed reality. 7th international conference, vamr* (Vol. 9179, 535–543). Springer.

Yiwen, G., Li, H., & Luo, Y. (2015). An empirical study of wearable technology acceptance in healthcare. *Industrial Management & Data Systems*, *115*(9), 1704–1723. doi:10.1108/IMDS-03-2015-0087

Yuen, S., Yaoyuneyong, G., & Johnson, E. (2011). Augmented reality: An overview and five directions for AR in education. *Journal of Educational Technology Development and Exchange*, *4*(1), 119–140.

APPENDIX 1

Table 4. Construct composition and reliability measures

Construct (Source)	Items	Alpha	AVE	C.R.
Google Brand Attitude (adapted from Park, MacInnis, Priester, Eisingerich, & Iacobucci, 2010)	I think Google is a good brand.	.94	.85	.95
	I like the brand Google.			
	I have a positive attitude towards Google.			
Familiarity	I did not know the information at all - I already knew the information.	n/a		
When Some Day Many People Use Google Glass, ...				
Societal Safety Improvement Potential	...crimes can be prosecuted more effectively.	.94	.75	.86
	...security in our country will increase.			
	...disaster relief forces can work more efficiently.			
	...lives can be saved in emergency cases, as disaster relief forces can be coordinated in a better way.			
Societal Progress Potential	...the world will partly grow together.	.83	.41	.68
	...our society will become more modern.			
	...jobs will be created.			
Societal Usefulness	For society, Google Glass is very useful. For society, Google Glass is a useful invention.	.92	.86	.92
Negative consequences for	...the world will become an antisocial planet.	.89	.66	.89
	...life quality of people will go down.			
	...mankind will be unhappier.			
	... people will have a lot less real friends.			
Negative consequences by loss of awareness	... many people will be distracted and put others at risk. ... people will be less cautious in their daily lives.	.90	.81	.90
Negative consequences for society level privacy	...freedom rights of people will be violated more often. ... there will be no more privacy.	.73	.58	.74
Desired Success for GG	I hope, Google Glass will prevail in the market.	.94	.88	.94
	I would be happy if Google Glass soon would become an established product.			
Anticipated Success for GG	I expect Google Glass to prevail in the market.	.94	.89	.94
	I expect Google Glass to become an established product soon.			

Note: CFA measures: χ^2=261,97; p<.001, df=173; CFI=.98; TLI=.98; NFI=.95; RMSEA=.038

APPENDIX 2

Table 5. Sample description

Gender	
Male	43.1%
Female	56.9%
Age	
Under 20	20.1%
21-22	31.8%
23-25	35.7%
26-30	12.3%
Average age (SD) in years	22,6 (2,4)
Degree	
Undergraduate	61.7%
Graduate	33.6%
n/a	.8%
Monthly Income	
below 500€	26.4%
501€ - 1.000 €	41.5%
1.001€ - 1.500€	8.0%
above 1.500€	10.4%
n/a	13.7%
Note: student sample with n = 364 respondents	

Chapter 2
Educational Augmented Reality (AR) Applications and Development Process

Muzaffer Özdemir
Çanakkale Onsekiz Mart University, Turkey

ABSTRACT

In recent years, presenting the useful information in an effective way has become a great necessity for educators. The opportunities provided by the AR technologies offer practical ways to meet this need of educators. By integrating the digital objects with real-world assets simultaneously, AR helps to concretize abstract concepts, and enhances the sense of reality, which in turn is a huge contribution to learning. In this chapter, it was presented the various limitations and advantages of AR revealed by some empirical studies in the literature. In addition, it was given information about AR development tools/ programs, add-on packages and presented development stages for an exemplary AR book page. The use of the Unity and Vuforia was explained as the development tools. It is believed that this information would be useful for those who will develop AR application which can be easily displayed by mobile or desktop PCs.

INTRODUCTION

Various information and communication technologies (ICT) including interactive technologies like online sharing of information, mobile technologies, and Augmented Reality (AR) have accomplished to come to our classrooms to facilitate traditional teaching (Chen & Wang, 2015). In this sense, it has been a great necessity to understand how that technology would be beneficial for learning in what context and with what kind of content use (Richards & Taylor, 2015). Especially, a competitive market has emerged with the rapid spread of mobile phones, and wireless network communications have largely facilitated the accessibility of these technologies. This has created important advantages for people in terms of technology-assisted learning. In recent years, research on learning technologies has focused on emerging technologies (e.g., adaptable learning, digital badges, mobile learning, open content, virtual and remote laboratories, learning analytics, augmented reality, volumetric and holographic displays,

DOI: 10.4018/978-1-5225-2110-5.ch002

virtual assistants, serious games, etc.) in terms of impact on learning (Johnson, Adams Becker, Estrada, & Freeman, 2015). These studies have contributed to the development of personalization processes placing students at the center of learning activities. AR technologies have taken important steps recently to prove the value of these options among these research trends. It can be said that AR technology, which can be used with mobile or non-mobile devices, represents an indispensable technology for educational environments in the near future.

Augmented Reality

AR is a technology blending context-based digital information and real-world environments dynamically (Sommerauer & Müller, 2014). A more formal definition is: "AR is a variation of Virtual Environments (VE), or Virtual Reality (VR) as it is more commonly called" (Azuma, 1997, p.355). "An AR system supplements the real world with virtual objects that appear to coexist in the same space as the real World." (Azuma et al., 2001, p.34). AR integrates 3D virtual objects with real environments in real time.

In these contexts, AR systems have the following features (Azuma et al., 2001):

1. Combine real and virtual objects in a real environment;
2. Run interactively, and in real time; and
3. Register (align) real and virtual objects with each other.

AR increases the sense of reality by adding virtual information into the user's real environment (Bokyung, 2009). We cannot restrict AR only to visualization technologies like head-mounted display or with our sense of sight (Azuma, 1997), because AR is a technology that has the capacity to affect all our senses. The AR applications of today have been improved with integrated cameras, GPS sensors, and mobile internet access that enable to overlay the real-world environments with dynamic, context-based, and interactive digital content (Sommerauer & Müller, 2014). AR systems employed in certain professional fields such as the military for 50 years have also become available to the general public with the spread of mobile technologies (Sommerauer & Müller, 2014). 3D models can also be used in addition to texts, images, videos, and animations in AR applications. AR applications, having such rich content support, have brought an innovative approach to multimedia-assisted learning environments, which are widely used at present. In particular, the use of 3D virtual worlds in learning environments offers interesting, attractive, and original learning experiences (Richards & Taylor, 2015).

The Use of AR in Education and Its Potential Benefits

Besides having several inspiring features for education environments, AR provides various opportunities for teaching designers and academicians to think more deeply about students' context and status (Haag, 2013). AR could enable (1) learning content in 3D perspectives; (2) ubiquitous, collaborative, and situated learning; (3) learners' senses of presence, immediacy, and immersion; (4) visualizing the invisible; and (5) bridging formal and informal learning (Wu, Lee, Chang, & Liang, 2013). In several studies, it has been found that elements of AR can increase the motivation, participation, and satisfaction of students while performing learning activities (Ibáñez, Di-Serio, Villarán-Molina, & Delgado-Kloos, 2016). AR technologies can take any situation, location, environment, or experience to a whole new level of meaning and understanding (Haag, 2013). In case AR is designed in conjunction with several technology types like mixed and enhanced reality, it has been claimed that the AR's potential and affordances can be further increased, and by this means it can have compelling features for achieving educational goals (Wu et al., 2013).

AR applications developed for learning environments are directly connected with some of multimedia design principles suggested by Mayer (2002) (Santos et al., 2014, p.48). For example, spatial contiguity principle state that people learn more deeply from a multimedia message when corresponding items of information are presented near rather than far from each other on the page or screen (Mayer, 2005) while temporal contiguity principle indicate that people learn better when corresponding elements of information are presented simultaneously rather than successively (Mayer, 2008). AR is in positive harmony with both of these principles as AR eliminates the need to search the required image for the desired information due to its ability to place digital images on a real-world item. Owing to this reason, AR is uniquely changing the way people learn with mobile devices (Haag, 2013). The modality principle, which is another one of the multimedia design principles, entails that learning increases in case the textual information is presented in a narrative format along with a related image instead of being presented visually (Moreno, 2006). AR can implement the modality principle by playing spoken text, instead of displaying printed text, when recognizing a stimulating event (Sommerauer & Müller, 2014). Apart from these, according to the signaling principle; "people learn better when cues that highlight the organization of the essential material are added" (Mayer, 2002) to a learning environment. AR can implement signaling by directing and guiding people through learning environments using geographic location information and visual triggers (Sommerauer & Müller, 2014).

The first attempts to analyze the potential benefits of AR have been realized by implementing AR for learning environments (Ibáñez et al., 2014). In these investigations, some evidence related to the effectiveness of AR in learning environments has been found, and these instances have been increased in subsequent studies. When experimental studies on the use of AR in education are examined, especially those conducted in recent years, it is proven that AR-assisted learning environments increase the learning performance (e.g., Wei, Weng, Liu, & Wang, 2015; Ibáñez et al., 2014; Sommerauer & Müller, 2014; Zhang, Sung, Hou, & Chang, 2014; Chang et al., 2014; Lin, Duh, Li, Wang, & Tsai, 2013; Chen & Tsai, 2012; Chiang, Yang, & Hwang, 2014; Chen & Wang, 2015; Hsiao, Chang, Lin, & Wang, 2013; Ibáñez, Di-Serio, Villarán-Molina, & Delgado-Kloos, 2016) and motivation (e.g., Wei et al., 2015; Di Serio, Ibáñez, & Kloos, 2013; Chiang et al., 2014; Hsiao et al., 2013) of students, create a positive emotional influence on them (e.g., Huang, Chen, & Chou, 2016; Chang, Hou, Pan, Sung, & Chang, 2015), influence their behavior patterns positively, (e.g., Wang, Duh, Li, Lin, & Tsai, 2014) and facilitate students to adopt a positive attitude for these applications (e.g., Hsiao, Chen, & Huang, 2012). However, in these experimental studies, the effectiveness of AR has been examined in the context of only certain education fields. The first of these major education fields is *Natural sciences, mathematics and statistics.* Most of the studies conducted in this education field are mainly related to *Physical Sciences* [*Physics* (e.g., Ibáñez et al., 2014; Zhang et al., 2014; Lin et al., 2013; Chen, & Wang, 2015; Ibáñez et al., 2016), *Biological and related sciences* (e.g., Chiang et al., 2014), and *Environment* (e.g., Huang et al., 2016; Hsiao et al., 2013)] subfield. Another major education field having evidence regarding the efficiency of AR is *Social sciences, journalism and information.* Studies conducted on this major education field are mostly related to the *Library, information and archival studies* [*Museum studies* (e.g., Sommerauer & Müller, 2014; Chang et al., 2014); *Library studies* (e.g., Chen & Tsai, 2012)] subfield. In addition, evidence has also been found that AR is efficient in *Arts and humanities* (e.g., Wei et al., 2015; Di Serio et al., 2013] and *Engineering, manufacturing and construction* (e.g., Chang et al., 2015) fields.

In literature, several benefits with regard to the use of AR in learning environments have been reported. Some of these benefits are as follows: Increasing attention and concentration (e.g., Chiang et al., 2014; Di Serio et al., 2013), helping to understand the topic (e.g., Ibáñez et al., 2014; Ke & Hsu, 2015; Hsiao et al., 2013), increasing the willingness to learn (e.g., Huang et al., 2016; Wei et al., 2015; Chang et al., 2014; Chen & Tsai, 2012), increasing the focus on tasks (e.g., Chiang et al., 2014; Chang et al., 2014), increasing the sense of entertainment (e.g., Huang et al., 2016; Chiang et al., 2014), helping to structure the information (e.g., Lin et al., 2013; Yoon, Elinich, Wang, Steinmeier, & Tucker, 2012), strengthening collaborative working (e.g., Ke & Hsu, 2015; Wang et al., 2014), enabling interactive participation (e.g., Han, Jo, Hyun, & So, 2015), providing convenience in student-centered activities (e.g., Di Serio et al., 2013), improving application skills (e.g., Chen & Tsai, 2012), and enhancing memory (e.g., Di Serio et al., 2013).

Examples of AR and Its Effectiveness in Education

It is possible to come across many studies investigating the effectiveness of AR in educational settings. Most of these studies have compared AR based learning with traditional or other technology based learning environments. Four examples selected from such studies were summarized below and they were addressed from the following aspects; AR learning applications/ systems which were addressed in these studies; other learning environments/ systems which were compared with the AR learning applications/ systems; dependent variables on which the effects of AR were discussed; used AR types, AR contents and display devices of these AR contents. These studies were chosen because they had four different types of AR, which are marker-based, location-based, robot- mediated and digital augmentation. Huang et al. (2016) developed an AR-based learning system that allows ecological explorations, and they investigated the effect of this AR system on the students' emotions and learning outcomes. The AR application they used was a geological location based one and its learning contents were interactive 2D objects and explanatory texts, which were related to ecological areas. Students who participated in the study used tablets and mobile phones to display AR objects in ecological areas. Researchers found that, compared to the human-guidance-only model, the AR-based learning model was quite effective in stimulating positive emotions among students and in improving their learning outcomes.

A study conducted by Ibáñez et al. (2014) assessed the effect of an AR learning application on students' enjoyment levels and learning outcomes. AR-based and web-based applications were compared in this study. With the AR-based application, 3D virtual models could be placed on marker-based tangible elements (physical paper circuit elements) in order to visualize the electromagnetic forces, electric fields, and magnetic fields and also in order to explore the behaviors of electrical circuit elements. Researchers found that their augmented reality approach was more effective in increasing students' knowledge and perceptions about the electromagnetic subject than a web-based application. They also indicated that participants reached higher flow experience levels with the AR-based application in comparison to the web-based application.

Another experimental study (Han et al., 2015) focused on enhancing children's engagement in dramatic play activities with the help of an AR-infused robot system. In this study, which used a marker-based approach, participants were pre-school children aged 5-6 years. Two AR-mediated dramatic play conditions (robot and computer mediated) created by researchers were compared in terms of the children's perception levels in relation to the variables of satisfaction, sensory immersion, and media recognition. The AR-based robot system used in this study utilized certain features to perform interactions with the children such as marker-based AR technology, facial expressions, physical actions, sound effects, text to speech and AR scene monitor. The study showed that children in the robot-mediated condition had significantly higher perceptional values than those in the computer-mediated condition regarding the above-mentioned aspects.

In another study (Hsiao et al., 2012) that came out with the suggestion of "learning while exercising", researchers developed a system called "Ecosystems Augmented Reality Learning". In this AR learning system, webcam data were used to capture the students' movements from the real world; and the information captured from student's movements was transferred to the system in combination with virtual images. The students' virtual interactions with AR learning system were displayed on a screen with the help of a projector. In the study, researchers compared the learning with this AR system with the existing keyboard/mouse-based computer assisted instruction, as well as with traditional face-to-face teaching approaches in terms of two dependent variables (learning achievement and attitude). The main findings of this study were as follows: learning while exercising groups (experimental group) did not display less learning achievement than the keyboard/mouse-based computer assisted instruction or traditional groups (control groups). However, they did more exercise without reducing their learning achievement quality. Besides, the students in the experimental group displayed more learning attitudes than the ones in the control groups, and the students in the group of keyboard/mouse-based computer assisted instruction attained the highest scores for the anxiety variable in their learning environments.

Challenges Encountered in the Use of AR in Education

While AR offers new learning opportunities, it also creates new challenges for educators (Wu et al., 2013) and learners. The marker recognition problem of an AR system is one of the main challenges that arise during the use of AR in learning environments. In fact, a study (Ibáñez et al., 2014) indicated the marker recognition problem as the most common problem they had encountered. On the other hand, other main problems reported by studies conducted on marker-based AR are loss of AR content while shaking the tablet (Sommerauer & Müller, 2014; Chang et al., 2014; Ke & Hsu, 2015), photosensitivity problem of cameras (Chang et al., 2015), and the necessity of shaking the tablets stopping the AR experience (Sommerauer & Müller, 2014).

The carrying requirement of tablets, which are heavier compared with mobile phones, while managing digital elements in AR applications is one of the most significant challenges experienced while using this technology. Especially in some of the Museum Studies (e.g., Sommerauer & Müller, 2014; Chang et al., 2014) related to AR, it has been noted that users have to carry the tablets with them while visiting the museum and to hold the tablets up and down in order to recognize the markers. Similar challenges can also be experienced in location-based AR application (e.g., Harley, Poitras, Jarrell, Duffy, & Lajoie, 2016) such that AR devices can be a burden during strolling for users who want to see digital AR assets overlapped on real assets in the environment. Problems related to the GPS accuracy of mobile devices also create similar challenges in terms of location-based AR applications as the locations of the AR objects cannot be determined outdoors in their real coordinates. As it is the case in classic measuring methods, there are several factors (environmental factors, the number of satellites accessed, distance from stable receiver, the quality of the GPS receiver, etc.) influencing the accuracy of GPS (Yalcin, Inal, Kalayci, & Tarihi, 2008). Thus, while designing AR applications the sizes of learning objects and the distances among them need to be taken into consideration (Chiang et al., 2014).

On the other hand, other problems encountered in AR applications are as follows: the necessity to use both the tablets and the physical elements at the same time (Ibáñez et al., 2014), the fact that students using the AR application do not have tablets having similar features, the fact that devices/ equipment used to visualize the AR content are expensive (Huang et al., 2016), the fact that students encounter problems while devices are prepared for AR environments (Chiang et al., 2014), parental resistance encountered against the use of such products in learning, and the fact that students, who are accustomed to use tablets at school or at home, are not accustomed to using these devices outdoors (Huang et al., 2016).

DEVELOPMENT PROCESS OF AN AR APPLICATION AT BASIC LEVEL

In this section, besides introducing the required tools to develop an educational AR application, the steps to develop an exemplary AR book with the help of the currently most preferred development tools will be described.

Tools Required for Developing an AR Application

Today, numerous AR application development tools are available. While commercial licenses are required to use these tools, some of them can be used as free trials. Information about five of these developing tools is provided in Table 1.

ARPA SDKs

ARPA SDK® is software that can be used to develop interactive AR applications for geographical applications, advertising, marketing campaigns, museums, games, and so on. Besides supporting various 3D file formats with multimedia content such as text, audio, video, 2D and static/ animated 3D content, the ARPA software platform enables the development of various tracking applications (high-precision

Table 1. AR development tools that can be used to develop applications for smartphones, tablets, and smart goggles (Roukounaki, 2015)

SDKs (Software Development Kit)	Company	License	Supported Platforms
ARPA SDKs	Arpa Solutions	Commercial*	Android, iOS (ARPA SDKs), Google Glass (ARPA GLASS SDK), Android, iOS, Windows PC (ARPA Unity Plugin)
ARLab SDKs	ARLab	Commercial	Android, iOS
DroidAR	--	Free and Commercial	Android
Metaio SDK	Metaio	Free and Commercial	Android, iOS, Windows PC, Google Glass, Epson Moverio BT-200, Vuzix M-100, Unity
Vuforia SDK	(NASDAQ: PTC)	Free and Commercial	Android, iOS, Unity
Wikitude SDK	Wikitude GmbH	Commercial*	Android, iOS, Google Glass, Epson Moverio, Vuzix M-100, Optinvent ORA1, PhoneGap, Titanium, Xamarin

* A free trial version is also available

marker, 2D/ 3D markerless, 3D face, infrared, sensor & mechanical, step motors, and GPS). ARPA SDK's professional version has image recognition, image detection and tracking, multidetection, and multitracking features. Furthermore, it enables the development of applications for Android, iOS, Google Glass, and Windows platforms. ARPA SDK®'s AR GPS SDK version supports the positioning feature along with GPS and sensor tracking. To use all functions of the ARPA SDK®, its commercial license must be purchased. However, there is also a free trial available for those who are not developing AR applications commercially.

ARLab SDKs

ARLab offers two different SDK (AR Browser and Image Matching) services. *AR Browser SDK* is software enabling the addition of the AR geolocation view feature to Android or IOS applications. This software, offering the opportunity to add or remove POIs (Point of Interests) in real time from scene, has a very simple interface API for developers. *Image Matching SDK* is real-time image recognition engine software to create apps for Android and iOS platforms. This SDK allows matching thousands of images within devices without an Internet connection. Besides these two SDKs, ARLab offers services like its image tracking engine, object tracking, virtual buttons, and 3D engine for AR application developers for Android and IOS.

DroidAR

DroidAR is a framework that allows the creation of location- and marker-based AR applications. DroidAR offers free SDK service to develop non-commercial AR applications, which enables the development of applications exclusively for Android platforms.

Metaio SDK

Metaio SDK allows the development of AR applications using 3D and SLAM (Simultaneous Localization and Mapping) tracking, markerless 3D tracking, bar code and QR code scanning features. Applications can be published offline for kiosks, desktops, or mobile devices via Metaio SDK. It supports Android, IOS, Windows PC, Google Glass, Epson Moverio BT-200, Vuzix M-100, and Unity platforms (Roukounaki, 2015).

Vuforia SDK

AR applications can be built for mobile devices and digital eyewear via Vuforia SDK. Developers can build these applications using Java and C++ programming languages with Android Studio and Unity. Vuforia SDK allows the addition of VR (Virtual Reality) features to existing AR applications. Vuforia SDK enables users to develop applications for Android, iOS, and Unity platforms, offering several kinds of target supports (e.g. Image Targets, Cylinder Targets, Multi Targets). On the other hand, Vuforia SDK has object, text, and cloud recognition features. Further, developers can create virtual touch-enabled AR applications by using the virtual button feature of Vuforia SDK. This SDK can also provide support for Microsoft Surface and HoloLens, which are currently the most innovative devices. Complexity in creating new 3D content, a labor-intensive development process, and prohibitive costs of the AR deployment on a large scale are important challenges to address. However, the company PTC (NASDAQ: PTC) announced that it has developed a new addition called *Vuforia Studio Enterprise* to meet these challenges. The company claims that Vuforia Studio Enterprise will enable users to quickly build an AR experience from existing 3D contents – whether created with PTC or third-party tools – and AR applications developed in this way can be published to the cloud in only a few steps.

Wikitude AR SDK

Developers can create vision-based and location-based augmented reality applications with Wikitude AR SDK. Location-based tracking, image recognition, 3D model rendering and video overlays are some of the features of this SDK. Platforms supported by Wikitude AR SDK are as follows: Android, IOS, Google Glass, Epson Moverio, Vuzix M-100, Optinvent ORA1, PhoneGap, Titanium, and Xamarin (Roukounaki, 2015).

AR Development Steps with Vuforia SDK and Unity

In this section, the steps to develop an educational AR application with Unity and Vuforia SDK will be presented.

Setting Up Unity

Unity is one of the most popular game development platforms preferred to develop real-time desktop or mobile games. Moreover, it enables the development of AR, VR (Virtual Reality), or MR (Mixed Reality) applications with the help of development packages like Vuforia SDK, Arpa SDK, and Metaio SDK. Unity also enables the user to choose from the C# or JavaScript languages to develop applications. Unity has personal, plus, professional and enterprise versions available. The commercial license must be purchased for all versions except the personal version. With Unity's free trial personal version, access is available to all engine features and continuous updates so as to develop non-commercial AR application for IOS, Android, or desktop PCs. All versions of Unity can be downloaded from its website (https://unity3d.com/) (Figure 1). One of the issues to consider here is choosing a Unity version that is compatible with the all features of the AR development packages (e.g., Vuforia) to be used together with Unity. For example, Vuforia SDK's 5.5.9 version launched at the time of preparation of this chapter was supported by Unity's 5.2.4 and lower versions.

Unity's free trial version requires free membership to have access to Unity's services and resources after installation. It is possible to develop applications offline using Unity (Figure 2).

Upon completing installation and establishing membership, Unity will be ready to develop AR applications with its 2D and 3D application development environment (scene), *game panel* (to be used to view AR applications built), *fundamental objects* (e.g., main camera, directional light etc.) and the *Inspectorpanel* showing the features of these objects (Figure 3).

Importing the AR Development SDK (Vuforia SDK) to Unity

Unity supports software platforms such as ARPA SDK, Metaio SDK, and Vuforia SDK to develop AR applications. In this section, the use of the Vuforia software platform enabling users to develop applica-

Figure 1. Download page for Unity

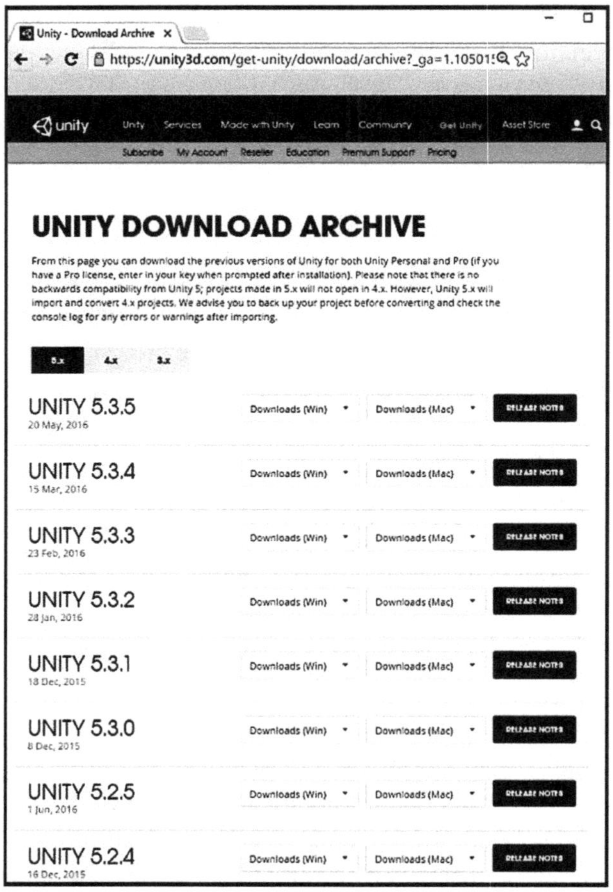

Figure 2. Unity account form

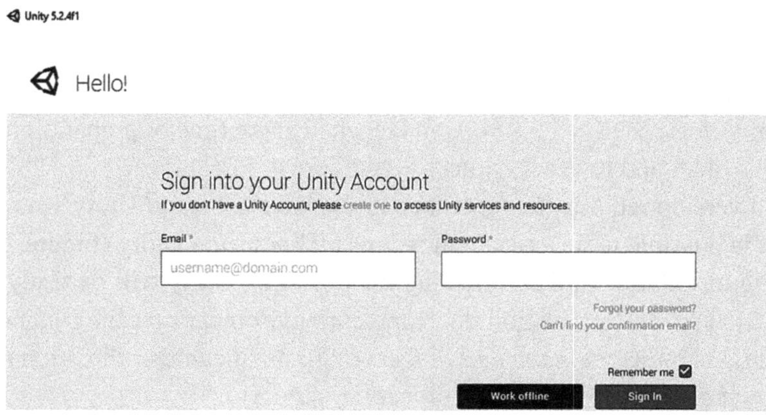

Figure 3. Unity's 2D and 3D application development environment, tools, and inspector panel

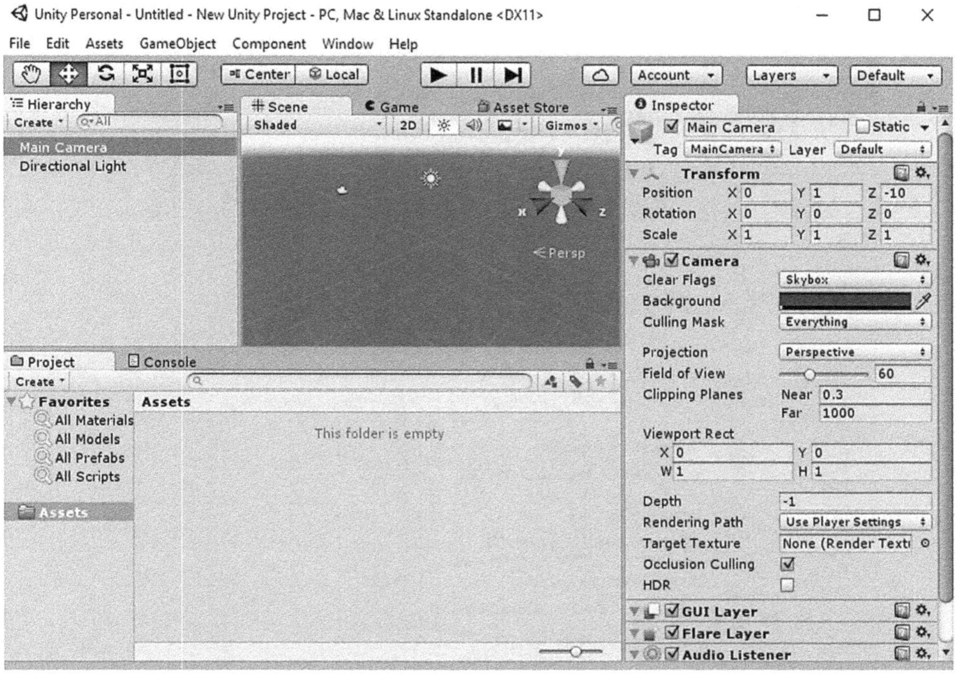

tions for innovative devices such as Microsoft Surface and HoloLens will be introduced. Vuforia SDK can be downloaded by visiting the site https://developer.vuforia.com and clicking the *download for Unity* link from the Vuforia developer portal (Figure 4). Before starting to develop an AR application, the downloaded Vuforia SDK package must be imported into Unity. To accomplish this, you can click *Import Package → Custom Package* options from the *Assets* menu in Unity and follow the steps in Figure 5.

After having imported the Vuforia SDK package into Unity, the AR application can now start to be developed. For this, some prefabs[1] (e.g. Image-Targets, Multi-Targets, Cylinder-Targets, text recognition, etc.) (Figure 6) required for recognition and tracking will be used with the AR Camera located within the Vuforia package. Vuforia enables the use of recognition and tracking features on various images and objects with the help of these prefabs. Image-Targets are 2D flat images (e.g., images like print media and product packages used in our daily lives). Vuforia's Image-Targets do not have to have special black and white regions or codes as in the traditional fiducial markers. Multi-Targets are structures combined in a geometrical arrangement and consist of more than one Image-Target (Figure 7a). Moreover, they can be arranged as smooth geometric shapes (e.g., Boxes). All surfaces of a Multi-Target can be tracked simultaneously. Cylinder-Targets enable the detection and tracking of images rolled around cylindrical and conical shapes (e.g., soda cans, coffee cups, mugs and beverage bottles) (Figure 8a, b). Text Recognition is used to develop applications that can recognize words in a dictionary covering approximately 100,000 English words. Finally, Object Recognition can be used to build apps that recognize and track rigid objects, as well.

Figure 4. Download page for Vuforia SDK

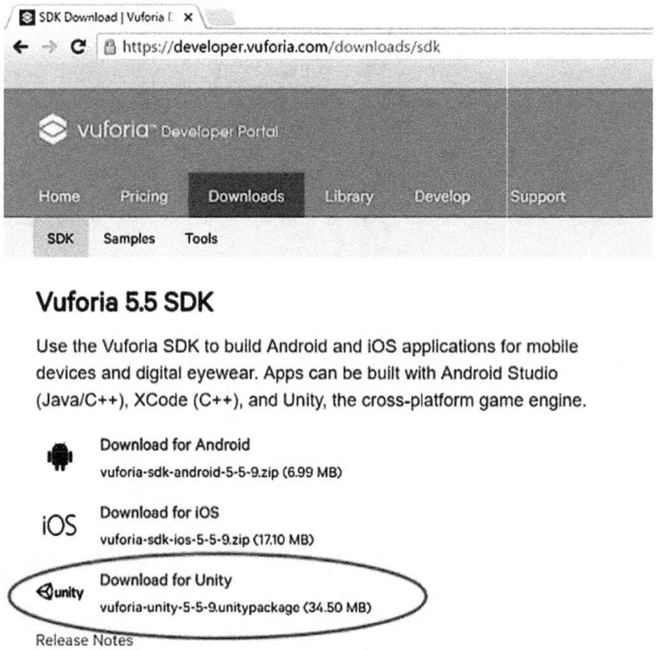

Figure 5. (a), (b) Steps to import Vuforia SDK into Unity and (c) Vuforia SDK assets

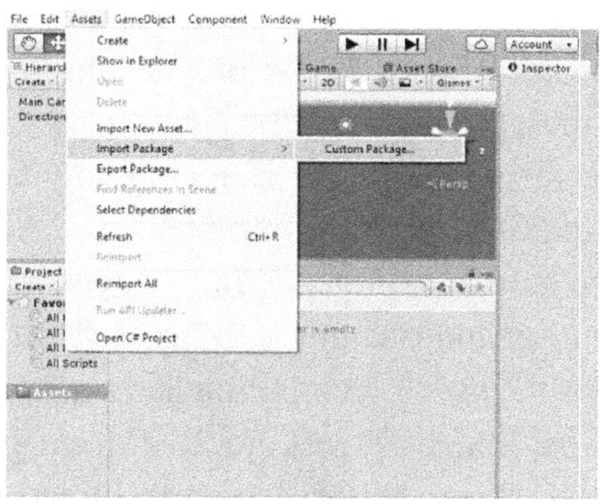

(a) Step 1 – Click Assets → Import Package → Custom Package

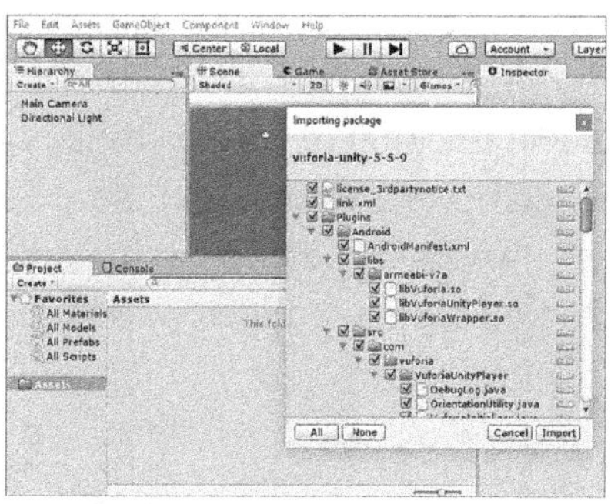

(b) Step 2 - Click import button

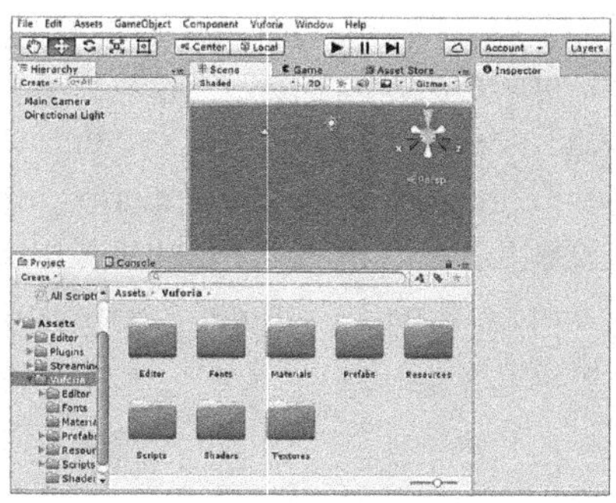

(c) Vuforia SDK assets

Figure 6. Prefabs available within the Vuforia package

Figure 7. (a) Image-Target sample and (b) 3D objects overlapped on Image-Target

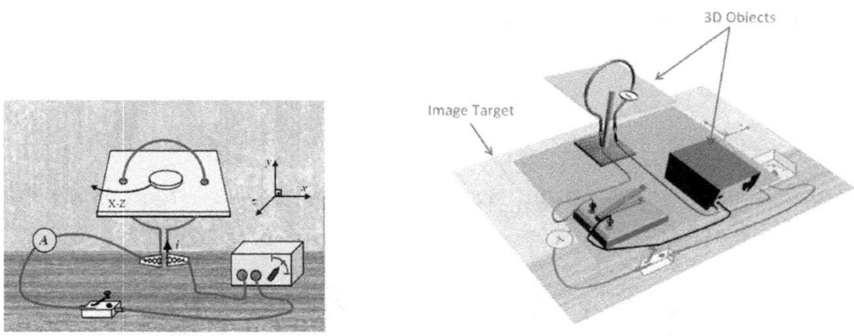

(a) (b)

Figure 8. (a) Multi-Target and (b) Cylinder-Target

(a) (b)

Creating an Exemplary Book Page for AR Application

Let us create an exemplary book page in order to examine the AR application to be prepared. Images from existing books can be used as Image-Targets, and markers (Figure 9) generated randomly from some websites (e.g., http://www.brosvision.com/ar-marker-generator/) providing service as a marker generator can also be used as Image-Targets. An image (Figure 11a) located on the right bottom corner of the sample page (Figure 10) to be used for this AR application will be tracked via AR imaging devices (web camera, mobile phone, etc.), and some 3D items (Figure 11b) will be viewed on these images.

Let Us Begin to Develop the AR Application

First, run the Unity program. Then, remove the *Main Camera* located as default on the scene and place the *ARCamera* in the *Prefabs* folder located within the Vuforia package to the *Hierarchy* panel. Procedures that should be followed to perform this action are as follows:

- Select the Main Camera on the Hierarchy panel
- Press the delete key on keyboard (Figure 12a)
- Select the ARCamera in the Vuforia's prefabs folder
- Drag and drop ARCamera to the Hierarchy panel situated on the Unity Window (Figure 12b)

After placing the *ARCamera* onto the scene, add the Image-Target located in the Vuforia's *Prefabs* folder onto the scene. Then, assign a name to this item to permit easy recognition at further stages of the application. Procedures to be performed for this action:

Figure 9. A marker randomly generated from a website (http://www.brosvision.com/ar-marker-generator/) providing service as marker generator

Figure 10. An exemplary book page for an AR application and an Image-Target located on the right bottom corner of the page

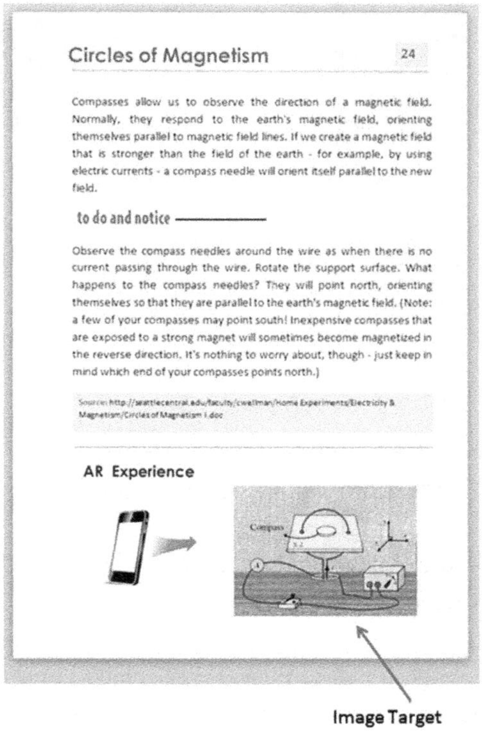

Circles of Magnetism 24

Compasses allow us to observe the direction of a magnetic field. Normally, they respond to the earth's magnetic field, orienting themselves parallel to magnetic field lines. If we create a magnetic field that is stronger than the field of the earth - for example, by using electric currents - a compass needle will orient itself parallel to the new field.

to do and notice

Observe the compass needles around the wire as when there is no current passing through the wire. Rotate the support surface. What happens to the compass needles? They will point north, orienting themselves so that they are parallel to the earth's magnetic field. (Note: a few of your compasses may point south! Inexpensive compasses that are exposed to a strong magnet will sometimes become magnetized in the reverse direction. It's nothing to worry about, though - just keep in mind which end of your compasses points north.)

Source: http://seattlecentral.edu/faculty/cwellman/Home Experiments/Electricity & Magnetism/Circles of Magnetism 1.doc

AR Experience

Figure 11. An Image-Target to be used in the AR application and (b) 3D items to be overlapped on this Image

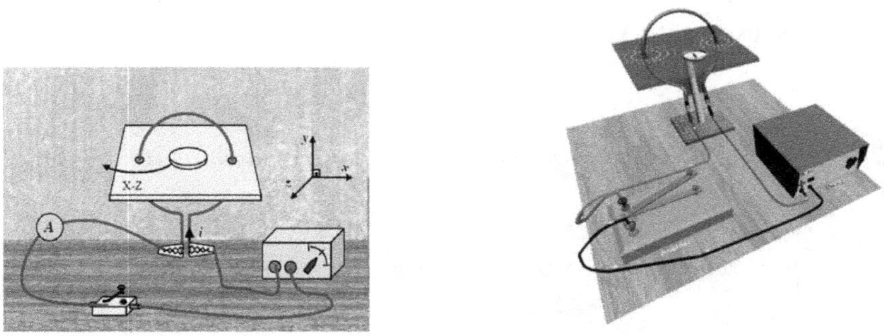

(a) (b)

Figure 12. Placement of the ARCamera on scene

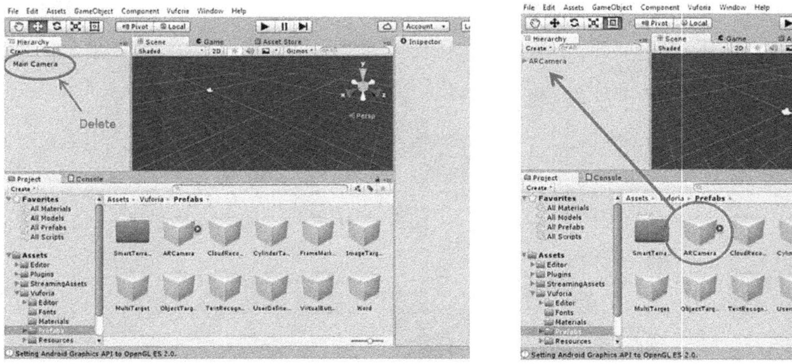

(a) (b)

- Select the Image-Target from the Vuforia's Prefabs folder,
- Drag and drop the Image-Target onto the Hierarchy panel (Figure 13a)
- Right click on the Image-Target (Figure 13b)
- Select Rename from the list and change the name of the Image-Target to "Circles_of_magnetism_Image_target"

Now it is time to define an image for the Image-Target placed onto the scene. As mentioned previously, this image will be the one indicated on our exemplary book page (Figure 10). Image-Targets can be generated in a web-based manner via the Vuforia Target Manager (Figure 14) located in the Vuforia Developer Portal. The generated Image-Targets can be updated using this manager. Vuforia Target Manager supports JPG or PNG images with a maximum size of 2MB in RGB or Grayscale.

The features of images imported to Vuforia Target Manager are stored in a database. This database is populated by the Target Manager and becomes ready to be downloaded. Then this database consisting of image features can be used with Vuforia by being imported into Unity.

Figure 13. (a) Placing the Image-Target onto the scene and (b) renaming it

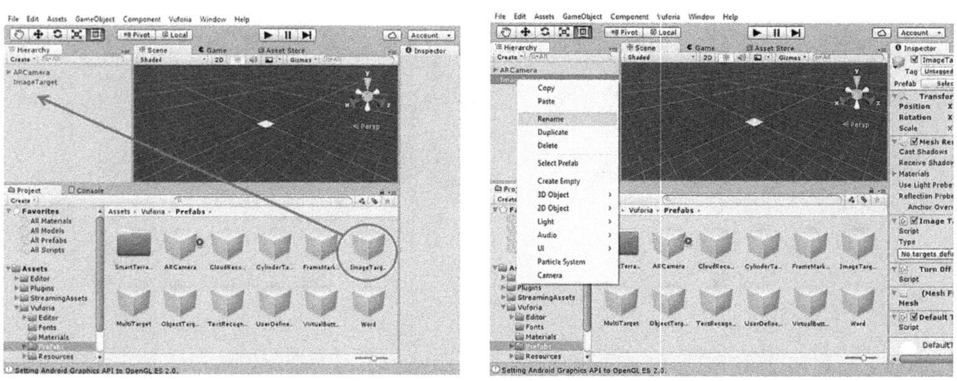

(a) (b)

Figure 14. Vuforia Target Manager

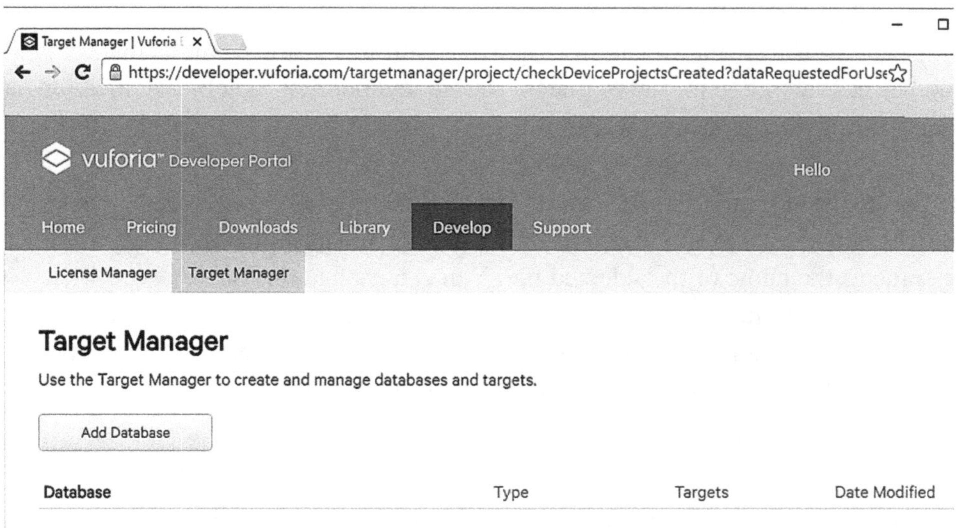

Generating an Image Target with Vuforia Target Manager

To begin working with the Target Manager, you will need a Vuforia developer account. If you do not have one, you can register for free at https://developer.vuforia.com. A database needs to be created before generating an Image-Target with Vuforia Target Manager. Procedures to be performed for this action:

- Click on the *Add Database* button on the Vuforia Target Manager page (Figure 14).
- Write the name of the database to be created from *Create Database* window (Figure 15a)
- Select *Device*[2*] as *Type* (*Cloud database* will not be used in this AR application).

After having created the database in Target Manager, you need to upload the image to be used as Image-Target into this database. Procedures to perform for this action:

Figure 15. (a) Creating a database in Target Manager and (b) the database created

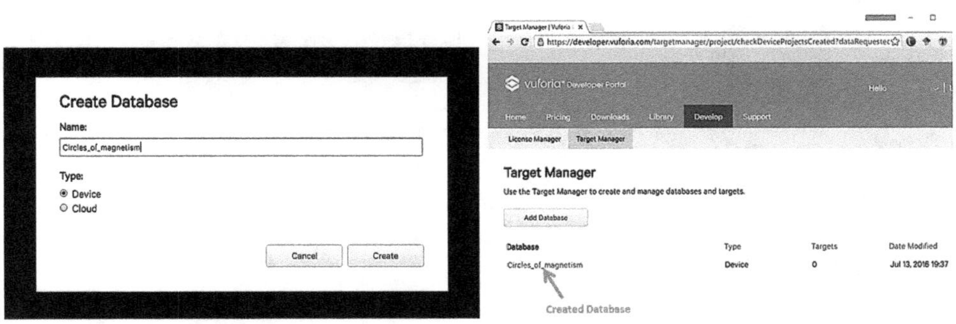

(a) (b)

- Click on the database created in the Target Manager ("Circles_of_magnetism" in Figure 15b). You will be directed to the page shown in Figure 16a.
- Click on the *Add Target* button on this page. You will see a form as indicated in Figure 16b.
- We can define the type (e.g., single image, cuboid, cylinder, or 3D object), location, and width of our target with the help of this form.
- *Single Image* will be defined as *type* for our application.
- Select the image (Figure 17a) defined as Image-Target by the computer by clicking the *Browse* button. The *Name* section located on the bottom of the form labeled *Add Target* will automatically be the same as the name of the selected file. You can change this name if you wish.
- In the scene units, write the width of the target within the *Width* box (let us define the width value as 20 for our application). This target's width size should be on the same scale as our augmented content. After the image has been uploaded, the target`s height size will be determined automatically by the Target Manager.

Figure 16. (a) The target generated with Target Manager and (b) Add Target form

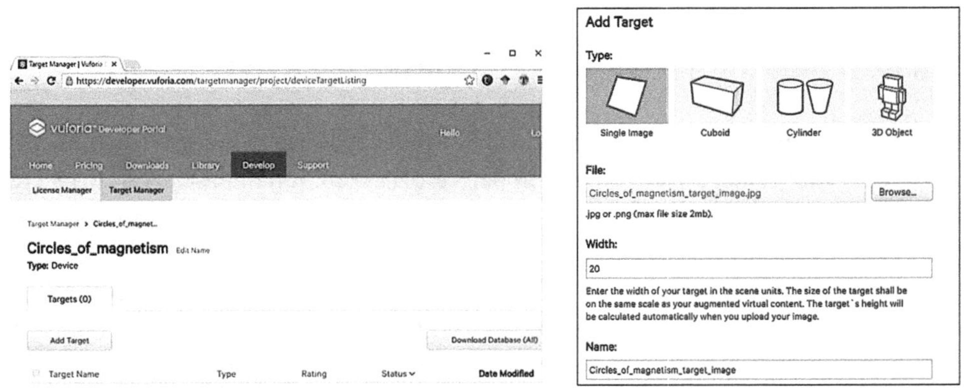

(a) (b)

Figure 17. (a) The image file defined as Image-Target in our AR application (jpeg file format and 118 KB in size) and (b) A target image generated with the Target Manager

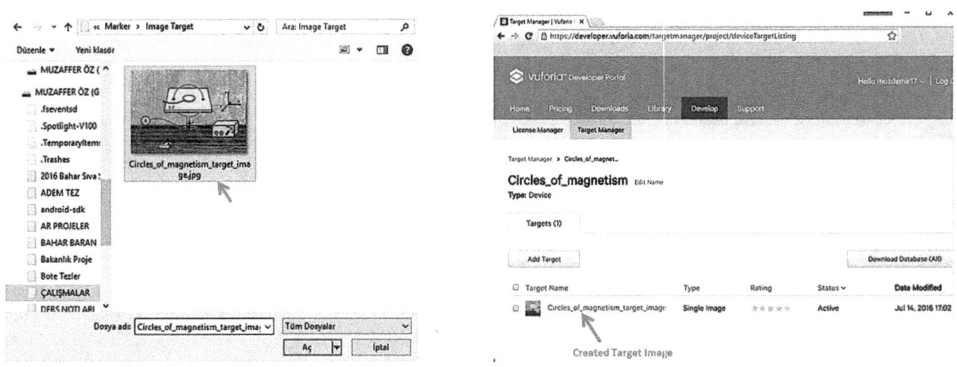

(a) (b)

As mentioned previously, the Target Manager packages the created image databases to be used in Unity and makes these packages available for download. Now let us package and then download the Image-Target (Figure 17b) we have just generated. Procedures to perform for this action:

- Select the generated Image-Target and click on the button *Download Database (All)* on the Target Manager window (Figure 18a)
- As shown in Figure 18b, the *Download Database* form will appear. Select the *Unity Editor* as the *development platform* and click the *Download* button.
- Let us keep the downloaded file named "Circles_of_magnetism.unitypackage" to be used in the later stages of the application.

Generating a License Key with Vuforia License Manager

A license key consisting of an alpha-numeric code is required in order to build all applications with Vuforia.

If AR applications to be developed will be published for more than one operating system (e.g., Android and iOS), you can use the same License Key for all of these applications. Vuforia offers three license options: Starter, classic, and cloud. With the starter license, without paying any fee, you can have access to all Vuforia platforms and develop demo applications by using just a single watermark (Figure 19). The other two licenses involve charges. You can obtain more information about other license options from the Vuforia developer portal (https://developer.vuforia.com/).

Figure 18. (a) The image file defined as Image-Target in our AR application and (b) Download Database form

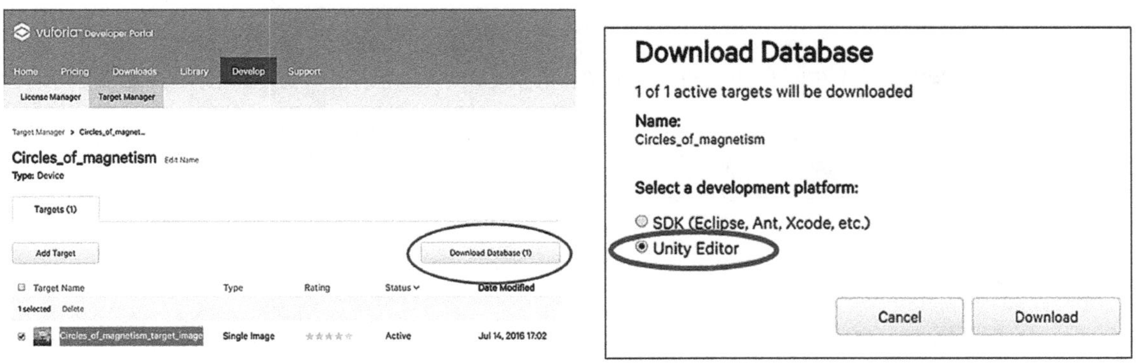

(a) (b)

Figure 19. Sample watermark used by Vuforia on the AR App preview

License manager (Figure 20) in Vuforia developer portal can be used to generate and manage the License Keys. Procedures to perform for this action:

- Click on the *Add License Key* button in the Vuforia License Manager window (Figure 20).
- Write an application name for the *License Key* to be obtained on the *Add License Key* form (Figure 21a). The name can be changed later.
- Select the device to be used for the application (select *mobile* for our example).
- Select a *license key* (select *starter* for our example).
- Click the *Next* button.
- Read and confirm the *terms and conditions of the Vuforia Developer Agreement* on the *Confirm License Key* form (Figure 21b). At the end of these processes, your license key will be created successfully as shown in Figure 22a and b.

Figure 20. Vuforia License Manager

Figure 21. Add License Key and Confirm License Key forms

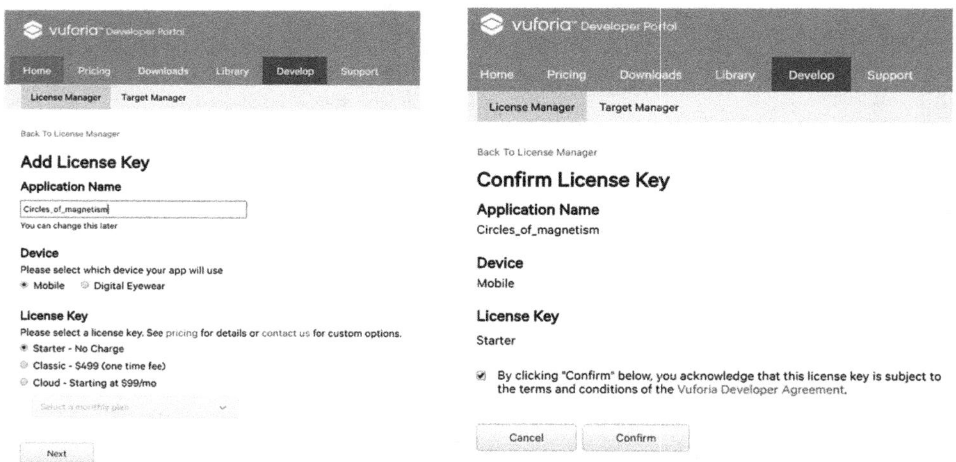

(a) (b)

Figure 22. (a) A license key generated with the Vuforia License Manager and (b) license key code

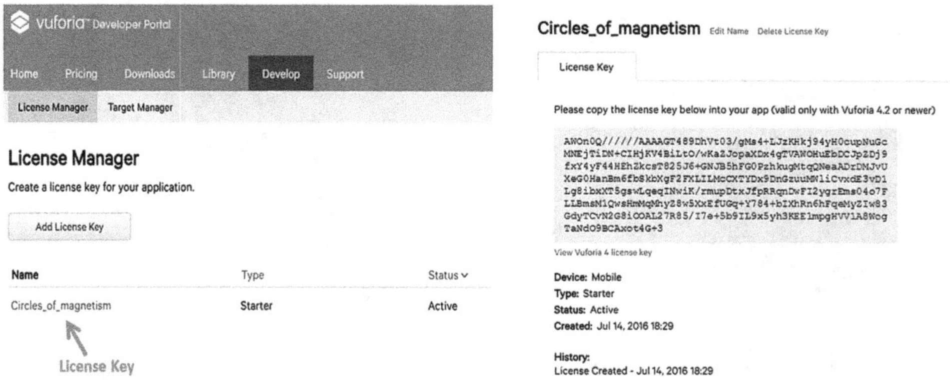

(a) (b)

Let Us Continue with Unity

We have previously placed the Image Target found in the Vuforia *Prefabs* folder onto the scene and named it as "Circles_of_magnetism_Image_target" so that we could easily recognize it. Now we need to define an image database for this Image-Target. To do this, we need to import the *Circles_of_magnetism_Image_target* file into Unity. Procedures to perform for this action:

- Click *Import Package → Custom Package* options from the *Assets* menu in Unity (Figure 23a).
- Select the *Circles_of_magnetism_Image_target* file.
- Click the *import* button on the *importing package* dialog box (Figure 23b).

Figure 23. Steps to import the image database to Unity

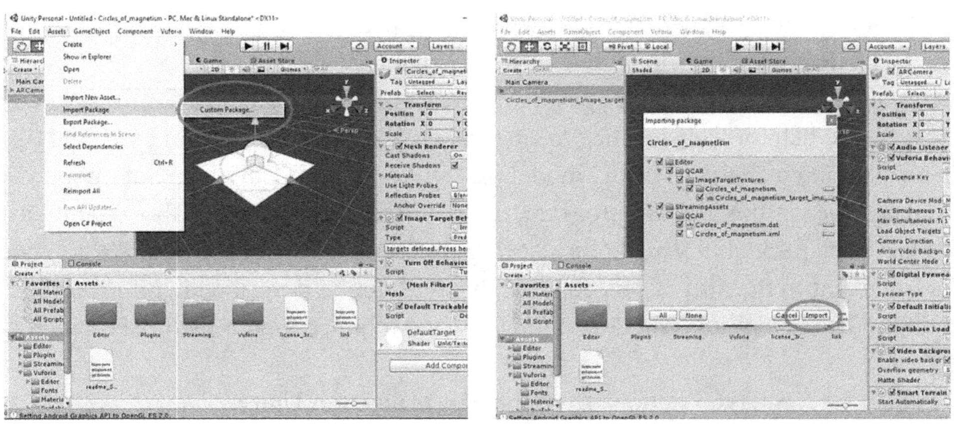

(a) (b)

Figure 24. Assigning an image to the Image-Target

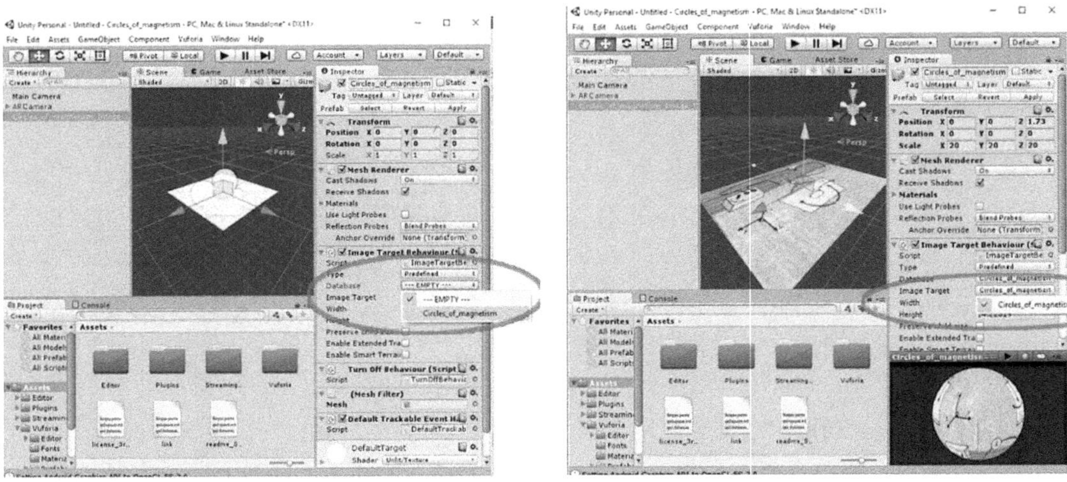

(a) (b)

Now, let us define the image database imported to Unity to assign an image to the *Circles_of_magnetism_Image_target* located in the scene. Procedures to perform for this action:

- Select *Circles_of_magnetism_Image_target* in Unity.
- Select *Circles_of_magnetism* from the combo box named *database* in the *Image Target Behaviour* section situated on Unity's *Inspector* panel (Figure 24a).
- Select *Circles_of_magnetism_target_image* from the combo box named *database* in the *Image Target Behaviour* section located on Unity's *Inspector* panel (Figure 24b).

Placing 3D Objects on Image Target

We have previously defined a database and image for the Image-Target to be tracked. Now it is time to import the 3D objects into Unity. These 3D objects will appear in the result of tracking the Image-Target on our exemplary AR book page via mobile devices (Figure 25a).

3D objects can be created via various programs such as 3DMax, Blender, Softimage, ZModeler, and SketchUp or with the help of websites like assetstore.unity3d.com, turbosquid.com, and tf3dm.com for free or with an associated charge. The 3D model that will be used in our exemplary book page has been created by our author via the 3DMax program, the required animations for this model have been prepared, and it has been packed in *fbx* file format (Figure 25b) and then exported. As the model creation stages are not relevant to the aim of this chapter, this will not be described here. Procedures needed to import the 3D model shown in Figure 25a into Unity and to place it onto the image target in the scene:

- Click *Import* → *New Asset* options from the *Assets* menu in Unity.
- Select the *fbx* file of the 3D model to be used for the AR application.[3*]
- As shown in Figure 26d, our 3D model will be placed on the Image-Target.
- Finally, adjust the size and coordinates of the 3D object according to the Image-Target's size and position. You can use the *scale* and *position* tools of Unity to perform this action.

Figure 25. (a) 3D model and Image-Target and (b) The fbx file of the 3D model and its associated files

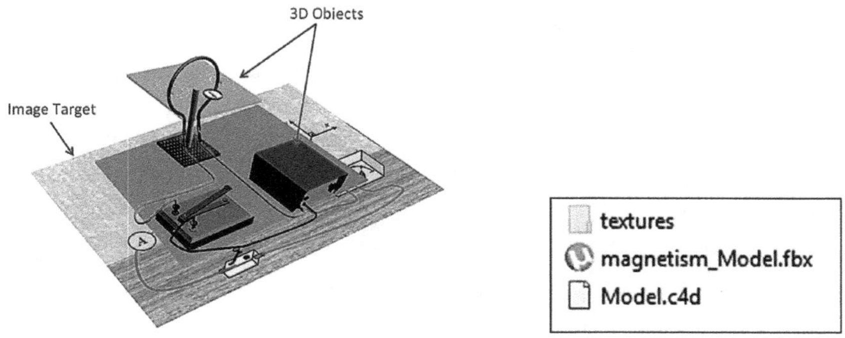

(a) (b)

Figure 26. Steps to import the 3D model into Unity

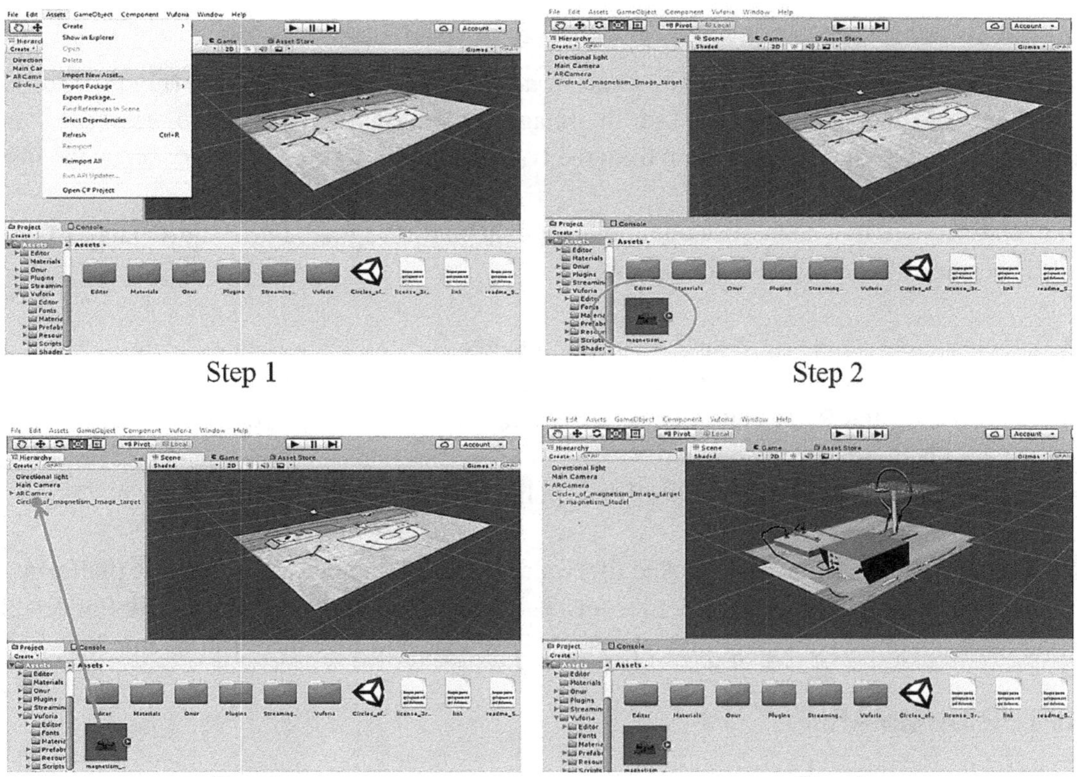

Step 1 Step 2

Step 3 Step 4

Adjusting ARCamera

Now let us adjust the settings of the ARCamera placed onto the scene. First of all, it is essential to note that you can place as many Image-Targets as the ARCamera can display. However, you need to increase the value in the *Max Simultaneous Tracked Images* box from the ARCamera's *Inspector* panel (Figure 27a). Another setting for the ARCamera is to check the confirmation dialog boxes *Load Circles_of_magne…*

Figure 27. ARCamera settings

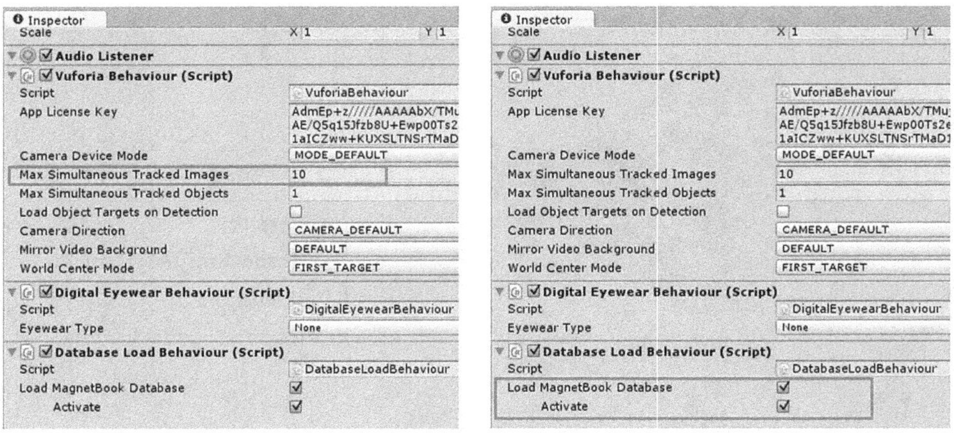

(a) (b)

Database and *Activate* located in the *Database Load Behaviour (Script)* section (Figure 27b). Consequently, the image database is assigned for the Image-Target. Finally, you need to select the camera to be used to overview your AR application from the *Web Cam Behaviour (Script)* section on the *Inspector* panel of ARCamera.

Now let us explain how to publish our application on platforms (e.g., IOS, Android, etc.) supporting the AR. Procedures to perform for this action:

- Click on *Build Settings* from the *File* menu in Unity.
- Select the platform (IOS or Android) to be published from the *Platform* section.
- Before clicking the *Build* button, press the *Player Setting* button to perform the required settings of the application platform.
- Finally, click the *Build* button to pack our application to be published.

Now, we can try our application packed for Android or IOS on devices using these platforms. After having installed the application on a mobile device, hold these devices on the image defined as Image-Target on our exemplary book page and observe the placement of our 3D object on this image.

CHAPTER SUMMARY

The increasing use of the internet and mobile technologies has accelerated the sharing of information in daily life. This facility has also created a pile of both useful and useless information that can easily be accessed. Today, filtering, modifying, and presenting the useful information in an effective way among this huge pile has become a great necessity for educators. In this sense, the opportunities that AR technologies present reveal practical ways to meet those needs of educators. It is possible to say that AR will become an indispensable technology for educational contexts in the future, since this technology not only presents an innovational approach for multimedia-supported learning environments but it can also be used with both mobile and non-mobile devices. Also, by integrating the digital information with real-world assets simultaneously, AR helps to concretize abstract concepts, enables the use of all senses, and enhances the sense of reality, which in turn is a huge contribution to learning.

Besides its contribution to education, this chapter presents the benefits and limitations of AR presented by empirical studies. These studies presented evidence that AR is not only useful in enhancing attention and concentration but also it provides a sense of enjoyment, helps in constructing the information, supports cooperative learning, enables interactive participation, facilitates learner-centered activities, and enhances memory.

These empirical studies also revealed various limitations of AR along with its advantages. In most of the studies, marker recognition is reported as the major problem. It is apparent that this problem can be solved by the improvements in image processing techniques. Although there has been remarkable recent progress in image processing techniques, there is a need for further improvements for innovative technologies such as AR and VR. Also, the practicality problem of AR imaging devices, the high costs of these devices, and the fact that students are not accustomed to using these devices outdoors are among the major problems reported. Although they provide a wider display area compared with other mobile devices, tablets are heavy and large in size and hence they represent the main challenge in terms of the keeping and carrying problems. In the near future, the downsizing of head-mounted virtual imaging devices such as Hololens and the widespread of use of them could be a solution both for keeping and carrying problems.

This chapter also presents information about AR development tools, along with steps for developing an educational AR application. Today, program developers face many problems since AR development tools are relatively new to them. Incompatibilities between developing programs (e.g., Unity) and add-on packages (e.g., Vuforia) are the major problems among them because these programs and add-on packages are usually produced by different companies. In order to overcome this problem, AR development environments could be designed to be compatible.

As an example, this chapter presents steps for developing an AR application. Unity was used as the development program, while Vuforia was utilized as the add-on pack. These two development tools are among the most popular instances of AR development software today. The AR application developed as a demo is based on a single page of a book, which can be generalized to the whole book. Furthermore, this AR application can be used by mobile devices and desktop computers, and it can be presented for the use of other users via markets.

REFERENCES

Azuma, R., Baillot, Y., Behringer, R., Feiner, S., Julier, S., & MacIntyre, B. (2001). Recent advances in augmented reality. *IEEE Computer Graphics and Applications*, *21*(6), 34–47. doi:10.1109/38.963459

Azuma, R. T. (1997). A survey of augmented reality. *Presence (Cambridge, Mass.)*, *6*(4), 355–385. doi:10.1162/pres.1997.6.4.355

Bokyung, K. (2009). Investigation on the relationships among media characteristics, presence, flow, and learning effects in augmented reality based learning. In P. A. Bruck (Ed.), *Multimedia and E-Content Trends* (pp. 21–37). Wiesbaden: Vieweg Teubner; doi:10.1007/978-3-8348-9313-0_3

Chang, K. E., Chang, C. T., Hou, H. T., Sung, Y. T., Chao, H. L., & Lee, C. M. (2014). Development and behavioral pattern analysis of a mobile guide system with augmented reality for painting appreciation instruction in an art museum. *Computers & Education, 71*, 185–197. doi:10.1016/j.compedu.2013.09.022

Chang, Y. L., Hou, H. T., Pan, C. Y., Sung, Y. T., & Chang, K. E. (2015). Apply an augmented reality in a mobile guidance to increase sense of place for heritage places. *Journal of Educational Technology & Society, 18*(2), 166–178.

Chen, C. M., & Tsai, Y. N. (2012). Interactive augmented reality system for enhancing library instruction in elementary schools. *Computers & Education, 59*(2), 638–652. doi:10.1016/j.compedu.2012.03.001

Chen, C. P., & Wang, C. H. (2015). Employing augmented-reality-embedded instruction to disperse the imparities of individual differences in earth science learning. *Journal of Science Education and Technology, 24*(6), 835–847. doi:10.1007/s10956-015-9567-3

Chiang, T. H., Yang, S. J., & Hwang, G. J. (2014). Students online interactive patterns in augmented reality-based inquiry activities. *Computers & Education, 78*, 97–108. doi:10.1016/j.compedu.2014.05.006

Di Serio, Á., Ibáñez, M. B., & Kloos, C. D. (2013). Impact of an augmented reality system on students motivation for a visual art course. *Computers & Education, 68*, 586–596. doi:10.1016/j.compedu.2012.03.002

Haag, J. (2013). Using augmented reality for contextual mobile learning. *Learning Solutions Magazine.* Retrieved January 7, 2016, from https://www.learningsolutionsmag.com/articles/1310/using-augmented-reality-for-contextual-mobile-learning

Han, J., Jo, M., Hyun, E., & So, H. J. (2015). Examining young childrens perception toward augmented reality-infused dramatic play. *Educational Technology Research and Development, 63*(3), 455–474. doi:10.1007/s11423-015-9374-9

Harley, J. M., Poitras, E. G., Jarrell, A., Duffy, M. C., & Lajoie, S. P. (2016). Comparing virtual and location-based augmented reality mobile learning: Emotions and learning outcomes. *Educational Technology Research and Development, 64*(3), 359–388. doi:10.1007/s11423-015-9420-7

Hsiao, H. S., Chang, C. S., Lin, C. Y., & Wang, Y. Z. (2016). Weather observers: A manipulative augmented reality system for weather simulations at home, in the classroom, and at a museum. *Interactive Learning Environments, 24*(1), 205–223. doi:10.1080/10494820.2013.834829

Hsiao, K. F., Chen, N. S., & Huang, S. Y. (2012). Learning while exercising for science education in augmented reality among adolescents. *Interactive Learning Environments, 20*(4), 331–349. doi:10.1080/10494820.2010.486682

Huang, T. C., Chen, C. C., & Chou, Y. W. (2016). Animating eco-education: To see, feel, and discover in an augmented reality-based experiential learning environment. *Computers & Education, 96*, 72–82. doi:10.1016/j.compedu.2016.02.008

Ibáñez, M. B., Di Serio, Á., Villarán-Molina, D., & Delgado-Kloos, C. (2014). Experimenting with electromagnetism using augmented reality: Impact on flow student experience and educational effectiveness. *Computers & Education, 71*, 1–13. doi:10.1016/j.compedu.2013.09.004

Ibáñez, M. B., Di-Serio, Á., Villarán-Molina, D., & Delgado-Kloos, C. (2016). Support for Augmented Reality Simulation Systems: The Effects of Scaffolding on Learning Outcomes and Behavior Patterns. *IEEE Transactions on Learning Technologies*, *9*(1), 46–56. doi:10.1109/TLT.2015.2445761

Johnson, L., Adams Becker, S., Estrada, V., & Freeman, A. (2015). NMC Horizon Report: 2015 Library Edition. Austin, TX: The New Media Consortium. Retrieved from http://www.nmc.org/publication/nmc-horizon-report-2015-library-edition/

Ke, F., & Hsu, Y. C. (2015). Mobile augmented-reality artifact creation as a component of mobile computer-supported collaborative learning. *The Internet and Higher Education*, *26*, 33–41. doi:10.1016/j.iheduc.2015.04.003

Lin, T. J., Duh, H. B. L., Li, N., Wang, H. Y., & Tsai, C. C. (2013). An investigation of learners collaborative knowledge construction performances and behavior patterns in an augmented reality simulation system. *Computers & Education*, *68*, 314–321. doi:10.1016/j.compedu.2013.05.011

Mayer, R. E. (2002). Multimedia learning. *Psychology of Learning and Motivation*, *41*, 85–139. doi:10.1016/S0079-7421(02)80005-6

Mayer, R. E. (2005). Principles for Reducing Extraneous Processing in Multimedia Learning: Coherence, Signaling, Redundancy, Spatial Contiguity, and Temporal Contiguity Principles. In R. Mayer (Ed.), *The Cambridge Handbook of Multimedia Learning* (pp. 183–200). Cambridge, UK: Cambridge University Press. doi:10.1017/CBO9780511816819.013

Mayer, R. E. (2008). Applying the science of learning: Evidence-based principles for the design of multimedia instruction. *The American Psychologist*, *63*(8), 760–769. doi:10.1037/0003-066X.63.8.760 PMID:19014238

Moreno, R. (2006). Does the modality principle hold for different media? A Test of the method-affects-learning hypothesis. *Journal of Computer Assisted Learning*, *22*(3), 149–158. doi:10.1111/j.1365-2729.2006.00170.x

PTC Inc. (2016). *Vuforia Target Manager,* Retrieved August 18, 2016 from https://library.vuforia.com/articles/Training/Getting-Started-with-the-Vuforia-Target-Manager

Richards, D., & Taylor, M. (2015). A Comparison of learning gains when using a 2D simulation tool versus a 3D virtual world: An experiment to find the right representation involving the Marginal Value Theorem. *Computers & Education*, *86*, 157–171. doi:10.1016/j.compedu.2015.03.009

Roukounaki, K. (2014, Mar 16). *Top 5 Tools for Augmented Reality in Mobile Apps* [Weblog]. Retrieved August 18, 2016 from https://www.developereconomics.com/top-5-tools-for-augmented-reality-in-mobile-apps/

Santos, M. E. C., Chen, A., Taketomi, T., Yamamoto, G., Miyazaki, J., & Kato, H. (2014). Augmented reality learning experiences: Survey of prototype design and evaluation. *Learning Technologies. IEEE Transactions on*, *7*(1), 38–56.

Sommerauer, P., & Müller, O. (2014). Augmented reality in informal learning environments: A field experiment in a mathematics exhibition. *Computers & Education, 79*, 59–68. doi:10.1016/j.compedu.2014.07.013

Steuer, J. (1992). Defining virtual reality: Dimensions determining telepresence. *Journal of Communication, 42*(4), 73–93. doi:10.1111/j.1460-2466.1992.tb00812.x

Unity Technologies. (2016). *Prefabs*. Publication 5.4-U. Retrieved August 18, 2016, from http://docs.unity3d.com/Manual/Prefabs.html?_ga=1.244525509.1541532845.1453019544

Wang, H. Y., Duh, H. B. L., Li, N., Lin, T. J., & Tsai, C. C. (2014). An investigation of university students collaborative inquiry learning behaviors in an augmented reality simulation and a traditional simulation. *Journal of Science Education and Technology, 23*(5), 682–691. doi:10.1007/s10956-014-9494-8

Wei, X., Weng, D., Liu, Y., & Wang, Y. (2015). Teaching based on augmented reality for a technical creative design course. *Computers & Education, 81*, 221–234. doi:10.1016/j.compedu.2014.10.017

Wu, H. K., Lee, S. W. Y., Chang, H. Y., & Liang, J. C. (2013). Current status, opportunities and challenges of augmented reality in education. *Computers & Education, 62*, 41–49. doi:10.1016/j.compedu.2012.10.024

Yalcin, B., Inal, C., Kalayci, I., & Tarihi, M. G. (2008). GPS ölçü süresinin nokta konum doğruluğu ile ilişkisi. *J. Fac. Eng. Arch. Selcuk Univ, 23*(1).

Yoon, S. A., Elinich, K., Wang, J., Steinmeier, C., & Tucker, S. (2012). Using augmented reality and knowledge-building scaffolds to improve learning in a science museum. *International Journal of Computer-Supported Collaborative Learning, 7*(4), 519–541. doi:10.1007/s11412-012-9156-x

Zhang, J., Sung, Y. T., Hou, H. T., & Chang, K. E. (2014). The development and evaluation of an augmented reality-based armillary sphere for astronomical observation instruction. *Computers & Education, 73*, 178–188. doi:10.1016/j.compedu.2014.01.003

KEY TERMS AND DEFINITIONS

ARLab SDK: The ARLab SDK is software that offers the opportunity to add or remove POIs (Point of Interests) in real time from AR application scenes.

ARPA SDK: The ARPA SDK® is a software that can be used to develop interactive AR applications for geographical applications, advertising, marketing campaigns, museums, games, and so on.

Augmented Reality: The Augmented Reality is an innovative technology that enables the coexistence and real-time interactions of real-world beings with three-dimensional virtual objects (Azuma et al., 2001).

DroidAR: The DroidAR is a framework that allows the creation of location- and marker-based AR applications.

Metaio SDK: The Metaio SDK is software that can be used to develop AR applications using 3D and SLAM (Simultaneous Localization and Mapping) tracking, markerless 3D tracking, bar code and QR code scanning features.

Unity: The Unity is one of the most popular game development platforms largely preferred to develop real-time desktop or mobile games.

Virtual Reality: "Virtual reality is a real or simulated environment in which a perceiver experiences telepresence" (Steuer, 1992).

Vuforia SDK: The Vuforia SDK is software that can be used to build AR applications for mobile devices and digital eyewear.

Wikitude AR SDK: The Wikitude AR SDK is a software that can be used to create vision-based and location-based augmented reality applications.

ENDNOTES

1. Prefabs in Unity are preconfigured game objects created in the scene and stored in any Unity project (Unity Technologies 2016).
2. *Vuforia Target Manager supports the creation of two database types, Device and Cloud (PTC Inc, 2016); • Device Databases are local databases that are stored on the user's device. • Cloud Databases are stored online and queried over the internet when the user runs the AR application.
3. *You can also use other file formats (e.g., max, blend, obj, etc.) supported by Unity.

Chapter 3
Augmented Reality:
Opportunity for Developing Spatial Visualization and Learning Calculus

Patricia Salinas
Tecnologico de Monterrey, Mexico

ABSTRACT

This chapter intends to advance a sociocultural perspective to frame the integration of augmented reality technology for the learning of mathematics. It is discussed the design of an application already produced with this technology. The purpose of the application is to promote the development of spatial ability while also reviewing the visualization of some calculus topics at college. The author contends the benefits of focusing in the graphical representation to approach mathematics knowledge, especially because digital devices are part of the modern culture. Students today are familiar to explore in digital media, and this should be an opportunity to inquire the power of digital design to engage in mathematical thinking.

INTRODUCTION

Mathematics is the science par excellence, keeping us in a profound relation with our modern world. We take advantage of its potential, which allows us to transform our reality, even though Mathematics is acknowledged for its level of abstraction. The interaction between this science and our minds must be carried out through symbols. Any mathematical representation is a symbol, and the proper handling of these symbols is an important issue related to the comprehension of mathematics.

While performing a mathematical task, three representations may be required: the numeric, the algebraic, and the graphical representation. They appear as numbers, formulas (combining numbers and letters) and also graphics, which relate to both numbers and formulas. Among those representations, the graphical is the one that offers more scope for visual perception, but at the same time, it requires much insight to capture the rich information it offers at first glance.

DOI: 10.4018/978-1-5225-2110-5.ch003

Up until now the teaching of mathematics takes place mainly through writing. In high school and college emphasis is placed in the practice of algebraic workouts that fit well with the classroom. To some extent, this has been the suitable kind of access to mathematics. Performing algebraic procedures on the blackboard results far easier than drawing mathematical figures. Written exercises have become the main test to certify the learning of students.

Today, one may wonder if emphasis on the algebraic representation of mathematical knowledge could be put aside to make room for graphical representations playing a key role in the learning process. The interaction with mathematics through emergent digital technologies as allies, allows for a new way to learn mathematics, introducing visual aspects. The access to graphical content enhancing the visual perception should improve the engagement of students with mathematics.

The present chapter is related with the conceptualization of a whole strategy integrating digital technologies to foster the students' interaction with mathematics and favor their engagement with the learning process.

In order to deepen the way these new technologies impact the learning of Mathematics, the author has been adopting a sociocultural perspective. It considers the incorporation of technology in education as part of a modern culture that impacts the cognitive process. This is attuned to the evolution of human consciousness presented by Donald (2001), who stresses the power of the representational capacity in humans' minds. In turn, the contribution of Moreno-Armella (2014) stands the mathematical thinking in the evolution of the human species. Concerning Mathematics Education, this work agrees with the perspective presented by Moreno-Armella and Hegedus (2013) about the advent of new types of digital technologies changing the nature of Mathematics reference fields. The symbolic thinking has been evolving from static notations (in paper or digital) to digital dynamic inscriptions, where mathematical representations could be grasped through intentional dragging actions. The present work also addresses an ecological perspective from Gibson (1979) and its use in design by Norman (1993), in order to underpin the affordances of digital technologies' environments in human perception, and promote the continuous act of perceiving, involving the co-perceiving of the self.

Since 2012 the author's research agenda addressed the effort to provide students an opportunity to "see" mathematics in a realistic way, taking advantage of Augmented Reality (AR) technology. The AR application that is presented here considers the Calculus learning for undergraduate students. It is intended to promote the development of spatial visualization skills through the route of several themes in Calculus courses at College. As a result of the reflective thought supported by the sociocultural perspective, the AR application should be understood as an environment for the interaction of students and mathematics, which emphasizes the human perception through eyes and body.

There are three objectives for the present chapter; the first one is to describe the features of the AR emergent technology and its involvement in education to date, addressing the discussion to the development of spatial visualization skills. The second objective is discussing about a sociocultural perspective that could frame the use of AR technology for mathematics learning. As a third objective, the AR application is described, making emphasis on its features. Its use will be discussed according to the calculus themes, and this will give the opportunity to value it as a tool for the development of spatial visualization and the learning of calculus simultaneously. Ultimately, solutions and recommendations, future research directions, and a conclusion, should stress the chapter's aim sharing a novel use of AR impacting in the interaction with mathematical knowledge.

It is important to mention that from the technical point o view, for its production, the AR application in this work is the result of combining different software. Autodesk Maya 2015 (http://www.autodesk.mx/products/maya/overview) is used for the creation of the 3D models and the animations. To visualize the RA into the electronic device camera, Vuforia technology is used (https://developer.vuforia.com/). Finally to compile the necessary application for Android or IOS is used the cross-platform game engine Unity (https://unity3d.com/es).

BACKGROUND

Augmented Reality (AR) combines virtual and real environments; it is a computer-generated imagery that is overlaid directly onto the real world in real time. With AR technology it is possible to bridge the gap between reality and the virtual world. This connection implies great potential to engage and motivate learners in a variety of interactive ways never possible before. It is definitely a promising tool in learning because it can provide the user with a unique discovery path through rich content, built from computer-generated, three-dimensional models. Technologies that support AR are powerful and compact enough to deliver AR experiences through mobile devices, supporting the effort to apply AR in classrooms (Lee, 2012).

Wu, Lee, Chang, & Liang (2013) point out the ubiquitous learning allowed by AR technology, enhanced by computer simulations, physical models and 3D or virtual objects. These authors outline important issues related with the implementation of AR in education. They stress two educational purposes using this digital technology. First, helping learners engage in authentic exploration in the real world environment, where virtual objects have been included, such as texts, videos and pictures. The most prevalent method consists in annotating real spaces with an overlay of location-based information. The second purpose is that AR allows the integration of digital learning resources in real world. In this way, learners can have firsthand experiences with scientific phenomena not easily accessible in the real world, like chemical reactions or the movements of a solar system.

Meanwhile Dunleavy and Dede (2014) name two forms of AR currently available for educators: location-aware and vision-based. The first presents digital media to learners as they move through a physical area, and the second, in contrast, presents digital media to learners after they point the camera in their mobile device at an object. They give a good illustration of the difference in both forms through the scenario of a life science student that passes by an oak tree and there is a video on her smartphone showing animals found near the tree (location -aware), but when pointing the camera at the base of the tree, it triggers a 3-dimensional model that illustrates de anatomical structure of the oak (vision-based). Both forms of AR create *immersive* learning experiences, which means the subjective impression that one is participating in a comprehensive realistic experience. The authors provide a literature review that focuses about AR for learning that use mobile location-aware technologies, and state that there are few studies about vision-based form, which by the way could bring a greater degree of digital immersion.

From a psychological perspective, the potential of AR learning it is quite useful for mathematics, as Bujak, Radu, Catrambone, MacIntyre, Zheng & Golubski (2013) highlight. They recall the use of physical objects in the mathematics classroom and point out the glean insight of "manipulatives" to the research agenda in order to understand the benefits of physical and virtual objects. Physical objects afford certain actions and thus, offer immediate kinesthetic feedback. Virtual objects "can include additional instructional information built right into them, such as allowing or disallowing certain configurations of the objects" (Bujak et al., 2013, p. 537). The combination of virtual and physical objects in the same space and time adds to AR technology the power to emphasize the strengths of each one. AR manipulatives, considered as digital interactive visual representations of a dynamic object, bring the opportunity for a new kind of educational products. They afford a natural interaction because of their engagement with virtual content. The user can move around and change perspective or move closer or farther to change scale. It is natural to point to objects and reach out to touch and move. Research has found that physical movements, like those, can favor the learning of spatial content and the easy recall for information. Existing AR systems enable the exploration of 3D spaces, and this can lead to richer learning experiences, at least due to the physical affordances that this technology provides.

The role of spatial ability for the learning of STEM education (Science, Technology, Engineering and Mathematics) has been a focus of research lately, and its improvement should be a relevant academic goal. "Ideally, learning science, mathematics and social studies ought to be intensely spatial activities" (Newcombe, 2013, p. 27). Spatial ability is characterized as the ability to construct and maintain high-quality internal spatial representations (spatial images) and to accurately transform these representations. Animated diagrams and static diagrams can be considered as external visual-spatial representations. On the other hand, spatial ability relates to internal visualization ability. Seen this way, effects of spatial ability on learning from animations could be framed as the interplay between perception of external visualizations and internal visualization processes. Hegarty and Kriz (2008) suggest the possibility for helping the lack of internal spatial visualization ability through watching an external visualization, like an animation on a computer display. According to Diaz, Hincapié and Moreno (2015), the differences between static and dynamic content in educative AR applications should be consider. In their experience with augmented reality application for electronics course, they found that students perceive learning is easier when using dynamic AR contents comparing with static AR objects.

The focus on new visualization methods to improve the current teaching models can be seen as necessary because the students, in their ordinary life, are constantly interacting with a lot of graphic information provided by videogames, Internet or 3D movies. For Martín-Gutiérrez, Fabiani, Benesova, Meneses & Mora (2015), AR is a promising technology "a combination of real world elements captured through a camera with multimedia elements such as text, images, video, or 3D models and animations (p. 752). AR environments allows to see the real world with virtual objects merged, combining real and virtual, interacting in real time and allowing to visualize elements from different angles and perspectives. As these authors claim, they "have performed small virtual animations for viewing invisible abstract concepts in the real world (this is precisely one of the main strengths of the AR technology)" (p. 757).

It is possible to allocate a motivational impact with the use of AR technology in education at least for the "magical" feeling it arouses when looking at a virtual object embedded in the physical space that surrounds us. Motivation and achievement in high school students has been studied by Estapa & Nadolny (2015). They claim that technology in itself is highly motivating, but the primary goal for studies should be to give evidence of the acquisition of knowledge. Their own experience with AR has led them to state that students gained experience with the topic of dimensional analysis. However, because of the early stages of AR in the researchers' agenda, they stress the need of further research to indicate significant impact at both types of mathematical activity: conceptual and technical understanding.

According to Sommerauer and Müller (2014) there is a clear advantage for AR applications to become available for the broad public due to the advances in mobile technologies, like smartphones and tablets with built-in cameras. These authors contributed to the still emerging quantitative empirical evidence of the effectiveness of AR technology. The design of an experiment in the informal learning environment of a mathematics exhibition took place, and the results measured through the pretest and posttest effect size indicated a medium gain effect on learning performance, but still inconclusive. They discuss about their contribution situated in the informal environment of a museum where learning is not an intentional and organized process, but voluntary and self-directed, and also it was focused on short-term knowledge acquisition and retention.

An experiment to evaluate an AR-assisted learning approach of solid geometry integrated into teaching activities was performed by Lin, Chen and Chang (2015). One of the exploratory questions proposed in the study was to compare achievements in mathematics and spatial ability; another one was to assess the AR-supported programs improving students' spatial ability. The conclusions include a high positive correlation of students' performance with respect to spatial ability and their mathematics achievement. Regarding the system-assisted teaching, students performances from the experimental groups improved, however there were nor significantly differences between control and experimental groups.

If until now there is scarce numerical evidence in research of AR benefits in experiments, it is important to question if research should stay focused on looking to AR technology producing the same learning reflected in test and reported as quantitative studies. As Stake (2010) claims, science is the explanation of how things work in general, and scientific research is quantitative but also qualitative. Quantitative thinking relies heavily on linear attributes, measurements and statistical analysis, but there is also a qualitative thinking to work alongside, where personal experience, intuition and skepticism refine theories and experiments. AR technology deserves to be considered in education from a different perspective about its potential to integrate a broad perception into the learning process.

As a researcher in mathematics education, the author agrees about the important outcome that represents learning to think spatially, and that it has a great opportunity to be developed through an appropriate use of AR. The challenge could be focus in AR technology to allow a new kind of learning not possible before. For the learning of mathematics, the author believes in the usefulness of build animations with AR functioning as an external visual-spatial process in order to promote an internal visualization process in students. Necessarily this relates with a vision-based form of use of AR technology that integrates the benefits of interactive media to enable a higher degree of digital immersion. Mathematics is distinguished from other sciences because of the high level of abstractions that should be worked through their symbolic representations; among them we have algebraic and graphical representations. AR represents a great opportunity to teach mathematics in a visual way, giving priority to the graphical representation, accompanied with the development of spatial ability. The AR application in this work includes the production of a digitally enhanced setting to display animations of 3D models showing visually mathematical processes, performed at the time the student improves spatial visualization skills.

The author's position emerges from a teacher of mathematics that has been pondering the usual practice in classroom, and is convinced to inquire about the grounds for this practice. Teachers and students, with their own role, and following a conventional mathematics curriculum, necessarily get used to some ways to do things. Unfortunately, today is easy to find students and teachers not pleased with math classes, and the psychological phenomenon of anxiety towards mathematics holds the attention of researchers (Furner & Gonzalez-DeHass, 2011). It is time to focus in a different perspective to study the situation; in this work, the author wants to advance in that direction, giving basis for the discussion in order to transform the way things are working right now in math education. The change is possible if digital technology integrates to curriculum in the appropriate form.

Following Cole (2010), formal education (at school) was extended to the broad masses of children, sometimes in a compulsory way, during the 19th century, coinciding with the rise of the industrial revolution. Schools follow several common features, among them, their internal organization, including age grading and a sequentially organized curriculum based on level of difficulty, and their design of permanent buildings, for the practice of teaching and learning. It is important to be aware of the cultural nature of the practice of teaching mathematics. Transforming it in order to improve the learning of mathematics should be a main concern. The author's aim is the improvement of curriculum as a result of the research's agenda. This is a struggle that should be confronted from a wider perspective.

Formal schooling in the modern world is facing great challenges. Modern culture includes digital technologies, and it is imperative to face their potential to transform the interaction with mathematics. The author believes that conventional curriculum, classifying theory and practice, or conceptual and technical understandings, has avoided truly exploring the nature of mathematics. Integrating digital technology in curriculum should not be seen as a way to improve the learning of what already is fixed there. Instead, the integration of AR technology in this work seeks to produce a different sense for mathematics learning, moving away from a fixed practice and giving place to digital technology as an ally to understand why mathematics are an important subject to be learned at school.

Digital technology advances with the commitment to become entrenched in our culture by minimizing costs and simplifying ways of accessing its progress. It is a real fact today the ease of owning a digital device at hand, and this is inherited in the classroom. AR applications are distributed through the now common online stores where, at a minimal cost, they are downloaded and updated on the device you have available. This is an opportunity to get around the constant financial pressures that the school institution faces. The author's claim is the challenge for those applications to be designed to transform interaction with mathematics and not just an aggregate that repeats the same curriculum. New learning emerges from the use of digital technology, and mathematics education research should address the design and development of products that go beyond what is offered to students now.

AUGMENTED REALITY IN MATHEMATICS EDUCATION

This section is intended to raise a sociocultural perspective that could frame the integration of AR technology for the learning of mathematics. It represents an invitation to rethink the nature of mathematics that allows us to recognize it as a product of the human capacity to symbolize, and in close connection with our interaction with the real world. It is also an invitation to assess the ability of visual perception for the interaction with reality around us, and to intentionally involve that ability in the process of learning mathematics. The discussion of the AR application should be framed in this perspective as a digital product designed and developed to offer a different interaction with mathematical knowledge, fostering a spatial visualization skill as part of learning mathematics.

Thinking of Mathematics and Culture

As Cole (2010) argues, education "is intimately linked to the human capacity and need to live in a cultural environment" (p. 461). The notion of culture refers to the full range of socially inherited accomplishments of past human activities. Culture is a symbolic and material inheritance, and it serves as a crucial resource for the current life of a social group. Culture is embodied in language and in social practices.

Cole stresses that the conception of culture originates in reference to the process of helping things to grow. From this perspective, schools are institutionalized cultures for growing next generations. "For purposes of discussing culture and contemporary education, it is significant that culture applies to people engaged together in the same classroom" (Cole, 2010, p. 462).

The experience of teaching in classroom needs to be analyzed from a wider perspective. The usual behavior of students has changed because the culture is changing. Today, capturing the students' attention requires more effort from the teacher than before. We are living a digital culture where communication and access to information is always at your fingertips, and this is a challenge for educational institutions nowadays.

In some way, the situation described above might be expected from the perspective of Donald (2001) about the evolution of human cognition and culture. He proposes four stages in human evolution that could have bridged the huge evolutionary distance between the minds of modern humans and the one of distant ancestors. The first shift, from Episodic to Mimetic stage, brings a more precise and self-conscious control of human action. The second shift, from Mimetic to Mythic stage, brings humans a richer and faster accumulation of cultural knowledge in speech. Finally, the third shift, from Mythic to Theoretic stage, brings much more powerful and abstract reflective cultures, driven by language and symbolic representation.

According to the proposal of Donald (2001) the main cognitive driving force underlying the transition in stages is the externalization of memory. Without this, earlier humans had to depend entirely on their biology (their brain) to remember. The initial externalizing phase is set with mimetic capacity, with its emotional representation through gesture. Another externalizing phase is the arrival of language. With it, humans supported narratives. A key externalization of memory is set through writing; this resource allowed the externalization of thoughts. Human minds had in writing an external resource that lasts over time, and written texts became reflective instruments that could be exposed to systematic analysis. This revolutionized the way humans think; changing from speech to writing expanded the human cognitive capacities. The use of symbolic devices, as language, changed the human mind by an intimate process of interaction. Influence flows both ways, from thinker to symbols and from symbols to thinker. Symbolizing mentality emerged, and modern humans now create and employ a huge number of powerful external symbolic devices to store and retrieve cultural knowledge. Language and science are cultural products that distinguish the human species, and today the power of digital technologies adds to the interaction between human mind and digital devices.

Returning to the current situation in classrooms, the author wants to emphasize that students have a modern human mind with evolved brains that reflect three ways of representing reality. They are cultural beings with interaction power through a mimetic layer, an oral-linguistic layer and an external-symbolic layer. They have internalized the symbolic conventions of the saturated digital culture that surrounds them but at the same time, this modern culture contains within it, a trace of each stage of cognitive evolution. The point to be noted here is that the conventional practice of teaching and learning in classroom may be wasting the potential of those minds to have interaction with mathematics. Leaving the learning of mathematics as a manipulation of symbols, mainly in an algebraic fashion, denies the meaning of mathematics and reduce it as a symbolic skill, performed by hand in paper. Mathematics is a cultural product, one in service of the human species. It emerged in the interaction between brains and world, it is an instrument to understand the world, resulting from the symbolic capacity that the human brain possesses. How can this be understandable in the classroom?

As Donald (2001) states, symbolic technology impacts on the conscious process, and transforms the architecture of conscious mental activity. External symbols introduce more powerful memory media into human cognition and increase the number of ways we can represent reality. They are a stable, permanent, virtually unlimited memory record. Their availability radically changed the total memory storage capacity of humans, as individuals and as a species. The external memory field provides sharper and more durable mental representations that allow the conscious mind to reflect on thought itself. When using a powerful symbolic device, like books or laptops, awareness finds itself juxtaposed between two storage systems at the same time, one biological and the other technological. A mirror image of natural working memory becomes available, and it changes the reflective power of the conscious mind.

From this perspective, it is useful to think of digital technologies and digital devices as a source to rethink learning experiences for this conscious mind in students. In some ways, the design and production of an "external memory field" for mathematics could make students react cognitively and bring the opportunity to analyze mathematical representations. The author's idea is to take advantage of digital technologies and devices to design environments where students get in touch with mathematics as part of the digital culture they are used to. The work the author presents in this paper follows the grounds of Donald (2001) holding that, arranged in the right order, a variety of symbolic technologies can lead the user to specified states of mind. The scripted aspect of external symbols, now in the digital display of the new mobile devices, can give rise to novel cultural possibilities. AR gives the opportunity to create a powerful experience for students. It could become a natural mediator of mathematical thought.

According to Moreno-Armella (2014) culture is a constituent part of the human species. Any human born today, besides the accumulation of meaningful experiences interacting with the real world, adds the cultural experiences of the environment. There is the direct interaction with the real world through the senses (sight, hearing, smell, taste and touch) and at the same time, the interaction is with culture, because the experiences occur in a cultural world. In some way, senses are transformed in order to access culture. Humans listen to music, and listen to their mother's language before being born; music and language are part of that culture. Human experience is hybrid, like the real world. It involves matter, organic, biology, but at the same time it involves culture. When someone says something "makes sense" there is some meaning perceived, as if it was captured by their organic senses. Mathematics should "make sense" for students.

Once an experience is transferred to language or another symbolic representational system, it can be thought out and polished, deepened, and organized. Thus, the possibility to be captured in symbolic representations gives to the experience some level of stability. Once existing and accessed by the symbols, it can be refined, reconsidered; it can even create new meanings. This occurs with mathematics. Every symbol is born from an experience, with meaning. Thinking of the number five, the mind could have displayed the symbol 5; this symbol synthesizes a meaning. In the human mind, a generalization process is now available, symbolized with 5, as this is part of the mathematical culture.

The symbolic capacity of humans is their support in getting to know about the world, to have clear ideas, to organize them and to systematize. This is the history of mathematics. But, as Moreno-Armella (2014) stresses, the symbolic experience is so intense, and often, it looks completely independent from its original roots in human experience full of meaning. External symbols aren't the original source, but they represent an original intentionality to get to know about the real world. Symbolic representations have the power to support the establishment of relations between symbols, and extend the symbolic net. There the meaning persists, but it takes the risk of being left aside when focusing on the handling of the necessary symbols. This is something that mathematics education should take into account.

Mathematics curriculum presents a prearranged organization of mathematical objects and processes being selected, in some way, to be taught. Typically, they propose a definition of the mathematical object, for example a function, and it is expected from the student to handle a set of processes performed with the diversity of symbolic representations of that object. Usually there is a numerical representation (an arrangement of columns with numerical values), accompanied by a graphical representation (a curve in a coordinate plane); it is also possible to have an algebraic representation of the function (expressions with numbers and letters). It has been finely argued by Duval (2006) that the cognitive difficulties with the learning of mathematics is related to the use of the different symbolic representations, although it is understood that any mathematical activity requires the use of symbolic representations. Among those problems, the author wants to point out that the curriculum and the practices it leads in classrooms is also a cultural issue. The way mathematics is taught suggests a conception of mathematics and this affects the learning process.

As Moreno-Armella (2014) argues, it is not possible to be sure about having the totality of representations for one mathematical object; in this sense, mathematical objects are always under construction. In the classroom, the author has noticed how students express their thoughts about a function making a gesture with their hands. This is an informal symbolic representation, natural in students, but it should be considered as a way to access the mathematical idea underlying the object named as function. Curriculum suggests that a mathematical object is understood if it is represented in two or three symbolic systems already written in textbooks. But those representations do not have meaning for students, there is no perception that could allow them to capture the mathematical idea; it makes no sense. Teachers should be conscious that the mathematical object does not exist detached of the totality of its representations; it is not contained in the textbook. Maybe this is frightening, but the author's feeling is that, being conscious of this, it is possible to break free from a teaching practice that does not work. Curriculum has set aside intuition and meaning to focus in formalization and symbolic systems of representation. Teaching and learning practices without meaning, only showing symbolic representations, leads to "the lost gaze of the students" (Moreno-Armella, 2014, p. 25).

Nowadays there is a new challenge in mathematics education because of the emergent digital technologies begging for changes in curriculums, where they could make a valuable contribution. It is time to consider new ways to perceive mathematics, to make sense for the students. Mathematics comprehension must deal with different symbolic representations, and digital technologies enhance the graphical one. The author's claim is to take advantage of the visual expression of mathematics ideas through AR technology, and allow students to incorporate a visual and tangible perception when learning mathematics. AR is a suitable technology for this approach, it is possible to set up and execute a coordinated experience on a scale that was impossible before; this is part of the digital culture right now.

An Ecological Approach for Perception

James Jerome Gibson (1904-1979) conceived vision as a way of acquiring information about the environment directly, coming in contact with it by means of active explorations. He pointed out that perception involves action, and psychological processes could only be understood in terms of the dynamic coupling with environment. Gibson's contribution to the study of perception, through an ecological approach, provides a framework to help educational researchers concerned and sensitized towards visual perception. His theory of direct perception under the ecological specificity contends that the environment affords perception to our bodies. The environment presents the possible actions to do, actions within the environment. "Affordances" are these perceived features of the environment that are there for our acting bodies, like a chair in the floor affords to sit down. In this view, affordances are preconditions for activity. "The presence in a situation of a system that provides an affordance for some activity does not imply that the activity occurs, although it contributes to the possibility of that activity" (Greeno, 1994, p. 336).

For Gibson, the environment is at least as essential as the brain in the existence and exercise of the human mind. Knowledge begins with perception of the environment. He observed that, as one scans an environment in one direction, the surface patterns of the environment that go out of sight come back into sight upon reversing the scan pattern. The surfaces can be seen having a continued existence, even if not seen all at the same moment. Thus, perceiving is a kind of temporal exploration of environment. "If the pattern of an existing surface can be seen over time this way, as a unit, then perceiving extends over time" (Mace, 2005, p. 197). A core insight in Gibson's frame is the existence of "invariants", which capture the idea that not everything changes during change. Visual perception involves actions, and knowledge relates to identifying these invariants in the environment.

Hutchins (2010) highlights Gibson's psychology, taking a deeply ecological approach to cognition over 30 years ago. He intended to show that our perceptual systems evolve and become attuned to environments. The eye-head-brain-body systems register a set of properties of the environment that stay invariant under transformation of action. Even though he examined the environment's structure, Hutchins complains: "but, alas, not the cultural environment" (p. 708). Gibson's insights about visual perception as a form of action have provided inspiration for a part of the philosophy of embodied mind movement. For them, "perception is something we do, not something that happens to us".

Several research traditions, including ecological psychology, committed to consider a cognitive ecosystem as the correct unit of analysis for the development of cognitive science. There, the points of contact between organism and environment come to be seen as locations of essential processes, and the role of the body in thinking becomes the focus. Cognitive science returns to consider issues like culture, context, history and their effects. This has been eased by the progress in understanding how the human brain works, and also by the advances showing the involvement of body and world in the constitution of human mind (Hutchins, 2010).

The design community adopted the word "affordance" coined by Gibson. Norman (1999) whom originally hated the idea, finally adopts it as "perceived affordance", referring to the actions that the user perceives to be possible when interacting with a device. In that time, Norman was doing a lot of design on computer screens. There, he said, the range of possible activities is limited to typing on a keyboard, pointing with a mouse, and clicking on mouse and keyboard switches. "Soon we will add spoken words and visual gestures to the list of interactions. All of these actions are abstract and arbitrary compared to the real physical manipulation of objects, which is where the power of real and perceived affordances lies" (p. 41). At the present time, the development of digital technology offers much more than what was thought in 1999, and it is the author's aim to involve the new technologies and devices that currently offer the kind of interaction that the manipulation of "physical objects" affords.

In his well-known book *Things That Make Us Smart*, Norman (1993) argues the power of representations. He contends that it is by the invention of external aids that humans have increased memory, thought and reasoning. Of those external issues, some arise from cooperative social behavior and others taking advantage of information present in the environment. In addition, there are some that emerge through the development of tools of thought: cognitive artifacts. He says that those cognitive artifacts complement human abilities and strengthen mental power.

The ability to represent perceptions, experiences and thoughts in some external medium, could present the opportunity to focus on the essential and take away irrelevant details. This is an important idea that the author wants to highlight for the design of educational products for mathematics. As Norman expresses, if the representation and processes are just right represented, then new experiences, insights and creations can emerge. There are several features, which are ideal for the development of representations, such as capturing the important while ignoring the irrelevant, and enhancing the process of interpretation. It is also advisable to enhance the ability to make judgments and to discover relevant regularities and structures.

In terms of educational purposes, the author's interest is to strengthen the performance of a visual perception while using AR technology in some calculus topics. The design of the experience through AR should function as a cognitive artifact, motivating intentional actions in students in order to produce knowledge. The important point is that the representational capacity of students can be expanded upon through external structures supported by devices showing the AR technology. The AR application that is discussed next, aims to promote the visual perception of a dynamical scanning of mathematical surfaces. There, the affordance that AR technology brings consists in the opportunity to move around whilst the animation takes place; it is possible also to move the animation showing up and observe the effect through a mobile device. The later is possible because the product in this work for AR visualization includes a physical paddle where the virtual reality it is shown, thus, physical manipulation adds to the power of real and perceived affordances.

AR on Behalf of a Visual and Tangible Mathematics

The author's experience in mathematics education has always been grounded in the mathematics classroom at college. There, the students usually express difficulties with spatial visualization, for example, they do not "see" what the teacher already has internalized as a solid of revolution. As said before, it is a research concern to provide a solution to the limitations of formal education for the development of this ability. Using spatial visualization is a transversal issue to curriculums, certainly not only for mathematics curriculum, but nevertheless, its importance is crucial for the students' understanding of this science.

It is important to highlight that the spatial visualization allowed through the AR application is not restricted to perception of 3D objects in space. The design with AR technology shared here includes the transition from 2D to 3D, and the return from 3D to 2D. In some way, the idea to develop with AR is similar to the "go out of sight and come back into sight" idea of visual perception. With this in mind, the author identified some topics in Calculus I, II and III courses at college, where the AR technology affordances could be exploited and function as a cognitive ally, promoting the visual perception process.

The final structure of the application consisted in three sections named: From 2D to 3D, Solids of Revolution and 3D Surfaces. The navigation on the app works based on traditional UI conventions such as a side panel for the controls, different buttons that control the actions in the display in relationship with the content displayed, a window for a video explanation of what is happening in the 3D animation, and a window for the mathematical expression to be displayed. The student handles the tablet with the AR application and two acrylic paddles with the marks ready to display the virtual objects. The second paddle in Figure 1 displays a yellow tinted 3D plane that could be used as a kind of blade to cut the solid. This is an affordance that AR technology offers without precedent for visual and gestural interaction, and it is useful for promoting a reflective thought. This is the kind of externalization without precedent that is available now because of AR technology.

Figure 1. The AR application use includes a tablet and two acrylic paddles for interaction

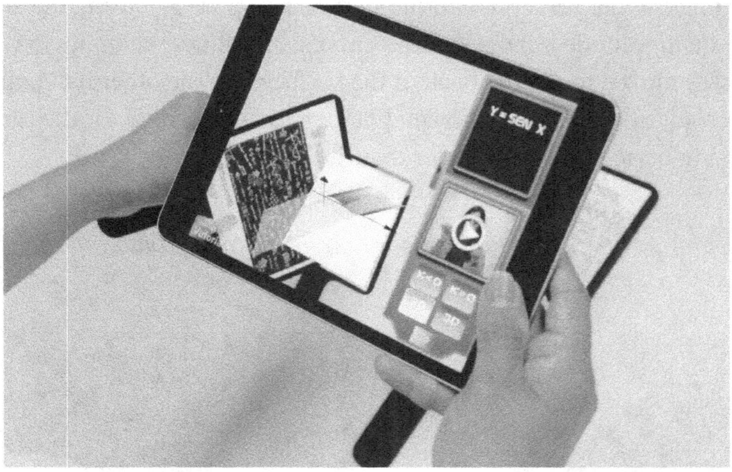

The Calculus I course considers the graphs of functions of a single real variable; there, the curves are contained in a plane. These curves, compelled in 2D dimensions, are taken as the starting point for the development of the spatial skill. The author's proposal is to produce a dynamic visualization of 2D curves in parallel planes, this way, the 3D visual perception is supported by the accumulation of curves. At the same time, the curves proposed comply certain pattern due to a parameter affecting the algebraic representation of the function. The curves in Figure 2 they appear as time passes fulfilling the behavior pattern set by the representation $y=\sin(x+k)$, where the parameter k (now negative) shows the shifting of the 2D curves to the right. While the shift is happening, at the same time the curves appear in parallel planes, behind the original curve. The complete 3D surface emerges as the final step in the dynamic external visualization with the accumulation of curves. The idea is to foster an internal visualization building a 3D surface in space.

Figure 2. Animation of 2D curves shifting to the right and promoting the visual perception of a 3D surface

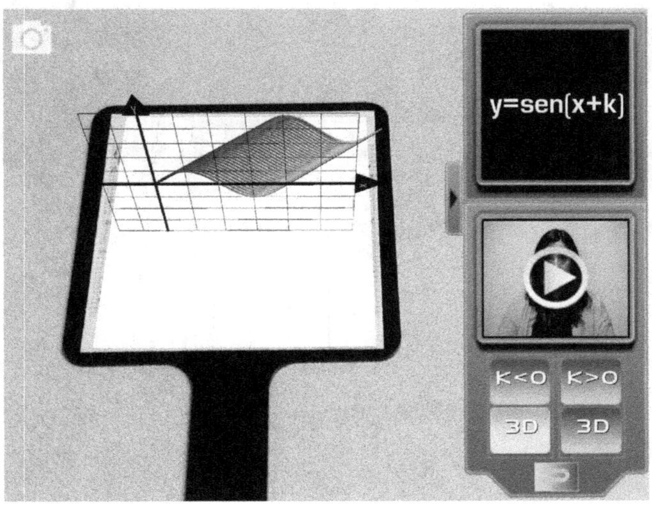

When the parameter k is affecting the function in the way y=ksin(x) represents, the visual animation from 2D to 3D gives place to the perception of the surface in Figure 3. During the animation for k greater than 1, it is noticed the amplitude changes in the curves; amplitude value increases as time passes. In Figure 3 the paddle was moved to see the back of the surface. On the other hand, the surface in Figure 4 represents the same result of the process of visual perception performed for the function y=sin(kx) that now shows a decreasing period being k greater than 1.

Figure 3. Animation of 2D curves showing bigger amplitude and promoting the visual perception of a 3D surface

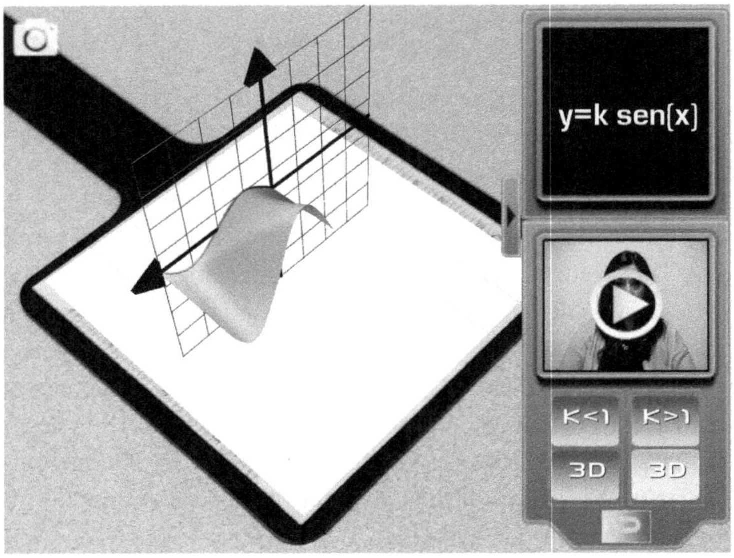

Figure 4. Animation of 2D curves showing smaller period and promoting the visual perception of a 3D surface

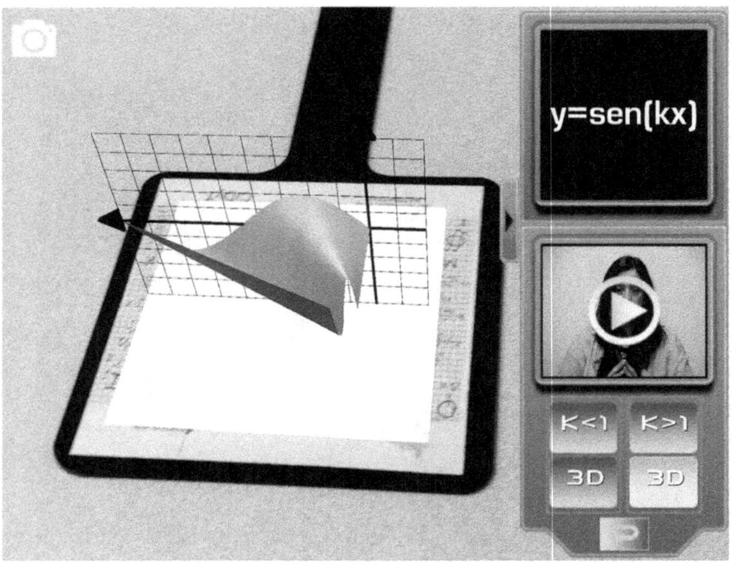

The design of AR technology, in the topic of solids of revolutions in Calculus II, includes several features in order to promote a generalization process. The author decided to consider four functions with graphs of combined increasing or decreasing behavior with concave upward or downward features. The approach presents a first step giving the natural visualization of the solid; this is the common way of spinning an area around an axis, like in Figure 5. For each of the four original curves the AR application includes both axes for rotation. There are a total of eight solids visualized through this animation in order to conceive different forms and volumes.

This first dynamic animation is followed by two more, corresponding with the methods for calculating the volume of the solid of revolution by means of a definite integral. An important step in these two visualizations is the generation of discs like in Figure 6, or the generation of shells, like in Figure 7; this evokes the visualization of the mathematical procedures according to the integral method of discs and shells applied to calculate the volume of the solid.

The fact that those visualizations are so different from the natural visualization of an area rotating around an axis is a key issue in the design. This is a perceived affordance introduced by the AR application design; it is intended to support the students' visual perception process. The animations showing the accumulation of disks in Figure 8, and the accumulation of cylindrical shells in Figure 9 are an important step for the interpretation of the integral concept, with its useful meaning associated to an accumulation process. Both animations finish proposing the corresponding definite integral associated with each of the two methods.

In Calculus III, the graphs of functions of two real variables relate to the visualization of a surface in a 3D space, like the one in Figure 10, corresponding to the function $z=y\sin(x)$. The author selected a diverse set of these functions, having different features in order to illustrate the kind of behavior that is possible to conceive in a 3D space.

Figure 5. Area under the decreasing and concave downward graph that rotates giving the natural visual perception of the solid of revolution

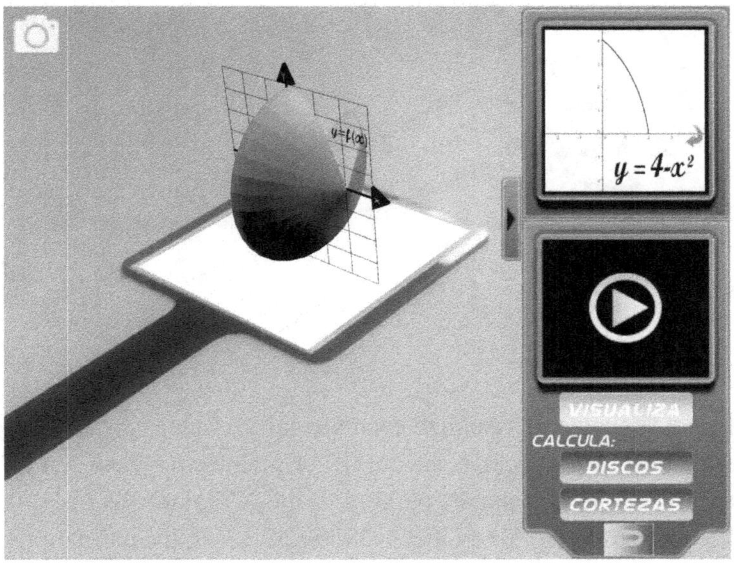

Figure 6. Animation of 3 discs generated by the rotation of a vertical segment limited by the decreasing and concave downward curve

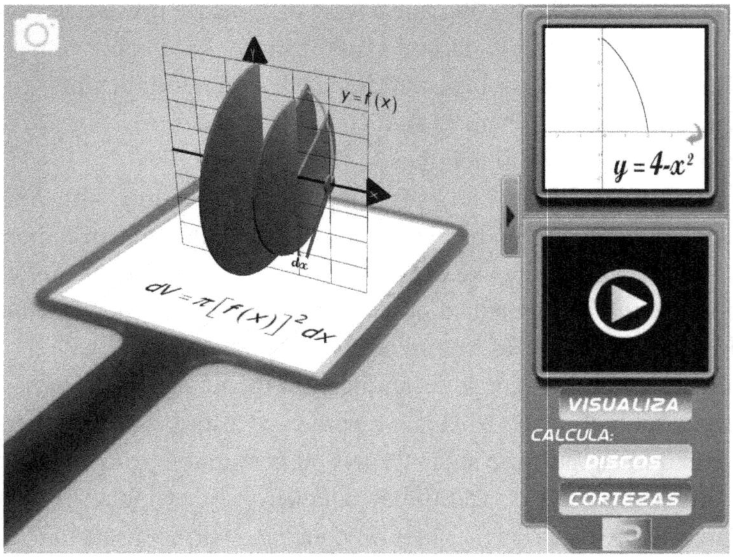

Figure 7. Animation of 3 shells generated by the rotation of a horizontal segment limited by the decreasing and concave downward curve

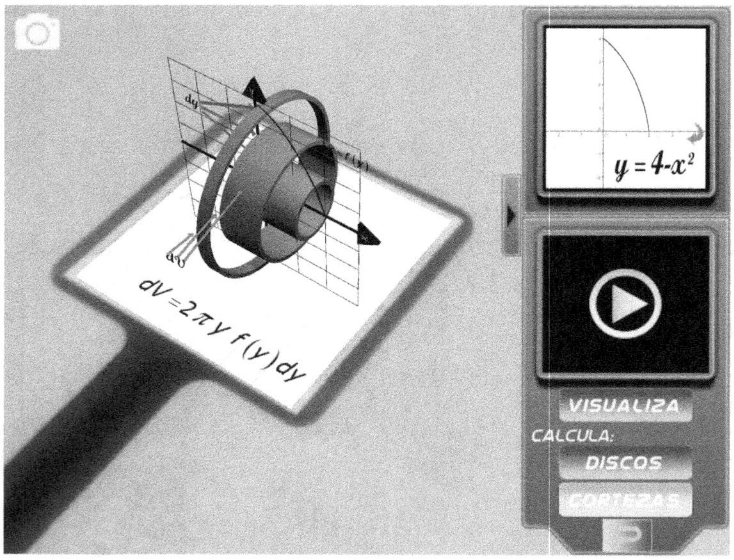

The design includes a first step to visualize each virtual surface as a static 3D AR image. The idea for the dynamical animation is to conceive the curves of intersection between planes and the original surface. The curves are generated by planes parallel to the XY plane, as in Figure 11, parallel to XZ plane, as in Figure 12, and parallel to YZ plane, as in Figure 13. In the animations, the evocation of the physical paddle with the virtual cutting effect by a yellow plane is used as an affordance. A dynamical

Figure 8. Accumulation of discs creating the solid through the discs method and the integral to calculate solid's volume

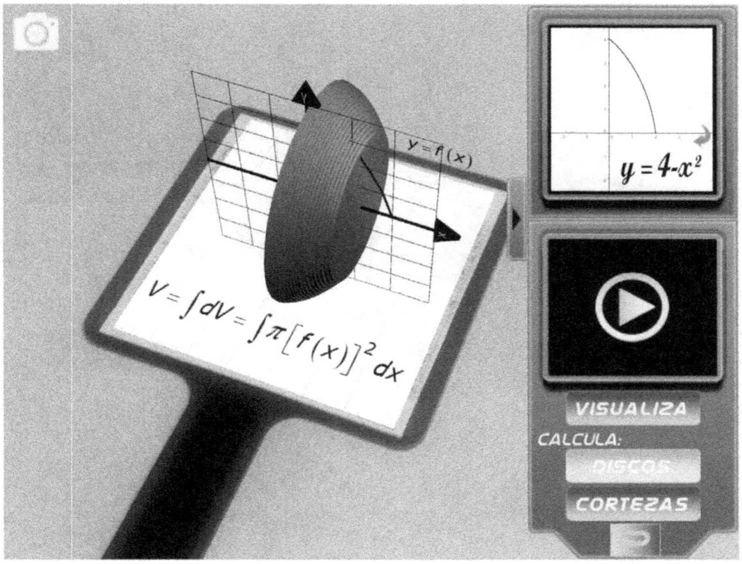

Figure 9. Accumulation of shells creating the solid through the shells method. The integral to calculate solid's volume

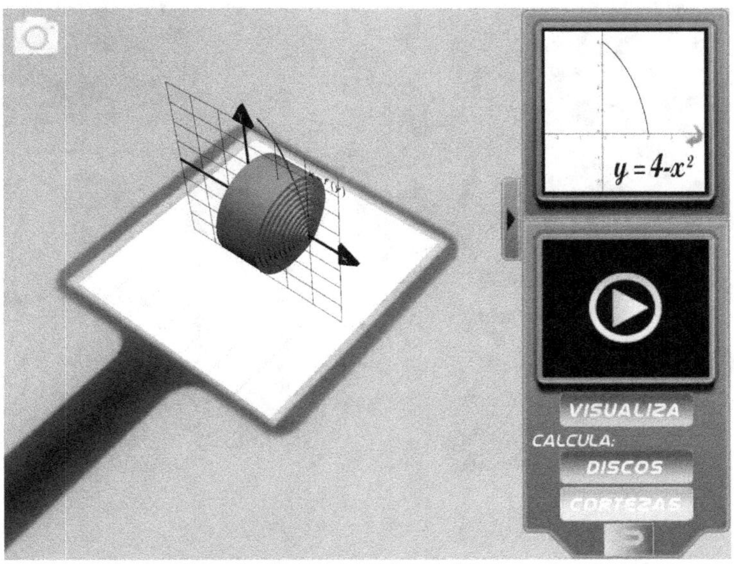

simulation of the intersections between the blade plane and the original surface allows for the visual perception of curves that evoke the recreation of the surface through a visual process. Static and dynamic visualizations are combined in the design to promote the reversing process from 3D to 2D.

The innovation introduced in this level promotes the perception of a visual process performed from 3D to 2D. The AR virtual model in 3D is the surface, the graph of the function of 2 variables, $z = y\sin(x)$.

Figure 10. Surface representing graphically a function of two variables

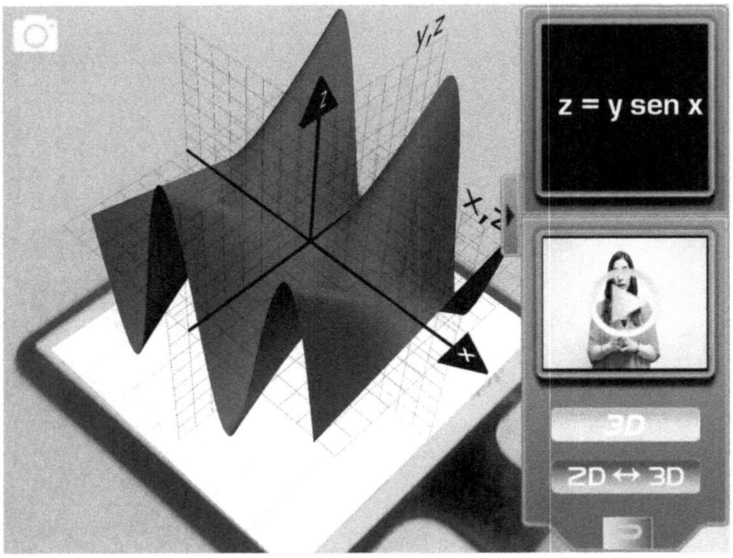

Figure 11. Animation of cutting the surface with planes parallel to XY plane

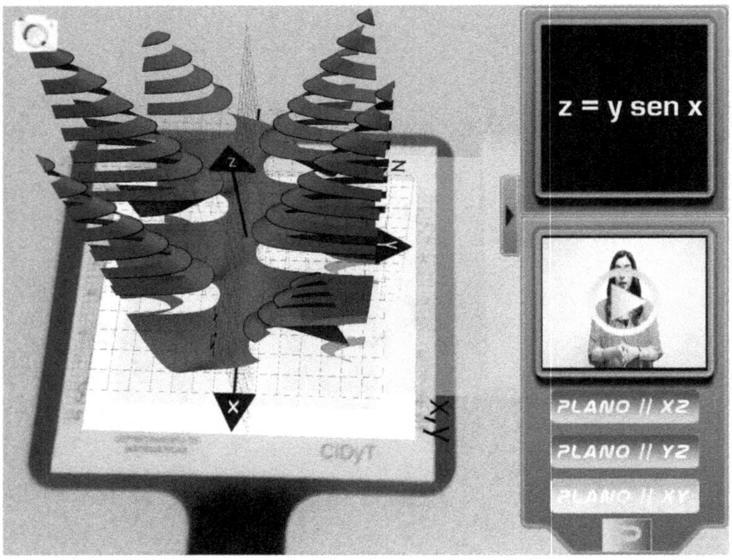

This continues with a sequence of yellowish planes that get in and get out the scene intersecting the surface. There is a visually return to 2D when the surface in 3D is affected by the intervention of planes parallel to the coordinate planes XY, YZ and XZ.

This visual external process through the AR application of cutting and rebuilding the mathematical surface promotes a cognitive evocation of a new way (like in Figure 14) of visualizing the original surface.

Figure 12. Animation of cutting the surface with planes parallel to XZ plane

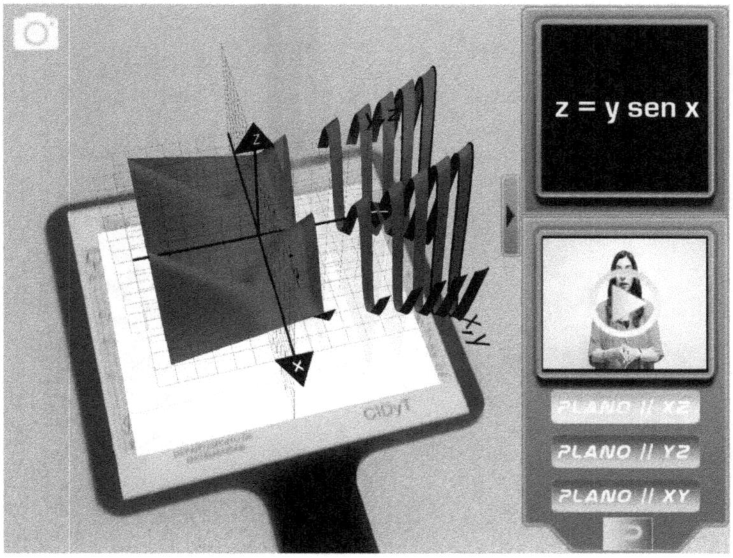

Figure 13. Animation of cutting the surface with planes parallel to YZ plane

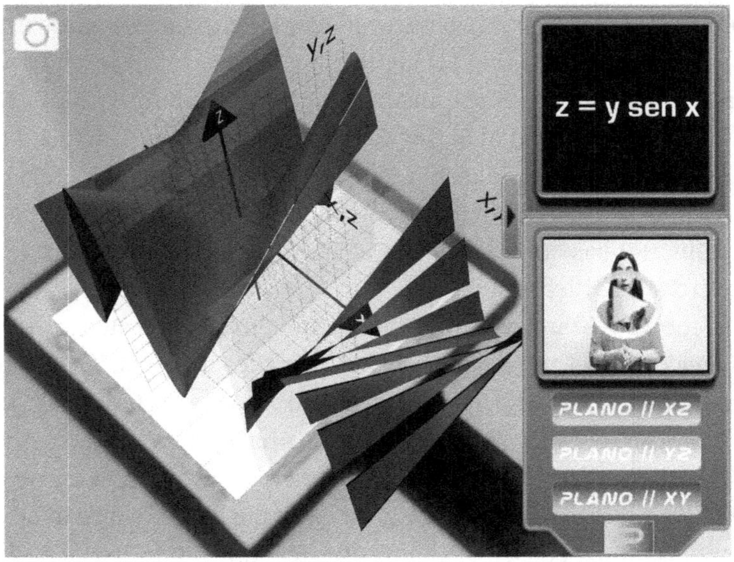

SOLUTIONS AND RECOMMENDATIONS

In this chapter the author discusses AR digital technology and its use in education, especially in mathematics education. It includes the description of the AR application designed for the development of spatial ability in students. The three sections presented are related to topics in Calculus courses at college level. The design of the application takes advantage of the affordances that AR technology offers. Students can move the paddle displaying a virtual object, or interact by slicing the virtual object with another paddle. This same physical action of slicing or cutting has been used in the dynamic 3D model animations that

Figure 14. Dynamic and static animations of curves in 2D that rebuild the static visualization of the surface in Figure 10

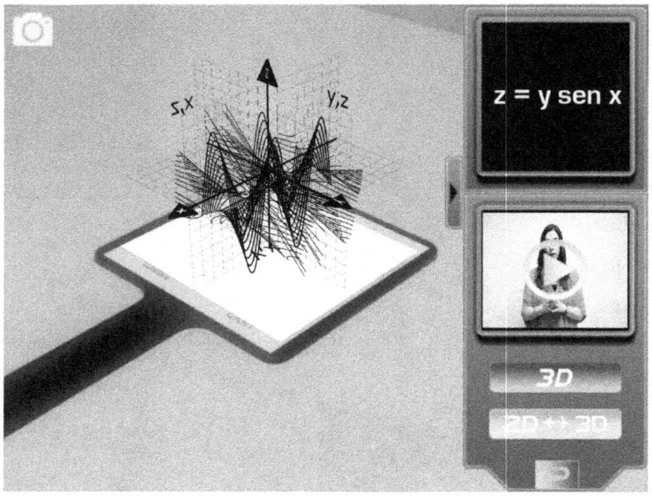

are displayed while the student is looking from different angles. The dynamic visualizations building the object promote an active visual perception when learning topics like solids of revolution or contour lines. There is a video element that could be use while looking to the 3D animations taking place.

The educational goal embraced in the AR application is to develop an ability that could be identified as part of a mathematical culture. The mathematical curriculum typically takes for granted that students have already developed a spatial ability for visualizing, but it is not, and this is precisely what has become a main concern in STEM education research. It should be noted that the AR application is not intended to teach a conventional topic of the calculus course. Even though it helps the teaching and learning for specific topics, it has been designed as a more general support tool for visualization that is possible only today, through AR technology.

As part of an educational research team for the learning of mathematics, the author knows that the way Mathematics is conceived influences the way Mathematics is taught. The Mathematics curriculum of today shows a theory that is already organized, and encourages a static point of view. To deal with this knowledge, students learn definitions, procedures, theorems and results, mostly in an algebraic fashion. The common practice for teachers and students is the performance of exercises as a meaningless, repetitive routine, the same used for grading purpose. It is the author's aim to advance a sociocultural perspective that allows becoming conscious of this practice; this is the first step to be ready and open to conceive a different interaction with mathematics knowledge. The integration of digital technology for the learning of mathematics is a key step to meet this objective.

FUTURE RESEARCH DIRECTIONS

The AR application in this work has been presented in several forums, there the evidences of use with students show the engagement it promotes (Salinas, González-Mendívil, Quintero, Ríos, Ramírez, & Morales, 2013; Salinas, Quintero, & González-Mendívil, 2014; Salinas, Quintero, Ramirez, & González-

Mendívil, 2015). They feel at ease using these kinds of devices, and expect more of this for mathematics. It is clear that technology is part of modern culture and it will not disappear; on the contrary, current developments attract the attention of any human mind. It is time for mathematics teachers to accept this and be open for the changes that should be made.

The experience with teachers and the AR application has shown their enthusiasm as well; the problem is that it doesn't currently fit in conventional curriculum. Teachers in formal education have the problem of having to follow curriculum that has not changed for a long time, and one that they are all used to. Nevertheless, new technological products are available outside school, and students know what is possible to do with digital devices. The growing stress in classrooms today begins to question the suitability of a curriculum arranged when technology did not yet play the important role it does today.

Educational research cannot stay in the same tradition, evaluating the learning of something that is not as important as before when technology was not part of our daily environment. This is not comparing achievement with and without technology where educational research should stay. Research should include the design of educational products for the learning of mathematics and assess the new kind of learning possible because of them. The perspective drawn in this work is not limited to the conventional mathematics curriculum but looking for a new curriculum where mathematics could be perceived as part of the culture of our modern world visualized through technology.

CONCLUSION

The science of mathematics is an evidence of the symbolic power of the human mind. It emerges from the interaction between the human brain and the social world. Its purpose has been to help humans to understand the different events perceived through their senses. For this to happen, the human species created a way to reflect in their own thinking: through the use of symbolic representations overlaid in the real world. Reflecting in a thought could seem difficult to do with mathematical symbols, but by evoking language in place of mathematics, the important role of symbols (like words) could be better understood. How to realize that learning math is as useful as it is to learn a language?

The difficulties with the learning of mathematics are widely recognized. Educational research reports problems with the use of the algebraic representation, the most common representation in mathematics teaching. But algebraic symbols are not the only way to access mathematics. Today digital technologies offer an approach to mathematics favoring visual perception. This is an increasingly suitable place for students to reflect on their own thinking. With the development of emergent technologies, new possibilities arise for mathematics education, highlighting graphical representations as a gateway to math. This is a crucial advantage because of the current permeance of digital devices in human social, and everyday, life.

REFERENCES

Bujak, K. R., Radu, I., Catrambone, R., MacIntyre, B., Zheng, R., & Golubski, G. (2013). A psychological perspective on augmented reality in the mathematics classroom. *Computers & Education*, *68*, 536–544. doi:10.1016/j.compedu.2013.02.017

Cole, M. (2010). Whats culture got to do with it? Educational research as a necessarily interdisciplinary enterprise. *Educational Researcher, 39*(6), 461–470. doi:10.3102/0013189X10380247

Diaz, C., Hincapié, M., & Moreno, G. (2015). How the Type of Content in Educative Augmented Reality Application Affects the Learning Experience. *Procedia Computer Science, 75*, 205–212.

Donald, M. (2001). *A mind so rare: The evolution of human consciousness*. New York: Norton.

Dunleavy, M., & Dede, C. (2014). Augmented reality teaching and learning. In J. Spector, M. Merrill, J. Elen, & M. Bishop (Eds.), *The handbook of research for educational communications and technology* (pp. 735–745). New York: Springer. doi:10.1007/978-1-4614-3185-5_59

Duval, R. (2006). A Cognitive Analysis of Problems of Comprehension in a Learning of Mathematics. *Educational Studies in Mathematics, 61*(1), 103–131. doi:10.1007/s10649-006-0400-z

Enyedy, N., Danish, J. A., Delacruz, G., & Kumar, M. (2012). Learning physics through play in an augmented reality environment. *International Journal of Computer-Supported Collaborative Learning, 7*(3), 347–378. doi:10.1007/s11412-012-9150-3

Estapa, A., & Nadolny, L. (2015). The Effect of an Augmented Reality Enhanced Mathematics. Lesson on Student Achievement and Motivation. *Journal of STEM Education, 16*(3), 40–47.

Furner, J., & Gonzalez-DeHass, A. (2011). How do Students' Mastery and Performance Goals Relate to Math Anxiety? *Eurasia Journal of Mathematics, Science & Technology Education, 7*(4), 227-242. Retrieved from http://www.ejmste.com/

Gibson, J. J. (1979). *The ecological approach to visual perception*. Boston: Houghton Mifflin.

Greeno, J. G. (1994). Gibsons affordances. *Psychological Review, 101*(2), 336–342. doi:10.1037/0033-295X.101.2.336 PMID:8022965

Hegarty, M., & Kriz, S. (2008). Effects of knowledge and spatial ability on learning from animation. *Learning with animation: Research implications for design*, 3-29.

Hutchins, E. (2010). Cognitive Ecology. *Topics in Cognitive Science, 2*(4), 705–715. doi:10.1111/j.1756-8765.2010.01089.x PMID:25164051

Lee, K. (2012). The Future of Learning and Training in Augmented Reality. *InSight: A Journal of Scholarly Teaching, 7*, 31-42.

Lin, H. C. K., Chen, M. C., & Chang, C. K. (2015). Assessing the effectiveness of learning solid geometry by using an augmented reality-assisted learning system. *Interactive Learning Environments, 23*(6), 799–810. doi:10.1080/10494820.2013.817435

Mace, W. M. (2005). James J. Gibson's Ecological Approach: Perceiving What Exists. *Ethics & the Environment, 10*(2), 195-216.

Martín-Gutiérrez, J., Fabiani, P., Benesova, W., Meneses, M. D., & Mora, C. E. (2015). Augmented reality to promote collaborative and autonomous learning in higher education. *Computers in Human Behavior, 51*, 752–761. doi:10.1016/j.chb.2014.11.093

Moreno-Armella, L. (2014). *Educación Matemática: del signo al pixel*. Bucaramanga, Colombia: Universidad Industrial de Santander.

Moreno-Armella, L., & Hegedus, S. J. (2013). From Static to Dynamic Mathematics: Historical and Representational Perspectives. In S. J. Hegedus & J. Roschelle (Eds.), *The SimCalc Vision and Contributions: Democratizing access to important mathematics* (pp. 27–46). Springer-Verlag; doi:10.1007/978-94-007-5696-0_3

Newcombe, N. S. (2013). Seeing Relationships: Using Spatial Thinking to Teach Science, Mathematics, and Social Studies. *American Educator, 37*(1), 26–31 and 40. Retrieved from http://silccenter.org/publications_pdfs/Newcombe final 3-4-13.pdf

Norman, D. A. (1993). *Things that make us smart: Defending human attributes in the age of the machine*. Reading, MA: Addison-Wesley Pub. Co.

Norman, D. A. (1999). Affordance, conventions, and design. *Interaction*, *6*(3), 38–43. doi:10.1145/301153.301168

Salinas, P., González-Mendívil, E., Quintero, E., Ríos, H., & Morales, S. (2013). 2013 International Conference on Virtual and Augmented Reality in Education. The Development of a Didactic Prototype for the Learning of Mathematics through Augmented Reality. *Procedia Computer Science, 25*, 62–70.

Salinas, P., Quintero, E., & González-Mendívil, E. (2014). How can Augmented Reality Favor the Learning of Calculus? *Proceedings of FECS 2014. World Academy of Science*. CSREA Press.

Salinas, P., Quintero, E., Ramirez, P. G., & Mendívil, E. G. (2015). Fostering spatial visualization through augmented reality in calculus learning. In *ASEE Annual Conference and Exposition, Conference Proceedings* (Vol. 122nd ASEE Annual Conference and Exposition: Making Value for Society). Retrieved from https://www.scopus.com/inward/record.uri?eid=2-s2.0-84941996184&partnerID=40&md5=6ab8d46079c7bf40467d03c0ba4bd417

Sommerauer, P., & Müller, O. (2014). Augmented reality in informal learning environments: A field experiment in a mathematics exhibition. *Computers & Education*, *79*, 59–68. doi:10.1016/j.compedu.2014.07.013

Stake, R. E. (2010). *Qualitative research: Studying how things work*. Guilford Publications.

Wu, H.-K., Lee, S. W.-Y., Chang, H.-Y., & Liang, J.-C. (2013). Current status, opportunities and challenges of augmented reality in education. *Computers & Education*, *62*, 41–49. doi:10.1016/j.compedu.2012.10.024

KEY TERMS AND DEFINITIONS

Affordances: Perceived features on things that promote the human action.

Augmented Reality: Is an emergent digital technology that integrates a virtual reality into the real world and allows the physical manipulation of virtual imagery.

Digital Culture: Is noticed by the wide use of technological devices in daily life that have changed the way to keep in communication and have access to information anytime and anywhere.

Dynamical Representation: A simulation or animation produced in a digital medium and performed by a digital device that includes changes in time, involving time to impact the visual perception.

Spatial Ability: It is related with the active visual perception of features occurring in a 3D environment, focusing in the global and also in the particular showing an intentional attitude.

Symbols: The representation of something else with an intentionality of communication.

Visualization: It implies the use of visual perception and gesture in order to infer useful information and be able to represent it to others.

Chapter 4
The Impact of Augmented Reality and Virtual Reality Study Material in the Future of Learning:
A Teamwork Experience

Giuliana Guazzaroni
Università Politecnica delle Marche, Italy

ABSTRACT

Mobile augmented reality offers important opportunities for learning. Moreover, it may represent new challenges for teachers and researchers. Implementing an augmented reality (AR) or a virtual reality (VR) learning experience involves the exploration of unusual pedagogical and technological boundaries. According to recent studies, it would be more productive to consider the augmented reality as a concept rather than an educational technology (Guazzaroni, 2015; Wu et al., 2013). This chapter is devoted to analyze a high school class of 23 students invited to use AR and VR tools to create their own study material. They are about 16-year-old attending Istituto Tecnico Tecnologico "Eustachio Divini" in San Severino Marche, Italy. The basic idea of the trial is to create a short printed document augmented with the technologies of AR and VR. The experience is evaluated using tests and direct observation. The aim is to observe the impact of augmented mobile learning and to demonstrate that AR and VR study material may represent a new communication object adequate to teach future students.

INTRODUCTION

Augmented reality (AR) technology has been increasingly applied to various subjects. AR allows offering additional information using physical operation as a medium, so that users can visually see the integration of the real world and virtual images (Chang et al., 2014). Consequently, AR preserves its users' connection with the real world. AR is generally considered an educational medium progressively

DOI: 10.4018/978-1-5225-2110-5.ch004

used by students and educators. Previous research has shown that AR technology has the potential to improve learning. Educational content can be available through different media. Digital learning experiences have been available in classrooms equipped with computers, interactive whiteboards, and more recently, learning experiences are increasingly reachable through pupils' personal devices such as smartphones or tablets. Consequently, the interaction with learning experiences is changing: people do not only use keyboards and mice to interact with screen content, but they can use their whole body to interact with educational study material that appears to exist in the physical world. This is possible through augmented reality technology (Radu, 2014). On the other hand, virtual reality (VR) technology is immersive. Users need headsets to block out the external world and literally "immerse" themselves in a digital realm. In VR, the concept of immersion is referred to user's involvement in the virtual environment. It causes a lack of awareness of time and of the real world, as well as a sense of "immersion". The "spatial immersion" in the virtual reality is a perception of being physically present in a non-physical world that provides a very absorbing environment (Freina & Ott, 2015). AR and VR are two closely related but very different technologies become widely popular. AR and VR rely on similar technology. The essential difference is that VR is immersive. In fact, the headsets block out the exterior world. VR main applications, therefore, are in video games and films, where users will lock themselves into computer-generated worlds. On the other hand, AR maintains its users' connection with the real world and a headset is not necessary. Students, using AR or VR, face a massive number of information and are immersed in a complex learning environment. AR and VR do not suggest a mechanical learning methodology. According to Manzelli (2013), AR and VR technologies offer a quantum correlation. They imply, in fact, the presence of correlations, as indeterminate sequences, chosen among the possible visions of the systems involved in the representation of images. These determine the non-local nature of perception using AR and VR. The quantum entanglement, or quantum correlation, is a metaphor that can be applied to the process of experiencing AR and VR objects. Mixing, in a real environment (e.g. The city, the school, the library etc.), physical elements with synthetic ones it is possible to observe, as in a quantum correlation, separate and, at the same time, interconnected systems. Moreover, using AR and VR technology in education means opening to informal learning environments. Furthermore, these technologies can favorite the transition from the "Learning Work" to the "Learning Adventure". In other words, a positive class atmosphere where students' activities are aimed to satisfy a personal desire of acquiring Knowledge (Leo et al., 2010). AR is an ideal bridge between the virtual world and the real one, offering an amplified perception (Bronack, 2011; Klopfer & Squire, 2008; Wu et al., 2013), visible to the user through mobile personal tools (e.g. smartphones, tablets etc.). The coexistence of virtual and real objects allows learners to visualize complex spatial relationships, quantum correlations or other abstract concepts. The student is allowed, for example, to experience and interact with three-dimensional synthetic objects to improve specific skills that could not be developed in traditional learning environments. These educational benefits are useful to see AR as an emerging educational technology (Squire & Klopfer, 2007; Wu et al., 2013, Guazzaroni, 2015).

This chapter is devoted to analyze a high school class of 23 students invited to use AR and VR to create their own learning content. They are about 16-year-old, mainly male, attending Istituto Tecnico Tecnologico "Eustachio Divini" in San Severino Marche, Italy. They generally use their smartphones to connect to the Internet, search information, communicate and play games. Nevertheless, when these students have to study a subject matter (e.g. History, Geography etc.) they chiefly use traditional paper books or school desktop computers. They need a clear model to improve their study method and to use new learning augmented and virtual medias. The basic idea of this learning experience is to create few paper notes (a short printed document) augmented with AR and VR. To access further content in AR, they can use their personal tools (e.g. smartphones plus apps) and to see VR they can use Google Cardboards or similar headsets. Google Cardboard is a VR and AR platform developed by Google used with a head mount for a mobile phone. Named for its foldout cardboard viewer, the platform is planned as a low-cost system to encourage interest and development in VR and AR applications. The subject matter selected to implement this immersive experience into school study material is Information Technology System Design. Students are encouraged to create their content using AR and VR and to collaborate in a team. The experience is fully evaluated using tests, autorating, and direct observation. The aim is to observe the impact of augmented mobile learning and to demonstrate that AR and VR study material may represent a new communication object adequate to teach neo-millennials and future students. The experience (Figure 1) will be described in this chapter.

BACKGROUND

Six Thinking Hat

According to De Bono (1985), in a positive team people should be different. *Six Thinking Hats* (1985) is a system by Edward De Bono that describes an instrument for group discussion and individual thinking involving six colored hats. This theory is helpful for teams to plan thinking processes in a detailed and cohesive way, and to think together successfully. People in a team should metaphorically wear six colored hats. Homogeneous teams are less brilliant than those with different opinions and attitudes. *Six*

Figure 1. A student enjoys an educational VR experience

Thinking Hats is a practical thinking framework to help people reflect plainly and directing their thinking attention in one direction at a time. The white hat represents facts, the green hat creativity, the yellow hat benefits, the black hat concerns, the red hat feelings, and the blue hat process.

A Teamwork Experience

Belbin (2003) observes that there are at least nine different profiles in a teamwork experience. In fact, by understanding the specific role within a particular team, people may develop their strengths and manage their weaknesses as team members, and consequently improve how they contribute to the group. The different roles are the following:

- **Plant:** Creative, imaginative, unorthodox. Solves difficult problems. Ignores incidentals. Too preoccupied to communicate effectively.
- **Resource Investigator:** Extrovert, enthusiastic, communicative. Explores opportunities. Develops contacts. Over-optimistic. Loses interest once initial enthusiasm has passed.
- **Coordinator:** Mature, confidant, a good chairperson. Clarifies goals, promotes decision-making, delegates well. Can be seen as manipulative.
- **Shaper:** Challenging, dynamic, thrives on pressure. The drive and courage to overcome obstacles. Prone to provocation.
- **Monitor:** sober, strategic and discerning member, who tries to see all options and judge accurately. Lacks drive and ability to inspire others.
- **Team Worker:** Co-operative, mild, perceptive and diplomatic. Listens, builds, averts friction. Indecisive in crunch situations.
- **Implementer:** Disciplined, reliable, conservative and efficient. Turns ideas into practical actions. Inflexible. Slow to respond to new possibilities.
- **Completer:** Painstaking, conscientious, anxious. Searches out errors and omissions. Delivers on time. Inclined to worry unduly. Reluctant to delegate.
- **Specialist:** Single-minded, self-starting, dedicated. Provides knowledge and skills. Contributes only on a narrow front. Dwells on details (Belbin, 2003).

Positive work in a team is essential to reach effective learning results in a cooperative class. Consequently in this research students were divided in groups of about five members. Each participant chooses his/her favorite role and reflects on it using Belbin's role model.

Teamwork Using AR and VR Technologies

In a traditional classroom learners work together better if they are focused on a shared workspace. In computer-based education, students generally work on separate computers. Group communication patterns are different in front of a computer. When users collaborate in front of a screen they sit side-by-side and their attention is on the computer desktop. On the other hand, when learners use an AR interface, they may sit around a table and the conversation seams similar to natural face-to-face collaboration (Kiyokawa et al., 2002; Billinghurst, 2002). In a collaborative environment students use the resources of the physical world to convey meaning. AR presents a deep relationship between virtual and physical objects. Real things may be enhanced by AR elements to provide dynamic information and interaction (Billinghurst,

2002). AR may be seen as a step between reality and VR, for the benefit of education (Pasareti et al., 2011). Although AR technology is not new, its impact in education has been discovered recently. It may be seen as a metaphor or as a bridge for transitioning between real and virtual worlds (Billinghurst, 2002). According to some authors, the neo-millenials generation has to cope with a completely different reality, and it is suggested to use ICT and especially AR to maintain their attention at school (Pasareti et al., 2011). The term AR is used in different ways by researchers. According to Wu et al. (2013), AR should be seen as a concept rather than a type of technology. Three principal categories of instructional approaches have been elaborated: approaches accentuating engaging learners into "roles", approaches highlighting learners' interactions with physical world and "locations", and approaches underlining the design of learning "tasks" (Wu et al., 2013). The approaches, highlighting engaging students into different roles in an AR setting, include role-playing and simulations. These methods point out the interactions and collaboration among students and are associated with mobile augmented learning and game based learning. Location-based learning is another approach; it emphasizes the interactions with physical environment using mobile devices and geological tracking of the participants. Through geological positioning systems and smartphones students' access information as they arrive in a certain location. Learning is authentic and users are dislocated in a physical environment (e.g. school campus). The third category underlines the tasks in an AR environment. Examples are: game-based, problem-based etc. Game-based learning is one of the most popular approach for AR. "A game provides digital information to one or more players, takes and processes input from players, and changes the digital information provided to the players based on the input" (Wu et al., 2013, p. 45).

However, the design of an AR experience for learners generally includes different methodologies including more than a single approach (Wu et al., 2013). The learning trial analyzed in this chapter mainly focuses on emphasizing the roles and the tasks. On the other hand, the locations are less important. In fact, only one group of students realized a virtual tour of their school. Research has shown that AR could be more effective than other technologies to develop crucial skills and knowledge (El Sayed et al., 2011; Wu et al., 2013). Students have the opportunity to manipulate interactive learning objects, to handle information in a lifelike way. An AR learning environment offers a sense of authenticity. Consequently, students can experience "dynamic models and complex causality" (Wu et al., 2013). Moreover, AR environments increase learners' motivation and interest in the topic (Guazzaroni, 2015). Immersive VR environments offer a great variety of experiences; students cannot normally practice such a significant selection of real situations at school. VR development can be found in head-mounted devices used by fighter pilots. The original aim was to place people in environments that barely contain the information they needed in order to interact in a natural way. "As a result of total immersion in a virtual world, participants report a very real sense of being in another place". This phenomenon is known as "cognitive presence" and is joined together with the "conviction that a virtual world is a valid, though different, form of reality". The first VR systems were designed to simplify the interface for airplane pilots. According to some authors, VR is immersive and the interface has been removed. In fact, the participants wear the computer and they can easily interact with the virtual environment. The subject-object distinction that exists between individuals and information in computers disappears when the learner use VR environments. Consequently, experiencing virtual worlds is a first-person experience. "First-person experiences are therefore natural, non-reflective, and predominant in our everyday interactions with the world [...] The interface creates a boundary around the computer and its information, and establishes the distinction between us – subject – and it – object". VR environments allow users to immerse in a virtual world without interfaces. The experiences in VR offer the same quality as experiences in the real world (Winn, 1993). VR should be designed in such a way that users would find it difficult to tell the difference from what is real and what is not. Though, VR is usually achieved by the wearing of a VR helmet or goggles (e.g. Google Cardboard or Oculus Rift). On the other hand, AR is the combination of VR and real life. Using AR, users are able to interact with virtual contents in the real world, and are able to distinguish between the two. Both virtual reality and augmented reality are related to the goal of immersing the participant, though both technologies do this in different ways. With AR, users continue to be in touch with the real world while interacting with virtual objects around them. With VR, the user is isolated from the real world while immersed in a world that is completely built. AR and VR both have the striking ability to alter human perception of the real world. Where they differ, is the perception of users' presence. VR transposes participants. In other words, VR brings learners elsewhere through visors. VR allows people to have virtual tours outside a room or a school and it allows students to interact with the world using a "natural semantic" of the world (Winn, 1993). AR however, adds something to the real world. It does not move users elsewhere, it barely augments the natural environment. Azuma (1997) defines AR as something that "allows the user to see the real world, with virtual objects superimposed upon or composited with the real world. Therefore, AR supplements reality, rather than completely replacing it". The result is a multisensory experience. Augmented reality offers an impression of sensory integration with the perception that the user has of the environment in which he/she is located. It is a superimposition of different levels of information integrated with real objects (Guazzaroni, 2013).

Experiential Technologies

Nowadays nomadic technologies are emergent. Moreover, many different social practices, supported by mobile devices, have reinvented the way people stay connected. In other words, mobile devices have changed "the rhythms of social time" and have altered the "uses of social spaces" (Alexander, 2004). New shared practices have been reinvented the reality through critical remix, mash-up, new contextualization and re-enactment of real and virtual objects. Griziotti (2011) describes the concept of "bio-hypermedia" to describe wearable technologies. He affirms that: "The bio-hypermedia is the birth of a new paradigm of internet-meets-body, prefigured by "always connected" devices like smartphones, tablets and others, invented in qualitatively and ontologically different ways to be infinitely more powerful than landline internet and the PC. This is the first media that interacts with our lives, freeing the exchange of information from the fixed time and place to which television and the PC are limited".

The exceptional growth of user-generated media (e.g. Facebook, Instagram etc.) can be considered as critical and disruptive for common users. In fact, "a new creation of the commons in all its forms – including revolt – are developing not only more and more rapidly and in vaster and vaster spaces, but are overwhelming the old world in ways and in locations that are quite unpredictable" such as Tunisia, Egypt, Brazil, Santiago del Chile, Tel Aviv etc. (Griziotti, 2011). What are the youth using their mobile phones to? An important characteristic is the adoption of joining mobile Internet. Consequently, people using smartphones are always connected to the web and they have the possibility to freely interact with peers, friends etc. Teenagers have launched the rules of a new mobile culture using mobile tools as a "lifeline" to share images, texts, and music in different social networks (Bressler, 2006). Damasio (2010) says that "the current digital revolution, the globalization of information and the arrival of the era of empathy represent the pressing examples that can lead to structural changes of the mind and the Self, attacking the cerebral processes that shape the mind and the Self". These powerful tools, capable of generating the "bio-hypermedia", and deeply rooted into youth's culture are the same that are used by educators to engage teens and young adults into formal and informal learning environments.

7E Learning Cycle

Huang et al. (2008), have proposed a concept for outdoor learning in the area of Natural Science for elementary schools. In this concept they have used the 7E learning cycle for technology enhancing outdoor learning (Eisenkraft, 2003; Huang et al., 2008). The 7E learning cycle seems to be suitable for mobile augmented learning. In fact, it allows students to gradually get acquainted with the complexity of a real-world learning environment. The seven phases can accompany the learners in a real setting (e.g. Museum, Library etc.), from the preparation of the experience to its evaluation. The student is firstly "engaged" with the cognitive contents, successively he/she starts to "explore" and discover the real locations, the tasks and the roles. After the exploration phase, the user has to be more active and elaborate new contents. At the end, the learner can "extend" the experience and the process is wholly evaluated.

The 7E learning cycle is divided in the following steps (Eisenkraft, 2003; Huang et al., 2008; Guazzaroni & Leo, 2011, Guazzaroni, 2012; Guazzaroni, 2013):

1. **Elicit:** Teachers prepare useful technologies (e.g. software, applications for different types of smartphones etc.) and appropriate didactic materials (e.g. videos, texts, pictures, tests etc.);
2. **Engage:** Teachers explain the experience they are going to have. Teachers divide students into several groups;
3. **Explore:** Students actively interact with AR and VR;
4. **Explain:** Groups start an active interaction with AR and VR. Groups create their contents;
5. **Elaborate:** Each group develops the final version of their outputs;
6. **Extend:** Students gather additional learning content or self assess their performances;
7. **Evaluate:** Teachers collate information to evaluate the experience.

The 7E learning cycle is useful to engage students with AR and VR learning environments to help learners to become more and more autonomous while interacting with new technologies. According to Wu et al. (2013), there are pedagogical issues that should be taken into consideration when AR is offered in a classroom. The use of AR could encounter constraints from schools or traditional teachers. In fact, AR and VR learning environments involve innovative methodologies such as teamwork, creative thinking, social co-construction of meaning, learning by doing (Dewey, 1916; Schank, 1995). It is learning in real life, which is the natural setting for learning something "by doing it" and situated learning (Lave & Wenger, 1991). Lave and Wenger (1991) affirm that learning is not the simple transmission of decontextualised knowledge from a teacher to a student, but a social process where knowledge is co-constructed; they suggest that learning is situated in a real context and embedded within a particular social and physical environment. When implementing new approaches the role of the teacher could not be conventional. The teacher can be seen as a facilitator and the classroom can be student-centered. For this reason teamwork is crucial. Moreover, the students' roles into groups should be appropriate to co-construct effective outputs. Mobile augmented learning is an educational environment, based on an accurate learning methodology, characterized by the use of handheld and smart devices, and augmented by the use of AR and VR facilities. It can represent a unique didactic opportunity if used in a proper manner. Mobile augmented learning is characterized by a deep change in learning scenarios. The facility of communication with teachers and peers offers the opportunity to design interactive and social activities to engage students in a real-world learning experience. Learning objects have changed, as well. For example, they can be embedded in an authentic environment to support situated learning (Lave & Wenger, 1991). Furthermore, mobile learning requires personal initiative. In fact, the participant has an active role in learning. The learner is no longer a passive consumer of predetermined educational resources, but he/she is a "prosumer", aware that today's knowledge is everywhere. Nowadays, peers interaction and networks appear to be decisive. The impact of new technologies in every field has determined a rethinking of knowledge, teaching and learning. New technologies and social media have profoundly influenced education (Leone & Guazzaroni, 2010).

Technologies of the Future

Artificial intelligence (AI) and robotics are technologies that will influence learning environments (Figure 2). Their impact will be profound and positive. China, India and other large countries generate an ample amount of data. All these records need AI's capabilities to be processed and used. Over the last decades, applications of AI have affected learning through specific tutor systems. Popular examples of AI for learning are an "Intelligent Tutor System" to track mental steps of students during problem solving tasks or a learning application useful to organize content delivery. AI and machine learning are rapidly developing to help students and teachers to organize their work. In fact, AI can play a crucial role in the field of learning analytics, evaluation, adaptive learning and mobile augmented learning. Using AI, it can be possible to create unique learning pathways for a student in MOOCs or in e-learning courses (Kurshan, 2016). A mobile augmented learning environment, as shown in the experience described in this chapter, are complex learning spaces. Consequently, specific applications of AI will be quite useful. Students using AR or VR to create their own study material need to master different ICT tools and software to develop proper results. An AI application can help learner in the process of creating their project and educators in the evaluation of the experience and in the assessment of pupils. Worldwide,

Figure 2. A student enjoys a learning experience using AR

people using video games spend 3 billion hours a week coping with them. Consequently, well-developed games require motivating and engaging elements to captivate players for hours on end. For this reason, educators should engage learners the way games engage players (Perry, 2014). AI will play a crucial role in gaming and in creating edutainment tools for individual learner. AI will strongly affect AR and VR game based learning as the content for the user can be properly selected using an intelligent agent application. Moreover, AR, VR and wearable computing have the potential to create a realistic learning environment using special glasses to receive augmented information while looking at real objects such as paintings or other real objects (Leue et al., 2015). Internet-of-Things (IoT) is the convergence of Internet with sensor and smart objects. IoT can be well-defined as "things belonging to the Internet" to supply and access real-world information. Billions of devices can be associated in networks and IoT is mainly considered a promise of technology today (Singh et al., 2014). The advancing of futuristic new technologies affects instructional design and teaching. New ICT tools can change the learning experience and for this reason it is important to study the effects of integrating technology in educational settings, and observe how such technologies can be maximized to improve learning (Santos et al., 2014).

A TEAMWORK EXPERIENCE USING AR AND VR LEARNING TOOLS

AR and VR Study Material

Participants in an augmented reality environment are immersed in a complex learning space. AR does not suggest a mechanical learning approach, implementing an AR environment at school metaphorically offers a quantum correlation. It implies, in fact, the presence of connections, as indeterminate sequences, chosen among the different possibilities (Manzelli, 2013). The concomitance of virtual and real objects lets students to manage complex spatial relationships, quantum correlations and abstract concepts. Learners experience and interact with three-dimensional artificial objects to develop specific skills. AR and VR learning environments may be considered as emerging educational scenarios (Squire & Klopfer, 2007; Wu et al., 2013, Guazzaroni, 2015). This chapter is dedicated to analyze a high school class of 23 students invited to use AR and VR to create their own study material. They are teens attending an ICT specialization in an Italian high school. They generally need a well-defined model to improve their study method and to use augmented and virtual medias. The aim of this learning experience is to create a short printed document augmented with AR and VR technologies. To access further content in AR or VR, students can use their smartphones and Google Cardboard, a low-cost platform to encourage mainstream development of VR. The subject matter selected to implement this immersive experience into school study material is Information Technology System Design. The experience is fully evaluated using tests and direct observation. The aim is to observe the impact of augmented mobile learning and to demonstrate that AR and VR objects may represent new communication medias adequate to teach future students. Two teachers of Information Technology System Design and an external expert proposed the activity to the students at the beginning of school year 2015/2016. After some lectures devoted to AR and VR implications, the educators and the specialist asked the students to divide into different groups and to choose an AR or VR tool to work with. The teachers to trial the impact of AR and VR environments followed the 7E learning cycle. The use of the 7E learning cycle was suggested by some previous experience with AR (Eisenkraft, 2003; Huang et al., 2008; Guazzaroni & Leo, 2011; Guazzaroni, 2012):

1. **Elicit:** Teachers prepare useful technologies, learning objects and lectures.
2. **Engage:** Teachers explain the experience they are going to have. Educators help students to divide into several groups and to choose a role.
3. **Explore:** Students actively interact with AR and VR.
4. **Explain:** Groups start an active interaction with AR and VR. Groups create their contents.
5. **Elaborate:** Each group develops the final version of their outputs.
6. **Extend:** Students swap outputs and self assess their performances.
7. **Evaluate:** Teachers organize information to evaluate the experience.

The groups were four and were divided as follow:

1. **Virtual Tour of an ICT Institute:** Six members composed the group, they were devoted to use Google Cardboard and Google Street View to create a virtual tour of their school.
2. **Inferno, Canto III – The Divine Comedy:** The group was formed by six members; they chose Aurasma, an AR application, to activate an image of Charon and animate this character.
3. **Martin Luther:** Five members formed the group; they selected Aurasma to activate an image of Martin Luther.
4. **3D School Logo:** Six members composed this group, they used Augment, a 3D augmented reality application, and a 3D printer to create a school logo visible using AR and printable in 3D.

The different members of the four groups were asked to choose a role using Belbin's team roles (Belbin, 2003). The Belbin team role model is designed to use the aptitudes and personalities of individual members working together in a group. Each team is composed by different roles to achieve the highest performances. The roles are nine and are basically complementary. A team role can be defined as the style of work of a team member and originates from the personality of the single. Teams work best when there is an equilibrium of crucial roles and when team members exactly know their roles, their strengths and weaknesses. To achieve the best balance, there should be: 1. A Coordinator or Shaper (not both) for leader; 2. A Plant to inspire ideas; 3. A Monitor to preserve clarity; 4. One or more Implementer, Team worker, Completer (Belbin, 2003). At the beginning of the experience, during the "engage" phase, students had to select their role in a chart showing Belbin's nine roles. The result was the following:

1. **Virtual Tour of an ICT Institute:** Six male members composed the group; they selected one Coordinator, a Monitor, a Specialist, two Completers and a Team worker.
2. **Inferno, Canto III – The Divine Comedy:** The group was formed by six members (four boys and two girls); they chose a Shaper; a Plant, a Monitor, an Implementer, a Specialist, a Resource investigator.
3. **Martin Luther:** Five members composed the group (four girls and one boy); they selected one Coordinator, a Monitor, a Specialist, an Implementer, and a Team worker.
4. **3D School Logo:** Six male members composed this group and they chose a Coordinator, a Plant, a Monitor, a Resource investigator, a Specialist, a Team worker.

Each student needed some time to select his/her specific role in the team. Additional time was spent explaining roles and how to work positively in a group. During this phase educators and specialist tried to engage students in the new activity (7E learning cycle: "engage"). Before this phase teachers and specialist offered some talks to elicit interest and engage learners in the social co-construction of study material using AR and VR environments (7E learning cycle: "elicit"). After these first stages, each group tried to explore AR and VR tools to select the right resources and to build a proper experience. The teachers helped each member. The specialist and the peers were facilitated to develop necessary skills to use AR and VR environments, combined with other useful applications (7E learning cycle: "explore"). AR may be used to enhance collaborative tasks, in AR there is a close relationship between synthetic and real world objects. "The physical objects can be enhanced in ways not normally possible such as providing dynamic information overlay, private and public data display, context sensitive visual appearance, and physically based interaction" (Billinghurst, 2002). Immersive VR environments offer first person and non-symbolic experiences that can be useful to help learners (Winn, 1993). The 23 students divided in four groups established a form of collaboration in the team using the assigned role. In the following phase (7E learning cycle: "explain"), learners developed their skills to create their specific content. The group: "Virtual Tour of an ICT Institute" composed by a Coordinator, a Monitor, a Specialist, two Completers and a Team worker was the most autonomous group. The Specialist was single-minded, self-starting and dedicated. He provided specialist knowledge and skills and the group creates a well-built tour of their Institute. The Monitor actively helped the others providing a logical eye. Their weakness was that they found it difficult to write a clear report of their activities. For example, they created a special tripod, using a carton, to take panoramic pictures. Unfortunately, they destroyed the object after having created their virtual tour. They used Google Street View, an app to take panoramas; Vieweet a web based software to build virtual tour; a QR-Code generator; Appypie to create an app for iOS and Android. Moreover, they elaborated a web site for their project. The use of a mix of different tools is a positive feature for education. In fact, an AR system should be designed by connecting different types of technologies (Wu et al., 2013) to create an experience. The group Inferno, Canto III – The Divine Comedy, formed by six members (four boys and two girls), a Shaper; a Plant, a Monitor, an Implementer, a Specialist, a Resource investigator was enough autonomous. They used an illustrated edition of The Divine Comedy to implement their activity. They found out the Divine Comedy, drawn by Lino Bianchi Barriviera, in their school Library. They selected a suggestive picture of Charon demon (trigger image) to activate a reading of Canto III, recorded using their voices.

To create their final output, they used Aurasma Studio, a web based application to generate AR experiences, Camtasia Studio to create a video of Canto III and TeamSpeak3, a software for the team internal communication. Moreover, they elaborate a web site for their project. The roles in the group were well balanced and they work agreeably in their team. The group Martin Luther was composed by five members (four girls and one boy), a coordinator, a monitor, a specialist, an implementer, and a team worker. This team needed scaffolding to build their output. To realize the final product they used Aurasma Studio and Movie Maker to animate Luther. The coordinator was not confident with her role and their task was redesigned with educators. Six male members, a coordinator, a plant, a monitor, a resource investigator, a specialist, and a team worker composed the group 3D School Logo. They used Augment, an AR software, to build a 3D logo of their school, they create the logo using Tinkercad. Their object was printed in 3D using CUBE 3. The group was enough independent and the roles were adequate to these students. Finding the proper role means to have better opportunities to succeed. Moreover, having a specific role means that each student knows how to contribute to the co-construction of educational material. Another important aspect is that the first phases of the 7E learning cycle are crucial. Students need clear instruction when they start a new activity, especially when they work with new medias. After having completed their group activities students swap outputs and self assess their performances. They fulfilled these tasks in an ICT Laboratory; each student could use a personal computer and one Interactive whiteboard (IWB) to show outputs to the other peers (7E learning phase: "extend").

The Autorating System

According to Brown (1995), it is difficult to "assess individual performance within teams from the results of team assessments". In fact, individual team members normally know who are the real contributors on their team and who are the "passengers". Consequently, if members had the power to influence over students' marks, assessment could be more accurate (Brown, 1995). For this reason, the two teachers and the specialist decided to offer team members the opportunity to assess themselves and their peers. The Autorating system has been used since 1991 in the Communication & Electronic Department and Computer Systems at the Royal Melbourne Institute of Technology University (RMIT), Australia. The rating category are: 1. Excellent (100%); 2. Very Good (87,5%); Satisfactory (75%); Ordinary (62,5%); Marginal (50%); Deficient (37,5%); Unsatisfactory (25%); Superficial (12,5%); No Show (0%). The students analyzed in the trial, mostly assessed themselves choosing the first three categories ("excellent", "very good" and "satisfactory"). Someone decides that two or three members were "unsatisfactory. The group: "Virtual Tour of an ICT Institute" resulted mainly "very good" students. They assigned to themselves a very good mark. In fact, it seemed to the teachers that the members of this team were very autonomous but they failed to write a complete report of how they realize their output. On the other hand, the app they created has been used for school guidance. The group Inferno, Canto III – The Divine Comedy self-assessed using high rates (mostly "very good", some "excellent" and one "ordinary"). Their output was complete, the report too short, but the activity they created using AR app Aurasma was fine and they worked autonomously. Educators considered the group Martin Luther the weaker one, they were not autonomous, and they needed lot of scaffolding and help to realize their AR experience. Surprisingly, their self-assessment was unanimously "excellent". This result was probably spoiled by the fact the team members need high marks to catch up inadequate previous marks. The group 3D School Logo used a wider range of marks to assess them. They mainly consider themselves "satisfactory" students or "very good" ones. They appear very honest, hardworking and interested in the activity. These marks were summed up with those assigned by educators to form the final mark. The autorating system facilitates effective teamwork and devolves power to student team members to reward deserving members in teams (Brown, 1995). In the trial, autorating helped educators to simplify the complexity of teams and the collaborative work. The last phase in the 7E learning cycle is devoted to evaluation; during this phase teachers collect elements to evaluate the experience. The whole experience resulted very innovative and needed new tools to be evaluated in a proper manner. Students affirmed to have appreciated the experience; only one out of 23 declared the activity was not useful for his future. Participants wrote their opinion about the use of AR and VR in education, based on their real experience. The result was satisfactory and educators indicated proper way to repeat the experience. The students had to answer to the following questions in a virtual environment:

1. After having completed your experience and realized your AR or VR output, do you think that these tools could be used in the school of the future?
2. After having completed your experience and realized your AR or VR output, do you think that you will continue to use these tools in the future for learning or working?

3. After having completed your experience and realized your AR or VR output, do you think that these tools are useful for learning or working?
4. After having completed your experience and realized your AR or VR output, write a comment about it.

All the students answered the first question. Twenty-two learners believed that AR and VR tools could be successfully used in education. Only one student thought that AR and VR tools would be hardly used in the future. In his opinion traditional schoolbooks were more efficient for learning. For the second questions, eighteen learners thought he/she would probably use AR and VR in the future for learning or working. Three students were convinced they would use these tools in a near future and only one would use AR and VR only if strictly necessary and for a specific aim. Most students thought AR and VR environments were useful for education and business (fifteen). Six learners believed AR and VR were very advantageous and only one student was persuaded that these tools were barely handy. The participants were interviewed and most of them affirmed that they did not know AR and VR before; they said these tools were very interesting for the future of education and business. Most of them described AR and VR environments as a discovery of a new interesting world. Others enthusiastically highlighted technological performances of AR and VR and their relevance in a near future. A few students strongly believed that AR and VR devices were funny tool, useful both for entertainment and learning. One girl pointed out the importance of team working to accomplish new tasks and using unknown technologies. In her opinion, working in a team could be a great way to achieve important goal.

ISSUES AND FINDINGS

The technology may promote the diffusion of a "Learning Adventure" approach. Using innovative technology to enhance learning may have a strong impact on learning organizations (Leo et al., 2010). In fact, such a use of technology could change the future education in institutions. For "Learning Adventure", Leo et al. (2010) mean learning activities directed to "satisfy a personal desire of acquiring Knowledge and/or aimed at providing proficiency in some competence domain". According to the above-mentioned authors there are three main categories that can be considered as "individual learning drivers": 1. Curiosity; 2. Desire of success; 3. Social/familiar pressure. During the trial students were curious to learn an innovative way to interact with peers using AR and VR environments. They really desired to finish their activities to prove themselves and others they could manage complex tools. Familiar and social pressure was an irrelevant factor; in fact families were not informed of this activity. After having produced their output, students expressed their satisfaction. During the "extend" phase of the 7E learning cycle, most learners were proud to show others their results with AR or VR. Moreover, they appreciated their involvement in the self-assessment process. The most important positive results emerging in the evaluation phase are:

1. Students could construct their learning material using VR and AR environments.
2. Students could construct their learning material working in a small group of peers.
3. Student could choose a specific role, useful to acquire an assertive behavior and develop autonomy.
4. Create an immersive object is fun.
5. Improve the creative mind (Gardner, 2006).

6. Develop new skills (e.g. use AR and VR tools).
7. Educators use learners' outputs to offer guidance to students with special needs.

During the experiment students appreciated working in small groups with a specific role. At the beginning in the "engage" phase (7E learning cycle), peers need lot of scaffolding and help to decide their specific role using Belbin's nine profiles scheme. But while working they realized the importance of roles, of their strengths and allowable weaknesses in the group. On the other hands, for negative aspects, the peers could manage with divisive behavior and, most of them, declared that they needed time to find out solutions accepted by all the members of the group. Working in a team is not easy and Belbin's material provided a handful tool. Most students appreciated using AR and VR applications. They could learn how to build an immersive learning object and how to deal with a 3D objects and eventually print it. In fact, AR and VR environments may offer "contextualization, authenticity, and engagement" (Wu et al., 2013) to all these students. They could feel they were participating to an important project for their future career and tried to elaborate excellent outputs. Additionally, some of them declared that the activity was fun. In fact, an "AR moulded-object" (Guazzaroni & Compagno, 2013) is a unique outcome resulting from real images mixed with audio, video, animations etc. The designer of the experience (or learner) has the power to "mould" real world objects and create educational dynamic performances (learning content). Students can improve their creativity and "moulded-objects" are open to be moulded in a countless way. The learner develop new skills for the future; they can design new activities and try a new way of learning. According to El Sayed et al. (2011), AR environments can help developing new skills and knowledge in a more effective way if compared to other technology enhanced learning environments. (Wu et al., 2013). Educators, can use the output of the trial to offer guidance to new students with special needs. The multimedia output seems to attract learners and to establish a more active interaction with the content of the "moulded-object" (e.g. Martin Luther).

SOLUTIONS AND RECOMMENDATIONS

Educators should explain roles and offer scaffolding. During the experiment, Belbin's role (2003) needed a solid explanation and individual conversation with students to help choosing the right role and to improve individual behavior while working in a group. Students required support to write reports about their activity in the group. They found it difficult to dispose a clear text containing detailed activity. For example, a group destroyed the special tripod they had create to take 360° pictures without taking clear documentation of their interesting tool. In fact, many students are not acquainted with writing and taking notes of their learning activity. On the other hand, educators should propose them, in alternative, a video explanation or a tutorial of the students' activities to complete their output. When asked about this alternative, students answered that it was quite easy for them creating video documentation instead of writing a text report.

FUTURE RESEARCH DIRECTIONS

AR and VR environments are emerging trends in education. Google cardboard, for instance, represents an economic platform to share a 3D world and an immersive educational experience. Moreover, integrating multiple technologies has a substantial potential to enhance learning and teaching. Using AR or VR environments is useful to consider the content, task or problems in different perspective, making invisible become visible. An inauthentic task can be transformed in an authentic one (Wu et al., 2013). VR allows first-person experience in virtual worlds or the possibility to visit special views, for instance, war settings or other places students could not visit. First-person non-symbolic experiences are natural are predominant in everyday interaction, using a VR headset student can experience a "natural semantic" of the world or of complex system (e.g. planets). VR promotes a strategy that allows pupils to learn from non-symbolic first-person experience. In fact, many students have low marks because they do not understand the symbol systems of subject matters. In these cases VR environments may offer a solution (Winn, 1993). AR books may represent a solution for the latest generation of students. For them it is highly recommended to make use of ICT to raise their attention. According to some author (Pasareti et al., 2011), a study has revealed that people prefer traditional books, but digital books offer interaction and collaboration between peers. Augmenting real books with AR and VR offers the advantage of both. With AR and VR facilities books can be enriched with different elements real and virtual. AR and VR unfold previously unexplored learning scenarios. The possibilities are still unknown, using mobile augmented environments students and teachers may expand their physical world and the potential for the future of education are enormous.

CONCLUSION

The described experiment analyzes a high school class of 23 students encouraged to use AR and VR instruments to create their own study material to raise and maintain their attention and to develop skills for the information technology system design. The basic idea of the trial was to create a short printed document, augmented with the technologies of AR and VR. The experience was evaluated using tests and direct observation. The aim was to observe the impact of augmented mobile learning and to demonstrate that AR and VR contents represent a new communication object adequate to teach future students. The results were positive and the educators considered the trial as a starting point for further developments. The most important results were: 1. Students could create their learning material using VR and AR environments; 2. They could construct their learning material working in a small group of peers; 3. They could choose a specific role, useful to acquire a self-confident behavior and develop autonomy; 4. They had fun while working; 5. They could improve their creative mind, a crucial attitude for the future (Gardner, 2006); 6. They developed new ICT skills; 7. Educators, after the trial, used learners' outputs to offer guidance to schoolchildren with special needs. This last result was important, in fact, it demonstrated that these new augmented learning objects may attract the last generation of learners and it may be flexible for different needs.

REFERENCES

Alexander, B. (2004). Going nomadic: Mobile learning in higher education. *EDUCAUSE Review*, *39*(5), 29–35.

Azuma, R. T. (1997). *A Survey of Augmented Reality*. Teleoperators and Virtual Environments.

Belbin, M. (2003). *Management Teams: Why They Succeed or Fail*. Oxford, UK: Butterworth-Heinemann.

Billinghurst, M. (2002). Augmented Reality in Education. In *New Horizons for Learning*. Seattle, WA: Technology in Education.

Bressler, D. (2006). Mobile Phones: A New Way to Engage Teenagers. In J. Trant & D. Bearman (Eds.), *Informal Science Learning. Proceedings of Museums and the Web 2006*. Toronto: Academic Press.

Bronack, S. C. (2011). The role of immersive media in online education. *Journal of Continuing Higher Education*, *59*(2), 113–117. doi:10.1080/07377363.2011.583186

Brown, R. W. (1995). Autorating: Getting individual marks from team marks and enhancing teamwork. In *Proceedings of Frontiers in Education Conference* (vol. 2, pp. 15-18). Atlanta, GA: IEEE Press. doi:10.1109/FIE.1995.483140

Chang, K. E., Chang, C. T., Hou, H. T., Sung, Y. T., Chao, H. L., & Lee, C. M. (2014). Development and behavioral pattern analysis of a mobile guide system with augmented reality for painting appreciation instruction in an art museum. *Computers & Education*, *71*, 185–197. doi:10.1016/j.compedu.2013.09.022

Damasio, A. R. (2010). *L'Autre Moi-même – Les Nouvelles Cartes du Cerveau, de la Conscience et des Emotions*. Paris, F: Odile Jacob.

De Bono, E. (1995). *Six thinking hats*. Boston: Little Brown and Company.

Dewey, J. (1966). Democracy and Education: An Introduction to the Philosophy of Education. New York: Free Press.

Eisenkraft, A. (2003). Expanding the 5E model. *Science Teacher (Normal, Ill.)*, *70*(6), 56–59.

El Sayed, N. A. M., Zayed, H. H., & Sharawy, M. I. (2011). ARSC: Augmented reality student card. An augmented reality solution for the education field. *Computers & Education*, *56*(4), 1045–1061. doi:10.1016/j.compedu.2010.10.019

Freina, L., & Ott, M. (2015). *A literature review on immersive virtual reality in education: State of the art and perspectives*. Paper presented at the conference of eLearning and Software for Education (eLSE), Bucharest, Romania.

Gardner, H. (2006). *Five minds for the future*. Boston: Harvard Business School Press.

Griziotti, G. (2011). *Murdoch and Berlusconi: the Fall of Two Media Empires and the Network Multitudes*. Retrieved April 13 2005, from http://www.opendemocracy.net/giorgio-griziotti/murdoch-and-berlusconi-fall-of-two-media-empires-and-network-multitudes

Guazzaroni, G. (2013). Emotional mapping of the archaeologist game. Advanced Human-Computer Interaction: Computers in Human Behavior, 29(2), 335-344. doi:10.1016/j.chb.2012.06.008

Guazzaroni, G. (2015). Realtà aumentata un'opportunità di apprendimento. In L. Salvucci (Ed.), Strumenti per la Didattica della Matematica. Ricerche, esperienze, buone pratiche (pp. 145-152). Milano, Italy: Franco Angeli.

Guazzaroni, G., & Compagno, M. (2013). AR moulded-objects performing Giuseppe Verdi's 200th birthday. *Archeomatica*, *4*, 38–41.

Guazzaroni, G., & Leo, T. (2011). Emotional mapping of a place of interest using mobile devices for learning. In I. Arnedillo Sánchez, & P. Isaías (Eds.), *Proceedings of IADIS International Conference on Mobile Learning* (pp. 277-281). IADIS.

Huang, K., Liu, T., Graf, S., & Lin, Y. (2008). Embedding Mobile Technology to Outdoor Natural Science Learning Based on The 7e Learning Cycle. In J. Luca, & E. Weippl (Eds.), *Proceedings of World Conference on Educational Multimedia, Hypermedia and Telecommunications* (pp. 2082-2086). Chesapeake, VA: Academic Press.

Kiyokawa, K., Billinghurst, M., Hayes, S., & Gupta, A., Sannohe & Y., Kato, H. (2002). Communication Behaviors of Co-Located Users in Collaborative AR Interfaces. In *Proceedings of the IEEE and ACM International Symposium on Mixed and Augmented Reality* (pp. 139-148). Los Alamitos, CA: IEEE Press. doi:10.1109/ISMAR.2002.1115083

Klopfer, E., & Squire, K. (2008). Environmental detectives: The development of an augmented reality platform for environmental simulations. *Educational Technology Research and Development*, *56*(2), 203–228. doi:10.1007/s11423-007-9037-6

Kurshan, B. (2016). *The future of artificial intelligence in education*. Retrieved December 8, 2016, from http://www.forbes.com/sites/barbarakurshan/2016/03/10/the-future-of-artificial-intelligence-in-education/#329b33a81e64

Lave, J., & Wenger, E. (1991). *Situated Learning: Legitimate Peripheral Participation*. Cambridge, UK: Cambridge University Press. doi:10.1017/CBO9780511815355

Leo, T., Manganello, F., & Chen, N. S. (2010). From the Learning Work to the Learning Adventure. In *Proceedings of EDEN 2010*. Academic Press.

Leone, S., & Guazzaroni, G. (2010). Pedagogical sustainability of interoperable formal and informal learning environments. In F. Lazarinis, S. Green, & E. Pearson (Eds.), *Developing and Utilizing E-Learning Applications*. doi:10.4018/978-1-61692-791-2.ch001

Leue, M. C., Jung, T., & Dieck, D. T. (2015). Google Glass Augmented Reality: Generic Learning Outcomes for Art Galleries. In I. Tussyadiah & A. Inversini (Eds.), *Information and Communication Technologies in Tourism* (pp. 463–476). Vienna: Springer Computer Science. doi:10.1007/978-3-319-14343-9_34

Manzelli, P. (2013). *Futuri orizzonti dell'arte quantistica e della realtà aumentata*. Paper presented at the meeting of Arte quantistica, scienza e realtà aumentata: sinergie d'innovazione, Pietrasanta, Italy.

Pasaréti, O., Hajdú, H., & Matuszka, T. Jámbor,i A., Molnár, I. & Turcsányi-Szabó, M. (2011). *Augmented reality in education*. Paper presented at INFODIDACT Informatika Szakmódszertani konferencia cd-kiadvány.

Perry, B. (2015). Gamifying French Language Learning: A Case Study Examining a Quest-based, Augmented Reality Mobile Learning-tool. *Procedia: Social and Behavioral Sciences*, *174*, 2308–2315. doi:10.1016/j.sbspro.2015.01.892

Radu, I. (2014). Augmented reality in education: A meta-review and cross-media analysis. *Personal and Ubiquitous Computing*, *18*(6), 1533–1543. doi:10.1007/s00779-013-0747-y

Santos, M. E. C., Chen, A., Taketomi, T., Yamamoto, G., Miyazaki, J., & Kato, H. (2014). Augmented reality learning experiences: Survey of prototype design and evaluation. IEEE Transactions on Learning Technologies, 7(1), 38-56.

Schank, R. C. (1995). What We Learn When We Learn by Doing. *Technical Report No. 60*. Northwestern University, Institute for Learning Sciences.

Singh, D., Tripathi, G., & Jara, A. J. (2014). *A survey of internet-of-things: Future vision, architecture, challenges and services*. Paper presented at the IEEE World forum on Internet of things (WF-IoT), Seoul, South Korea.

Squire, K., & Klopfer, E. (2007). Augmented reality simulations on handheld computers. *Journal of the Learning Sciences*, *16*(3), 371–413. doi:10.1080/10508400701413435

Winn, W. (1993). A conceptual basis for educational applications of virtual reality. In Human Interface Technology Laboratory. Washington Technology Center, University of Washington.

Wu, H. K., Lee, S. W.-L., Chang, H. Y., & Liang, J. C. (2013). Current status, opportunities and challenges of augmented reality in education. *Computers & Education*, *62*, 41–49. doi:10.1016/j.compedu.2012.10.024

KEY TERMS AND DEFINITIONS

Augmented Reality: The possibility to overlay computer graphics or computer audio onto the real world and augment, in this way, human perceptions.

Bio-Hypermedia: The bio-hypermedia is a new paradigm of internet-meets-body, characterized by "always connected" devices like smartphones, tablets and other smart devices.

Learning Adventure: The adventure of learning is directed to satisfy a personal desire of acquiring Knowledge and it is aimed at providing expertise in some specific domain.

Mobile Learning: Mobile learning is learning through exploration and conversation across multiple contexts amongst people and interactive technologies.

Moulded-Object: The designer of an AR or VR experience has the power to "mould" real world objects and create dynamic performances that may involve learners for educational purposes.

Prosumer: The learner is no longer a passive consumer of predetermined educational resources, but he/she is a "prosumer", aware that today's knowledge is everywhere.

Virtual Reality: The possibility to immerse in a completely virtual world using special goggles to visualize computer graphics or computer sound.

Chapter 5
Use of Augmented Reality in Mobile Devices for Educational Purposes

Bülent Gürsel Emiroğlu
Kırıkkale University, Turkey

Adile Aşkım Kurt
Anadolu University, Turkey

ABSTRACT

Use of technology in education has been widespread in the last decade, thanks to developments and improvements in information and communication technologies, especially in mobile devices. Among the fields in which mobile devices play important roles, education is one of the leading ones. Mobile devices help teachers and learners access educational resources when needed. To increase the reality of virtual learning environments on mobile devices, Augmented Reality (AR) technologies were introduced for mobile platforms, and the term Mobile Augmented Reality (MAR) arose. MAR opens a new door for educators and trainers to experience new methods of teaching for mobile learners. In this chapter, educational use of AR on mobile devices will be explained. Throughout the content of the chapter, readers will be informed about how AR applications changed people's teaching and learning styles.

INTRODUCTION

Use of technology in education has been widespread in the last decade, thanks to developments and improvements in information and communication technologies, especially in mobile devices. Mobile devices, mostly referred "smart phones", enabled us to reach information anywhere and anytime much more easily. Among the fields in which mobile devices play important roles, education is one of the leading ones. Mobile devices help teachers and learners access educational resources when needed. To increase the reality of virtual learning environments on mobile devices, Augmented Reality (AR) technology has been adopted with the help of high-speed cellular networks and high performance graphical and micro processing capabilities of the latest smart phones.

DOI: 10.4018/978-1-5225-2110-5.ch005

AR can be defined as visualising the artificial world with virtual effects of real subjects with the help of computer-generated graphics, sound, speech, text, still and/or motion pictures, etc. It enables users of its system to perceive objects of the real world and discover their functionalities with the help of computer-generated visualizations. AR is somehow different from Virtual Reality (VR). AR is used to enhance the digital view of the real world whereas VR is used to create a virtual world in a digital environment. AR is used in many fields including entertainment, education, training, sports, modelling and simulation. Educational use of AR (for both teaching and learning) is gaining popularity and attention especially amongst the Z generation. As the use of mobile devices for accessing and using educational resources increased dramatically during the last decade, educational AR applications are mostly moved to mobile platforms.

With the AR movement in mobile platforms, the term Mobile Augmented Reality (MAR) arose. MAR opens a new door for educators and trainers to experience new methods of teaching for mobile learners and to try new techniques in this context and setting. MAR it enables its learners to access online AR resources without the limitations of physical locations of the learning activities and formal connections of the processes needed. MAR is accessible anytime and anywhere, enabling users to experience a new way of thinking and understanding by experiencing the augmented reality of the concepts shown. Educational MAR attracts learners and raises their motivation by providing augmented reality for the objects of the real world in virtual settings on mobile devices.

In this chapter, educational use of AR on mobile devices will be explained. Throughout the chapter, you will be informed about how AR applications changed people's teaching and learning styles with the help of incremented reality graphical software.

BACKGROUND

Thanks to developments and improvements in information and communication technologies, use of technology in education has been widespread in the last decade. Many studies exist for using suitable technology in educational environments to increase students' success (Kulik, 2003; Wenglinsky, 2006), fun, effective and active learning process (Korkmaz, 2013; Sumadio & Rambli, 2010). Many new tools and platforms exist in the educational field due to the developments in technology, such as social networks, educational blogs and wikis, web 2.0 – 3.0 tools, podcasting, interactive whiteboards, and mobile devices. As a generic term, mobile device refers to a variety of devices that allow people to access data and information from wherever they are (Bucki, 2016). Handheld and portable devices are generally called as mobile devices due to their lightness and compactness. New display, processing and storage technologies of mobile devices provided many facilities similar to high-tech personal computers (technopedia, 2016). Our daily lives have been widely affected by these mobile devices, especially in interaction, sharing, and communication with others. In recent years, mobile devices are growing in complexity and diversity, leading to providing new interaction paradigms, modalities, shapes, and purposes (de Sa & Carriço, 2010). Mobile devices have some advantages compared to laptops for using in teaching and learning (Warschauer, 2011):

1. Lightness and orientation flexibility for better digital reading or accessing of content.
2. Instant readiness and fast switching among applications for rapid starting in learning activities.
3. Better user interaction with the help of large touch screen display.

4. Better mobility comparing to laptops, carrying is easy for collecting data and taking notes on.
5. Low-cost application development for mobile platforms, variety of free or very cheap educational applications for mobile devices.
6. Relatively long battery life comparing to laptops enabling more suitable for whole day activities.

Mobile devices, mostly referred to "smart phones" enabled us to reach information anywhere and anytime much more easily. Users can connect to the Internet and use various applications installed. Also, in addition to audio communication and text messaging, users are provided with video communication (Kaya & Koçyiğit, 2014). For the fields that mobile devices play important roles, education is one of the leading ones. Mobile devices help teachers and learners access educational resources when needed. To increase the reality of virtual learning environments on mobile devices, AR technology has been adopted with the help of high-speed cellular networks and high performance graphical and micro processing capabilities of the latest smart phones.

In this chapter, firstly AR and then educational use of AR on mobile devices will be explained. Throughout the chapter, you will be informed about how AR applications changed people's teaching and learning styles.

Augmented Reality (AR)

AR was first realised by Tom Caudell and David Mizell in 1992 for the overlaying of computerized information on the real world (Ghouaiel, Cieutat, & Jessel, 2014). AR can be defined as visualising the artificial world with virtual effects of real subjects with the help of computer-generated graphics, sound, speech, text, still and/or motion pictures, etc. In other words, "AR merges the physical world into virtual images, graphics and sound" (Delello, 2014). AR enables users of its system to perceive objects of the real world and discover their functionalities with the help of computer-generated visualizations. Considering these features, AR can be described as synchronous mixed reality environment created by adding virtual objects to the real world environment (Sırakaya & Seferoğlu, 2016). With AR you could see more than others see, hear more than others hear, and perhaps even touch, smell and taste things that others can not (van Krevelen & Poelman, 2010).

AR and VR concepts can be confused because of hosting virtual objects in their structure. AR is somehow different from VR. AR is used to enhance the digital view of the real world whereas VR is used to create a virtual world in a digital environment. In VR, the user is entirely in a virtual environment so that the user cannot see the real world whereas in AR the user can see both real world and virtual objects added on to the real world (Sırakaya & Seferoğlu, 2016). Reality-virtuality continuum concept, which is constituted by Milgram and Kishino (1994), sorted the transition between real and virtual environments (Figure 1).

At the left side of the diagram, real environment is located. Towards the right of the diagram, the amount of virtual objects increase and with reality, AR is derived. At the right side of the diagram, virtual environment is located and with addition of real objects to the virtual environments, augmented virtuality is derived. In each case of intersection of real and virtual environments is called mixed reality. Mixed reality is also called as AR or augmented virtuality according to the hosting rate of reality and virtuality (Sırakaya & Seferoğlu, 2016). Before explaining the use of AR in education, for more detailed information, looking at the history of AR will be useful.

Figure 1. Reality-virtuality continuum (Milgram & Kishino, 1994)

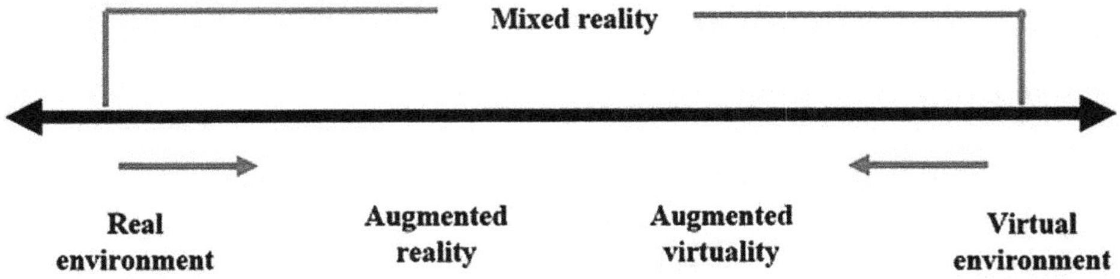

History of AR

An immersive, multi-sensory theatre experience, Sensoroma that was developed by Morton Heiling in late 1950s can be shown as the first example of AR. In 1962, Ivan Sutherland created the first computer graphic user interface called "Sketchpad". In 1975, Myron Kruger developed an artifical reality laboratory called videoplace for human computer interaction at the University of Connecticut. In 1980, Steve Mann developed wearable computers. In 1990, Tom Caudell introduced the term AR into the literature. In the following years, we see the AR technology being used in various fields. The Horizon Report 2005 listed AR as a key emerging technology for the five-year period. They predicted that as the interface became easier to use, AR would become common in education and training (New Media Consortium, 2005; University of Calgary, 2016). By 2010, the regularity and popularity of smart phones provided the incentive to put AR into the user's hands.

Significant developments in AR can be seen more clearly on the timeline presented in Figure 2.

Application Platforms Used in AR

All around the world, there are many AR applications used for educational purposes. These platforms can be listed as: Augmented Reality Development Lab (ARDL), Reliving the revoluation, PhysicsPlay-

Figure 2. Historical development of AR (Sırakaya, 2015; Yılmaz, 2014)

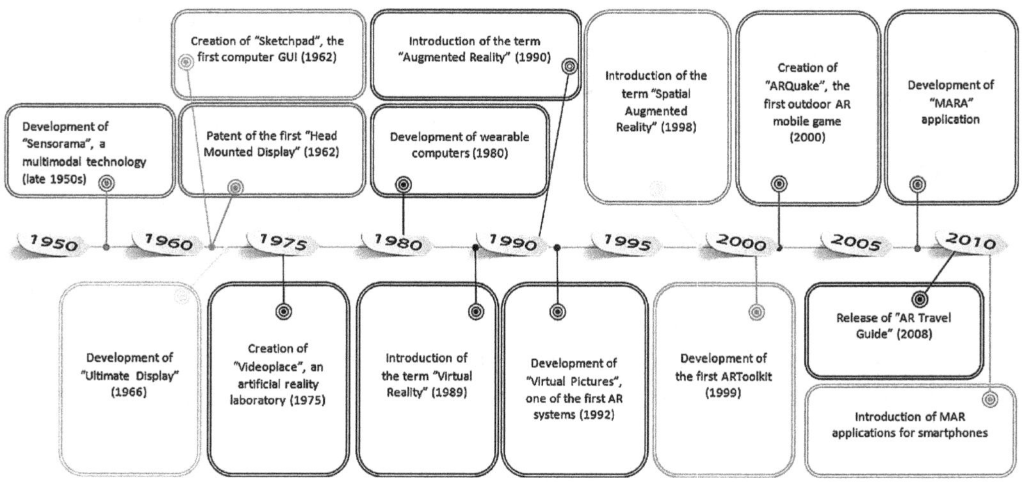

ground, MITAR games, New horizon, Occupational Safety Scaffolding, FETCH! Lunch Rush, Field trips, School in the park augmented reality experience, QR Code scavenger hunts, Driver's ed, Geotagging, Dow day, SciMorph, Imaginary words, Sky Map and Star Walk, Handled augmented reality project and Project glass.

AR allows teachers and students to increase visuality and reality on the digital real life environment seen on a high-resolution screen of a mobile device with the help of camera and running on Android or iOS platforms. On their first experience with AR, most users state incredible thoughts but having problems to think of it in classroom activities. Both teachers and instructors are aware that learning can be enhanced with reading and listening, and also with creation and interaction. Elements 4D, which is an AR application created by Daqri, allows students to manipulate and combine chemical elements on their mobile devices (IOS or Android) rather than just reading those subjects from course books. Another application created by Daqri, Anatomy 4D provides learners to discover human body parts and systems. Similary, Aurasma, another free application, offers users to be enrolled in and compose personal AR environments. Teachers and students can benefit from this free and open source application to carry their real-life experience to digital environment.

- You can get Aurasma application with a free account from its website.
- With Aurasma, you can identify the object whichever you want as a source and you can associate this object with an existing or prepared video by your own
- Firstly, the source (posters, pictures or questions as text), the video will be set to be played on the screen after the source is displayed
- After installing the application on the phone or tablet, video will start to run when held on the source.
- Thanks to the Aurasma, videos can be added to the clipboard. For example, when introducing a project made by your student, you can hang pictures and writings about the project on clipboard. If a video is identified to these pictures and writings, in the video, students can explain project steps, the challenges s/he faced when making project and tell what he has learned. Suppose you are a math teacher; you have prepared a worksheet on the subject of similarities in your geometry class for your students. If you recorded the solutions of the problems on a video and associated the video with Aurasma, your students can learn the solution of the problem, which forced them, by watching the video.
- Aurasma is an effective tool that can be used for music lesson. For example, you can hang musical notes on your class wall, when your students read musical notes with their smart devices they can hear the music created by notes. You can also post news about the choir on your clipboard, so the people who read the news can listen to the performance of the choir.

Quiver is another application that can be used to provide three-dimensional thinking especially in younger age groups. Quiver application has several painting papers on the website. Some of these painting papers are free. After printing out the desired painting you can color the painting with crayons you want, even you can add writings on your painting and complete it. When you look your painting with your loaded Quiver application tablet or smartphone, the picture becomes three-dimensional.

Comparison of AR application development platforms can be seen at Table 1.

Table 1. Comparison of AR application development platforms

Application	Operating System	Medium Run On	2D	3D	Video Playback	Social Media	External Web Page	Location Based
Alive	A/i/M	PC/SP/Tab	+	-	+	-	+	-
Augment	A/i	SP/Tab	+	+	+	-	+	-
Aurasma	A/i	SP/Tab	+	+	+	+	+	-
Blippar	A/i/M	SP/Tab/Glass	+	+	+	+	+	-
Junaino	A/i	SP/Tab/Glass	+	+	+	+	+	+
Layar	A/i	SP/Tab	+	+	+	+	+	-
Wikitude	A/i	SP/Tab/Glass	+	+	+	+	+	+

A: Android, i: iOS, M: Microsoft phone, PC: Computer, Sp: Smart Phone, Tab: Tablet, Glass: Google Glass

Use of AR in Education

AR is used in many areas including entertainment, education, training, sports, modelling and simulation. Educational use of AR (for both teaching and learning) is gaining popularity and attention especially amongst the Z generation. In the literature, there are lots of AR studies carried out in various education fields, such as maths education (İbili, 2013; Kaufmann, 2003; Kaufmann & Schmalstieg, 2003; Tomi & Rambli, 2013; Tosik-Gün, 2014), biology education (Cai, 2013; Özarslan, 2013; Perez-Lopez & Contero, 2013; Rosenbaum, Klopfer & Perry, 2006; Squire & Jan, 2007; Vilkoniene, 2009), physics education (Abdüsselam, 2014; Abdüsselam & Karal, 2012; Cai, Chiang & Wang, 2013; Enyedy, Danish, Delacruz & Kumar, 2012; Kaufmann & Meyer, 2008; Matcha & Rambli, 2013; Ibáñez, Di Serio, Villarán & Kloos, 2014), chemistry education (Boletsis & McCallum, 2013; Cai, Wang & Chiang, 2014; Fjeld & Voegtli, 2002; Núñez, Quirós, Núñez, Carda & Camahort, 2008), medical education (Botden & Jakimowicz, 2009; Hansen, Wieferich, Ritter, Rieder & Peitgen, 2010; Küçük, Kapakin & Göktaş, 2016), science education (Bressler & Bodzin, 2013; Kerawalla, Luckin, Seljeflot & Woolard, 2006; Brandt, Nielsen & Swensen, 2015; Yen, Tsai & Wang, 2012; Yen, Tsai & Wu, 2013). Some of the areas which AR technology is used for education can be listed as follows (Somyürek, 2014):

- Providing two-dimensional books with a third dimension
- Giving training for cognitive and psychomotor repair tasks
- Three-dimensional representations of concepts or carrying out experiments in areas such as physics, chemistry, biology
- Following videos or audios and conducting experiments about phenomenon in various topics in science museums
- Visualization of spatial concepts and relationships in mathematics and geometry lessons
- Visualizing concepts in geography education
- Providing a variety of knowledge and skills in health education, intervention guidance
- Gaining experience through meaningful and authentic tasks in military personnel training
- Classroom management experience in teacher education
- Gaining knowledge/skill about tools and materials in engineering education

AR technology also provides numerous educational benefits. The educational benefits of AR applications and the references listed at Table 2.

The views of students and teachers, who are the practitioners of the system, through AR is a key point. Students' views on using AR in education can be summarized as; courses' being fun, reducing cognitive load, increase in motivation and interest towards the course, increased opportunity to ask questions, increase in interaction between students, new opportunities for individual learning, concretising abstract concepts, rise of success. On the other hand, teachers' views on using AR in education can be summarized as; contribution to the development of creativity in students, ensuring effective participation of students to the course, students' being able to carry out the course with their own pace, increase in

Table 2. Educational Benefits of AR applications

Educational Benefits	References
Enhancing attention	Delello (2014); İbili and Şahin (2013); O'Brien and Toms (2005); Perez-Lopez and Contero (2013); Sumadio and Rambli (2010); Tomi and Rambli (2013)
Providing attractive and effective learning	Dünser and Hornecker (2007); Lester, Converse, Kahler, Barlow, Stone and Bhogal (1997); Oh and Woo (2008); Wojciechowski and Cellary (2013); Zhou, Cheok and Pan (2004)
Enhancing motivation	Delello (2014); Fleck and Simon (2013); İbili and Şahin (2013); Kerawalla, Luckin, Seljeflot and Woolard (2006); Küçük, Yılmaz and Yüksel (2014); O'Brien and Toms (2005); Perez-Lopez and Contero (2013); Sumadio and Rambli (2010); Serio, Ibanez and Kloos (2013); Taşkıran, Koral and Bozkurt (2015); Tomi and Rambli (2013); Yen, Tsai and Wang (2012)
Enriching interaction	Azuma (2004); Bujak, Radu, Catrambone, MacIntyre, Zheng and Golubski (2013); Kerawalla, Luckin, Seljeflot and Woolard (2006); Ivanova and Ivanov (2011); Wojciechowski and Cellary (2013); Wu, Lee, Chang and Liang (2013)
Facilitating learning	Abdüsselam (2014); Cai, Wang and Chiang (2014); Delello (2014); Ivanova and Ivanov (2011); Kaufmann (2003); Kerawalla, Luckin, Seljeflot and Woolard (2006); Núñez, Quiros, Núñez, Carda and Camahort (2008); Rosenbaum, Klopfer and Perry (2006); Shelton and Hedley (2002); Shelton and Stevens (2004); Tian, Endo, Urata, Mouri and Yasuda (2013); Yen, Tsai and Wu (2013)
Enhancing engagement	Abdüsselam and Karal (2012); Bai, Blackwell and Coulouris (2013); Bujak, Radu, Catrambone, MacIntyre, Zheng and Golubski (2013); Cai (2013); Delello (2014); Dunleavy, Dede and Mitchell (2009); Ivanova and Ivanov (2011); Yusoff and Dahlan (2013)
Providing cooperation	Yuen, Yaoyuneyong and Johnson (2011)
Triggering creativity	Klopfer and Yoon (2004); Yuen, Yaoyuneyong and Johnson (2011); Zhou, Cheok and Pan (2004)
Developing imagination	İbili and Şahin (2013); Kaufmann (2003); Klopfer and Yoon (2004); Núñez, Quirós, Núñez, Carda and Camahort (2008); Shelton and Hedley (2002), Yuen, Yaoyuneyong and Johnson (2011)
Enhancing spatial ability	Bujak, Radu, Catrambone, MacIntyre, Zheng and Golubski (2013); Chen and Tsai (2012); Kaufmann (2003); Medicherla, Chang and Morreale (2010); Shelton and Stevens (2004); Wojciechowski and Cellary (2013); Yen, Tsai and Wu (2013)
Teaching of media objects that cannot be created and conditions that cannot be achieved in real-world	Kerawalla, Luckin, Seljeflot and Woolard (2006); Shelton and Hedley (2002); Wojciechowski and Cellary (2013); Wu, Lee, Chang and Liang (2013); Yuen, Yaoyuneyong and Johnson (2011)
Helping to remove misconceptions	Fleck, Hachet and Bastien (2015); Fleck and Simon (2013); Rosenbaum, Klopfer and Perry (2006); Shelton and Hedley (2002); Tian, Endo, Urata, Mouri and Yasuda (2014); Yen, Tsai and Wang (2012)
Making learning fun	Rambli, Matcha and Sulaiman (2013); Taşkıran, Koral and Bozkurt (2015); Tomi and Rambli (2013); Zarzuela, Pernas, Martínez, Ortega and Rodríguez (2013)
Concretising abstract concepts	Abdüsselam (2014); Abdüsselam and Karal (2012); Özarslan (2013); Taşkıran, Koral and Bozkurt (2015); Tosik-Gün (2014)
Performing dangerous experiments in a safe environment	Eursch (2007); Wojciechowski and Cellary (2013)

interaction between students, ease of use in classroom. There are many AR applications used in educational environments. They all have both advantages and disadvantages according to different criteria. In the following paragraphs, advantages and disadvantages of educational AR applications are discussed:

Advantages of Augmented Reality

It can be said that the most notable advantage of AR is that it supports commonly used learning approaches such as situated learning (Clarkson, 2014; Dunleavy, Dede & Mitchell, 2009; Taşkıran, Koral & Bozkurt, 2015), constructivist learning (Abdoli-Sejzi, 2015; Delello, 2014; Gardner & Elliott, 2014), authentic learning (Wu, Lee, Chang & Liang, 2013). Bujak, Radu, Catrambone, MacIntyre, Zheng and Golubski (2013) present a framework for understanding AR learning from physical, cognitive, and contextual perspectives. On the physical dimension, they argue that physical manipulation affords natural interactions, in other words, more learner control through natural interaction. On the cognitive dimension, they discuss that facilitation of the understanding of difficult issues by reducing the cognitive load results in improved understanding of abstract concepts. On the contextual dimension, they argue that AR creates possibilities for collaborative learning around virtual content and the ease of use of AR increases by face-to-face communication and motivation.

By integrating AR into education (especially in lectures), teachers capture the attention of learners. It allows the students to participate into the lessons because students are able to access models on their own devices via AR applications and so the students can gain a better understanding of the concepts they are studying. Learners will retain more knowledge for a longer period and will be excited by new ideas and think critically about the world around them (augment.com, 2016). In other words, learners could improve their enthusiasm for learning and increase their educational realism-based practices with AR (Abdoli-Sejzi, 2015). Another benefit with AR learning is there are no costs for making mistakes and errors, as they are not real (Wang, 2012). The other advantages of AR can be listed as follows (Sırakaya & Seferoğlu, 2016):

- Provides more easily understanding by visualizing the complex nature of the issues
- Makes learning fun and attractive
- Increases the motivation of learners
- Improves the spatial skills of learners
- Helps to remove misconceptions
- Increases the participation of the course
- Provides learners to learn in accordance with their learning styles
- Allows the concretization of abstract concepts
- Provides the implementation of a teaching situation which is not possible to observe in real-life
- Allows implementation of dangerous experiments securely
- Allows students to establish natural experience with virtual objects without departing from the reality of the classroom
- Allows learners to control their learning process

Disadvantages of Augmented Reality

Apart from the numerous advantages of AR in educational settings, there are also some limitations to overcome (Sırakaya & Seferoğlu, 2016):

- Difficulty of developing content for AR. Development of 3D content and applications requires technical knowledge.
- Physical requirements of the class, hardware requirements, and individual differences between students in a class limit the use of AR applications.
- Educators are prejudiced about using AR in the classroom and they need time to accept this.
- AR holds multimedia elements that support permanent learning, if these elements are not used in suitable amount, it will cause excessive cognitive load.
- The availability of desktop computers, portable computers, portable devices, augmented reality glasses, where the AR applications will run, will be an economical limitation
- Expensive to develop and maintain
- Lack of data portability between AR environments
- Lack of privacy

Mobile Augmented Reality (MAR)

After the technology revolution during the mid 2000s, as nature and use of mobile devices fundamentally changed and widened, easily accessible high tech mobile platforms enabled movement of educational AR applications to portable devices. As a result, MAR term come to the point. MAR created a new era for both teachers and learners for applying new practices of teaching in the area of mobile learners and experiencing new methods in the context and environment of them. The main advantage of MAR is that it removes the physical location barriers for learners by accessing online AR resources and related activities and processes. Due to its nature, MAR is always accessible whenever and wherever the users want to experience its benefits as it provides a new way of discovering and recognising by facing the augmented reality of the real concepts. Use of MAR in education offers new opportunities for learners and increase their motivation by supplying enhanced reality for the real life objects on digital environment of mobile platforms. Looking at the components of MAR will be more useful before using MAR in education. The components of MAR are briefly described below.

Components of MAR

The components of MAR can be listed as; computerised infrastructure, display, associations, relations, and connections. *Computerised infrastructure;* to generate and manage the virtual materials, which will be placed on real physical environment, processing tracker information and controlling the screen of the device. *Display;* to show the digitised virtual material within the physical world. To increase the visual sensing experience, high-density head-up displays, head-worn displays, mobile hand-held displays are used. Additionaly, other senses (audio, haptic, etc.) can also be integrated. *Associations and relations;* to establish the relationships and connections between real life objects and virtual elements. Virtual and audial relations, dimensions, locations and attributes of the virtual model in the environment should be defined and associated. *Connections;* to connect mobile device to the wireless network and Internet, to communicate with other users and learners, update information of materials on the system and run online applications.

Wagner (2009; cited in Specht, Ternier & Greller, 2011) stated that following technical components are important in MAR systems:

- Display systems such as head-mounted displays, mobile phones with built-in camera and hand-held projectors. Display technologies are becoming more flexible and less costly produced.
- Sensor systems such as gyroscopes, GPS, electronic compasses, cameras, and microphones, as well as indoor location tracking systems.
- Wireless connection and related protocols for real time augmentation settings and enabling multi-user real-time interaction.
- Computerised Mobile phones for real time visualization of 3D objects and overlays on a stand-alone device.
- Tagging and tracking technologies with multi-marker tracking and hybrid tracking systems for coping with the registration problem (Bimber & Raskar, 2005), which is the challenge of linking the real-world perception of a mobile AR user and the presentation of the augmentation layer.
- Linking of location-based AR information in storytelling and gaming used for learning support to link AR experiences with instructional designs or task structures and sequencing approaches.
- Flexible layer-based AR browsers with social media integration for enabling AR systems to build on existing information channels and present existing information to users in a new kind of user interface.

Use of Mobile Augmented Reality (MAR) in Education

The development and rapid increase in mobile phone usage have made MAR possible (Azuma, Baillot, Behringer, Feiner, Julier & MacIntyre, 2001; Papagiannakis, Singh, & Magnenat-Thalmann, 2008) and spread. MAR can be applied in various educational domains. It can help learners to have better understanding, experience embedded learning content in real world environment, or explore content related to their current profile or environmental context. Most important examples support exploration of the physical environment with different subject domains, including history, arts, technology, biology, astronomy, and so, by enriching produced figures in the real world environment with AR techniques (Specht, Ternier, & Greller, 2011). In this section, we tried to summarize related work of educational use of MAR in the literature.

Küçük, Kapakin and Göktaş (2016) conducted a research, which aimed to determine the effects of learning anatomy via MAR on medical students' achievement and cognitive load. 70 second-year undergraduate medical students enrolled the research. 34 students in experimental group, which used MAR applications, reported higher achievement and lower cognitive load. As a result, the mobile learning approach helped students learn better by exerting less cognitive effort because MAR applications contributed to the formation of an effective learning environment.

In another study, Küçük, Kapakin and Göktaş (2015) aimed to determine the medical faculty students' views on anatomy learning via MAR technology. 34 sophomore students, studying medicine enrolled the study. Questionnaire and interview form were used as data collection tools and descriptive analysis were used for analysing the data. At the end of the study, it was found that students' views were highly positive towards MAR. Students emphasized that MAR-based learning created a sense of reality, increased interest in lesson, provided flexible learning environment. In conclusion, the authors reported that incorporating MAR applications in anatomy education would be beneficial.

The aim of the research conducted by Shirazi and Behzadan (2015) was to assess the effectiveness of a collaborative context-aware mobile augmented reality tool (CAM-ART). Sample of the study were construction and civil engineering students, who were randomly assigned to control and test groups. Students' pre-performance and post-performance data was collected through a designed learning tool and a feedback questionnaire was used to measure the perceptions of students' using MAR based tools. The results showed that CAM-ART had a positive impact on students' learning and students found CAM-ART to be an interesting, helpful, and motivational approach in the classroom.

Castillo, Sánchez, and Villegas (2015) introduced a new software, which provides a detailed view of the subsystems, and tasks that encompass the creation of a MAR application. The application was used to observe the impact of an AR experience in Mexican undergraduate students. Data was collected trough a questionnaire. Analysis results showed that using MAR can enhance teaching-learning process and motivates students. The authors suggested that MAR could be an alternative technology to change the learning paradigm in the future.

Albrecht, Folta-Schoofs, Behrends and von Jan (2013) aimed to compare the impact of the heightened realism of a self-developed MAR blended learning environment (mARble) on learners to textbook material. Profile of Mood States Questionnaire was used to measure initial emotional status and 10-item single-choice was used to measure previous knowledge of 10 third-year medical students. During the 30-minute learning period, the students were randomized into two groups: the first group equipped with an iPhone with a preinstalled copy of mARble, while the second group was provided with textbook material. The results showed that mARble group performed considerably better regarding learning efficiency.

Martín, Díaz, Cáceres, Gago and Gibert (2012) introduced an educational AR application called EnredaMadrid, which was built on mobile device based on geolocalisation, to cope with the complexity of history. A questionnaire was used to evaluate the technology used in EnredaMadrid. The results showed that AR is the most positive element in EnredaMadrid, making learning fun and motivating students. Moreover, the results indicate that the students believe that AR is the most appropriate tool to learn the history.

In their experimental study, Tarng and Ou (2012) used mobile phones with AR technologies as assistive tools to learn butterfly ecology by breeding their virtual caterpillars on host plants and so learn butterfly's life cycle. Research results reveal that learning effectiveness was better in experimental group and thus AR butterfly ecology learning system could enhance students' learning.

A MAR game, ARGreenet presented by Juan, Alem and Cano (2011) in their study, aimed to increase the awareness about recycling. The research was conducted with 38 children, compared ARGreenet, and basic mobile game on recycling through students' knowledge of recyling, level of student engagement, fun and ease of use of the games, precieved willingness to change behaviour of students. According to the results of the study, 69,4% of the students preffered ARGreenet game because of ease of use and fun of the game and found the game more engaging while no significant differences found for other variables. Besides, both games influenced students' willingness to change behaviour.

Huizenga, Admiraal, Akkerman and Dann (2009), incorporated Frequency 1550, which is a MAR game about Medieval Amsterdam, into their research, which investigates students' engagement, historical knowledge and motivation for history. In the study, 10 of the students played the mobile game while the other 10 of them took regular project-based lessons. Results show that students who played mobile game were more engaged and gained more historical knowledge than the regular group while no significant differences were found for motivation for history.

In their study, Ardito, Buono, Costabile, Lanzilotti and Piccinno (2009) introduced a MAR enhanced game called Explore! The game aimed to assist middle school students through their visit to archaeological sites in Italy. The game shows a map of the site on cell phones, which each group of students given two, and requires students to explore important places in the site giving hints. The study was carried out with 42 students. 19 of them played the game on paper-based version while 23 students played the mobile version. Results showed no significant difference in terms of learning while students found the game enjoyable.

In the Alien Contact! game, which focuses on learning maths, language arts and scientific literacy skills for middle and high school students, is a MAR game developed by Dunleavy, Dede and Mitchell (2009). Students play the game in teams of four and each member has different roles according to the scenario. In this setting, players interact with virtual characters and items to solve science, math and language problems. The study aimed to determine how teachers and students' views towards teaching and learning with an AR simulation. Results showed that students' engagement were high in all three case studies.

In the CONNECT project, using MAR and Head Mount Display (HMD) technologies, a system was developed for science learning in both formal and informal learning settings. In the system, students are able to interact with the learning environment phsysically and intellectually through hands-on experimentations and "minds on" reflection. In the study conducted with learners with disabilities, it was found that disabled students and able-bodied students using the system had similar results, which supports that the project have the potential for the education of disabled (Arvanitis et al., 2007).

Schmalstieg and Wagner (2007) conducted a case study with a group of high school students using two handheld AR enhanced games they have developed. The games, which are about discovery of historical artifacts, include puzzles and other tasks related to exhibits and show a map of the exhibition, which also list the solved and remaining tasks. In the first game, students in groups of two three students recieved one handheld while, in the second one, each student had their own. Data gathered from interviews, observations and logs were analyzed and results show that students were highly motivated and satisfied.

Environmental Detectives is an AR simulation developed in the collaboration of Squire and Klopfer (2007) and Environmental Science Faculty at MIT. The simulation in which students play environmental engineer roles aims to give students experiences in environmental investigations. The study revealed that the simulation assists students in scientific practice.

Wagner, Schmalstieg and Billinghurst (2006), developed Virtuoso, which is a handheld educational game-based collaborative AR application. This game-based environment has also two other versions, one of which was developed as a desktop application created in Macromedia Flash and the other as a paper-based material. In the game, players need to sort artworks in accordance to their creation dates. The players were tested in three different conditions (paper-based, PC- based, PDA-based). The study failed to find a significant difference for educational outcomes. However, paper and PDA versions were prefered over PC version for more effective collaboration. Besides, players report that PDA interface was the most enjoyable one.

Google Glass is a wearable augmented reality technology introduced by Google Company in 2012. It is a technology with a potential use in educational practice. With Google Glass, researchers, teachers, colleagues and students in different locations can share instant information and work collaboratively and simultaneously. Thanks to the audiovisual capabilities of Google Glass, audios, photos or videos can be recorded and used in lessons. With this feature, virtual trips can be organized, in and off classroom activities can be recorded and published. In addition, new documents can be created and reproduced using the document creation feature of Google Glass. Because of its wireless Internet connection, online facilities can be utilized; instant communication between teacher and student can be established through various social networks (Erbaş & Demirer, 2014).

SOLUTIONS AND RECOMMENDATIONS

Recently, especially during the last 5-6 years, educational use of AR applications has increased due to the benefits and opportunities they provide to the users. In these AR applications, virtual figures and characters were synchronously added to the real-life objects on the screen. MAR provide support to the instructors for attaching digitalised content having enhanced facilities together with geographic location of an object or an environment. With enhancements added to the educational environments, better interaction with students, configurable information for learners, enjoyful teaching, effective and attractive mediums can be provided (Kirkley & Kirkley, 2005). MAR applications increase the interest and motivation of students with the visualization of the objects used (Delello, 2014). Additionaly, MAR can be used for serving course content to the students more interactive and attractive with the help of Bring Your Own Device (BYOD) approach, which is a very popular trend in education. MAR can support the efficiency of teaching and learning in educational and academic environments by adding information at proper time and place and contribute with computer generated digitally enhanced 3D imagery content. From the constructivist point of view, MAR also provides self-control to the students for their own learning activities. Authentic teaching and learning activities can also be established with the help of MAR. Furthermore, MAR enables teaching media objects that cannot be constructed and settings that cannot be accomplished in real life and also execute threatening experiments with a safer way. According to the results of many studies about MAR, these systems can contribute for more motivated, entertaining and engaging environments for learning activities. Furthermore, educational MAR applications can be pleasing, inspiring and interesting for students. Appropriate MAR applications can be used to create effective and efficient learning environments that will provide students with minimum cognitive load and maximum achievement (Küçük, Yılmaz, & Göktaş, 2014).

Recently, many of the students in higher education have smart mobile phones, elementary and secondary students have tablet PCs given by the government within FATIH project. Considering this ownership, MAR applications can be used effectively for teaching-learning process in educational settings of Turkey. As they provide two and three-dimensional visual support, video playing, hyperlinks of web pages and location-based operating characteristics, educational MAR applications can be used for many purposes such as material development, orientation activities, etc. The critical points that should be considered for using educational MAR applications are as follows:

- Light level of the teaching and learning environment as lightening is important for monitoring the pointer during image-based MAR usage.
- Physical conditions of the educational settings as students should be placed and seated appropriately for MAR usage.
- A brief introduction and opening about MAR before the application start in case of any potential problems.

- Demostration of in-class materials and resouces used for improving MAR applications
- Printed materials explaining in-class rules, regulations and announcements about MAR applications.

FUTURE RESEARCH DIRECTIONS

This chapter presents a detailed systematic review of MAR as an actual, trendy and promising technology for supporting technology-enhanced learning. In this section, future research directions are discussed for educational settings. Further research and studies are required for the usability of educational MAR applications as well as describing principles and suggesting guidelines for interested authorities and participants of MAR platforms of instructional settings. Many inter-disciplinary studies have been conducted to establish comparisons between MAR learning experiences with respect to other platforms and tools used in teaching and learning environments. Accordingly, more studies need to be undertaken considering a large-scale evaluation and longitudinal evaluations in future researches.

Educators should take guidance, advice and support from domain experts and technical specialists for integrating MAR into lesson plans and curriculum. Moreover, new authoring tools are needed for educators to create 3D content and experience new methods for creating interactive MAR learning environments. However, nature and constraints of the institutional context should also be considered during the development process of educational MAR applications. Additionnaly, deep understanding and knowledge in the subject domain of the topics would be an advantage for educators to design MAR learning experiences according to the skills and competencies of learners. Learners should be capable of exploring educational MAR content and this process should be carefully constructed for improving learning opportunities. In educational activities, creating multisensory experiences with MAR and exploring their impact in the learning outcomes are very important.

To enable success efficiently, much more studies should be conducted to understand and discover user experiences and knowledge construction processes in MAR applications. Diversity amongs learners having different characteristics, backgrounds and socio-economic status should also have considered for addressing the special necessities and individual needs of the target group. Educators can easily adapt MAR content for adding - removing elements and changing the speed of animations with the help of flexible technology lying under MAR.

To increase the attention and motivation of learners, new learning environments can be created allowing individual use of MAR applications (Küçük, Yılmaz, & Göktaş, 2014). To do so, different frameworks should be designed and used for personalized MAR considering pedagogical and didactical components in order to guide the development of MAR-based educational settings.

Recently, a well-known anime, Pokemon, which was a passion of a generation before and now, was released in a mobile game format called Pokemon Go. Although it was released for a few countries at the beginning, the game unexpectedly succeeded to have the whole world under its spell. By the help of Augmented Reality (AR) technology, Pokemon Go combines the virtual world with real world.

The objective of this AR-based game is to go where the game targets you and catch virtual characters in real world using mobile cameras (Akkoyunlu, 2016). Pokemon Go attracted many people at any age in a short time and is the one of the most popular topic in social media lately. While it is so popular and current, discussion about its use in education also arises. Highlighting research and discovery, Pokemon Go may be useful in lessons such as history and geography with its geolocation feature. The game may be helpful for motor skills development of children who try to catch "Pikachu". As the game encourages people to walk, it can be useful in lessons that require physical activities as well. It is also beneficial for socializing and cooperating people while chasing Pokemons in the streets and waiting at "PokeStops". These being the case, it is not hard to say that studies about educational use of Pokemon Go would be carried out soon in the literature. On the other hand, common technological, physiological and psychological problems that mobile gaming bring out (such as virus threat, neck pain, game addiction etc.) should be taken into consideration while using Pokemon Go for educational purposes.

CONCLUSION

The possibility of the use of MAR in educational practice opened a new door for educators, students and researchers, providing them with access to online AR resources and communications ubiquitously. AR comes up as a powerful way of enhancing attention, motivation and engagement; providing cooperative, attractive, effective and entertaining learning practices; fostering creativity and imagination; and presenting conditions, practices and environments that cannot be created in real-world settings. Notable educational advantages of AR technologies can be regarded as supporting commonly used learning approaches, increasing attention and participation of students in learning activities, constituting better understanding of concepts, and freedom to make mistakes and errors with no cost. On the other hand, difficulty and cost of developing AR platforms, cost of physical requirements to utilise these platforms, and educators' acceptance of this technology can be regarded as disadvantages of AR technologies. MAR offers new opportunities for educational practice removing the locational barriers; utilising mobile devices such as smart phones and tablets, which are widespread among students, as educational environments; integrating audial, virtual and haptic responses and enabling real-time multi-user interaction. Several researches in this field reveal that educational use of MAR leads higher achievement, motivation, interest, satisfaction, engagement and joy, lower cognitive load and positive attitude towards MAR-based educational practices. Considering all, if economical disadvantages are overcomed by aiming commonly used mobile devices and calling on govermental support, MAR might open a new era in educational practices and paradigms.

REFERENCES

Abdoli-Sejzi, A. (2015). Augmented reality and virtual learning environment. *Journal of Applied Sciences Research*, *11*(8), 1–5.

Abdüsselam, M. S. (2014). Fizik öğretiminde artırılmış gerçeklik ortamlarının kullanımlarına ilişkin öğretmen ve öğrenci görüşleri: 11. Sınıf manyetizma konusu örneği. *Pegem Eğitim ve Öğretim Dergisi*, *4*(1), 59–74.

Abdüsselam, M. S., & Karal, H. (2012). Fizik öğretiminde artırılmış gerçeklik ortamlarının öğrenci akademik başarısı üzerine etkisi: 11. Sınıf manyetizma konusu örneği. *Eğitim ve Öğretim Araştırmaları Dergisi*, *1*(4), 170–181.

Akkoyunlu, B. (2016). *Son Zamanların Çılgınlığı "Pokemon Go" Oyunu Hakkında Neler Biliyoruz?* Retrieved 3 August, 2016 from http://www.bilimiletisimi.com/show.php?id=5103

Albrecht, U. V., Folta-Schoofs, K., Behrends, M., & von Jan, U. (2013). Effects of mobile augmented reality learning compared to textbook learning on medical students: Randomized controlled pilot study. *Journal of Medical Internet Research, 15*(8), e182. doi:10.2196/jmir.2497 PMID:23963306

Ardito, C., Buono, P., Costabile, M. F., Lanzilotti, R., & Piccinno, A. (2009). Enabling interactive exploration of cultural heritage: An experience of designing systems for mobile devices. *Knowledge, Technology & Policy*, *22*(1), 79–86. doi:10.1007/s12130-009-9079-7

Arvanitis, T. N., Petrou, A., Knight, J. F., Savas, S., Sotiriou, S., Gargalakos, M., & Gialouri, E. (2007). Human factors and qualitative pedagogical evaluation of a mobile augmented reality system for science education used by learners with physical disabilities. *Personal and Ubiquitous Computing*, *13*(3), 243–250. doi:10.1007/s00779-007-0187-7

augment.com. (2016). *5 reasons to use Augmented Reality in Education.* Retrieved 23 April, 2016 from http://www.augment.com/blog/5-reasons-use-augmented-reality-education/

Azuma, R., Baillot, Y., Behringer, R., Feiner, S., Julier, S., & MacIntyre, B. (2001, November/December). Recent advances in augmented reality. *IEEE Computer Graphics and Applications*, *21*(6), 34–47. doi:10.1109/38.963459

Azuma, R. T. (2004). Overview of augmented reality.*Proceeding of SIGGRAPH '04.*

Bai, Z., Blackwell, F., & Coulouris, G. (2013). Th rough the looking glass: Pretend play for children with autism.*IEEE International Symposium on Mixed and Augmented Reality 2013* (pp. 49–58). Adelaide, Australia: IEEE. doi:10.1109/ISMAR.2013.6671763

Bimber, O., & Raskar, R. (2005). *Spatial augmented reality: Merging real and virtual worlds.* Retrieved 11 July, 2016 from http://web.media.mit.edu/~raskar/SAR.pdf

Boletsis, C., & McCallum, S. (2013). The table mystery: An augmented reality collaborative game for chemistry education. In M. Ma, M.F. Oliveria, S. Petersen & J.B. Hauge (Eds.), Serious games development and applications (pp. 86-95). Springer Berlin Heidelberg.

Botden, S. M., & Jakimowicz, J. J. (2009). What is going on in augmented reality simulation in laparoscopic surgery? *Surgical Endoscopy*, *23*(8), 1693–1700. doi:10.1007/s00464-008-0144-1 PMID:18813987

Brandt, H., Nielsen, B. L., Georgsen, M., & Swensen, H. (2015). Augmented Reality for Science Education. Lecture Notes in Computer Science.

Bressler, D. M., & Bodzin, A. M. (2013). A mixed methods assessment of students flow experiences during a mobile augmented reality science game. *Journal of Computer Assisted Learning*, *29*(6), 505–517. doi:10.1111/jcal.12008

Bucki, J. (2016). *What is mobile computing?* Retrieved 14 May, 2016 from http://operationstech.about.com/od/glossary/g/Definition-Of-Mobile-Computing.htm

Bujak, K. R., Radu, I., Catrambone, R., MacIntyre, B., Zheng, R., & Golubski, G. (2013). A psychological perspective on augmented reality in the mathematics classroom. *Computers & Education*, *68*, 536–544. doi:10.1016/j.compedu.2013.02.017

Cai, H. (2013). *Using augmented reality as motivators for youth environmental education: An American Harts's tongue fern conservation project* (Unpublished Master Thesis). State University of New York College of Environmental Science and Forestry.

Cai, S., Chiang, F. K., & Wang, X. (2013). Using the augmented reality 3D technique for a convex imaging experiment in a physics course. *International Journal of Engineering Education*, *29*(4), 856–865.

Cai, S., Wang, X., & Chiang, F. K. (2014). A case study of augmented reality simulation system application in a chemistry course. *Computers in Human Behavior*, *37*, 31–40. doi:10.1016/j.chb.2014.04.018

Castillo, R. I. B., Sánchez, V. G. C., & Villegas, O. O. V. (2015). A pilot study on the use of mobile augmented reality for ınteractive experimentation in quadratic equations. *Mathematical Problems in Engineering*, *2015*, 1–13. doi:10.1155/2015/946034

Chen, C. M., & Tsai, Y. N. (2012). Interactive augmented reality system for enhancing library instruction in elementary schools. *Computers & Education*, *59*(2), 638–652. doi:10.1016/j.compedu.2012.03.001

Clarkson, J. (2014, February). *Exploring situated learning in augmented reality apps for adults with autism spectrum disorder.* Paper presented at Graduate Student Research Conference, Oshawa, ON.

de Sa, M., & Carriço, L. (2010). Designing and evaluating mobile interaction: Challenges and trends. *Foundations and Trends in Human-Computer Interaction*, *4*(3), 175–243. doi:10.1561/1100000025

Delello, J. A. (2014). Insights from pre-service teachers using science-based augmented reality. *Journal of Computers in Education*, *1*(4), 295–311. doi:10.1007/s40692-014-0021-y

Dunleavy, M., Dede, C., & Mitchell, R. (2009). Affordances and limitations of immersive participatory augmented reality simulations for teaching and learning. *Journal of Science Education and Technology*, *18*(1), 7–22. doi:10.1007/s10956-008-9119-1

Dünser, A., & Hornecker, E. (2007). An observational study of children interacting with an augmented story book. In *Proceedings of 2nd International Conference of E-Learning and Games (Edutainment 2007)*. CUHK. doi:10.1007/978-3-540-73011-8_31

Enyedy, N., Danish, J. A., Delacruz, G., & Kumar, M. (2012). Learning physics through play in an augmented reality environment. *International Journal of Computer-Supported Collaborative Learning*, *7*(3), 347–378. doi:10.1007/s11412-012-9150-3

Erbaş, Ç., & Demirer, V. (2014). Eğitimde artırılmış gerçeklik uygulamaları: Google Glass örneği. *Journal of Instructional Technologies & Teacher Education*, *3*(2), 8–16.

Eursch, A. (2007). *Increased safety for manual tasks in the field of nuclear science using the technology of augmented reality.* Paper presented at 2007 IEEE Nuclear Science Symposium Conference. doi:10.1109/NSSMIC.2007.4436557

Fjeld, M., & Voegtli, B. M. (2002). *Augmented chemistry: An interactive educational workbench.* Paper presented at International Symposium on Mixed and Augmented Reality, 2002. ISMAR 2002, Darmstadt, Germany. doi:10.1109/ISMAR.2002.1115100

Fleck, S., Hachet, M., & Bastien, J. M. (2015). *Marker-based augmented reality: Instructional-design to improve children interactions with astronomical concepts.* Paper presented at 14th International Conference on Interaction Design and Children, New York, NY. doi:10.1145/2771839.2771842

Fleck, S., & Simon, G. (2013). *An augmented reality environment for astronomy learning in elementary grades: An exploratory study.* Paper presented at 25ème conférence francophone sur l'Interaction Homme-Machine, IHM'13, Bordeaux, France. doi:10.1145/2534903.2534907

Gardner, M., & Elliott, J. (2014). The immersive education laboratory: Understanding affordances, structuring experiences, and creating constructivist, collaborative processes, in mixed-reality smart environments. *EAI Endorsed Transactions on Future Intelligent Educational Environments, 14*(1).

Ghouaiel, N., Cieutat, J. M., & Jessel, J. P. (2014). *Adaptive augmented reality: Plasticity of augmentations.* Paper presented at Virtual Reality International Conference, Laval, France.

Hansen, C., Wieferich, J., Ritter, F., Rieder, C., & Peitgen, H. O. (2010). Illustrative visualization of 3D planning models for augmented reality in liver surgery. *International Journal of Computer Assisted Radiology and Surgery*, *5*(2), 133–141. doi:10.1007/s11548-009-0365-3 PMID:20033519

Huizenga, J., Admiraal, W., Akkerman, S., & Dam, G. T. (2009). Mobile game-based learning in secondary education: Engagement, motivation and learning in a mobile city game. *Journal of Computer Assisted Learning*, *25*(4), 332–344. doi:10.1111/j.1365-2729.2009.00316.x

Ibáñez, M. B., Di Serio, A., Villarán, D., & Kloos, C. D. (2014). Experimenting with electromagnetism using augmented reality: Impact on flow student experience and educational effectiveness. *Computers & Education*, *71*, 1–13. doi:10.1016/j.compedu.2013.09.004

İbili, E. (2013). *Development, implementation and assessment of the effect augmented reality on geometry teaching materials for geometry classes* (Unpublished doctoral dissertation). Gazi University, Ankara, Turkey.

İbili, E., & Şahin, S. (2015). Investigation of the effects on computer attitudes and computer self-efficacy to use of augmented reality in geometry teaching. *Necatibey Faculty of Education Electronic Journal of Science and Mathematics Education*, *9*(1), 332–350.

İbili, E., & Şahin, S. (2015). Artırılmış gerçeklik ile interaktif 3D geometri kitabı yazılımın tasarımı ve geliştirilmesi: ARGE3D. *AKÜ Fen ve Mühendislik Bilimleri Dergisi*, *13*(1), 1–8.

Ivanova, M., & Ivanov, G. (2011). Enhancement of learning and teaching in computer graphics through marker augmented reality technology. *International Journal on New Computer Architectures and Their Applications*, *1*(1), 176–184.

Juan, M. C. F. D., Alem, L. A. P., & Cano, J. (2011). ARGreenet and Basic-Greenet: Two mobile games for learning how to recycle.*Proceedings of the 19th International Conference on Computer Graphics, Visualization and Computer Vision*, 25–32.

Kaufmann, H. (2003). *Collaborative augmented reality in education*. Institute of Software Technology and Interactive Systems, Vienna University of Technology.

Kaufmann, H., & Meyer, B. (2008). *Simulating educational physical experiments in augmented reality*. Interactive Media Systems Group, Institute of Software Technology and Interactive Systems, Vienna University of Technology. Retrieved, 10 June, 2016 from http://publik.tuwien.ac.at/files/PubDat_170658.pdf

Kaufmann, H., & Schmalstieg, D. (2003). Mathematics and geometry education with collaborative augmented reality. *Computers & Graphics*, *27*(3), 339–345. doi:10.1016/S0097-8493(03)00028-1

Kaya, M., & Koçyigit, A. (2014). *Mobil uygulamalarda vekil tabanlı kod taşıma yönteminin farklı seviyelerdeki bulut bilişim altyapılarının kullanılması durumundaki başarılarının karşılaştırılması*. Retrieved 20 June, 2016 from http://ceur-ws.org/Vol-1221/20_Bildiri.pdf

Kerawalla, L., Luckin, R., Seljeflot, S., & Woolard, A. (2006). Making it real: Exploring the potential of augmented reality for teaching primary school science. *Virtual Reality (Waltham Cross)*, *10*(3-4), 163–174. doi:10.1007/s10055-006-0036-4

Kirkley, S., & Kirkley, J. (2005). Creating next generation blended learning environments using mixed reality, video games and simulations. *TechTrends*, *49*(3), 42–53. doi:10.1007/BF02763646

Klopfer, E., & Yoon, S. (2004). Developing games and simulations for today and tomorrow's tech savvy youth. *TechTrends*, *49*(3), 41–49.

Korkmaz, Ö. (2013). İlk ve orta öğretimde öğretimsel amaçlı teknoloji kullanımı. In K. Çağıltay & Y. Göktaş (Ed.), Içinde Öğretim teknolojilerinin temelleri: Teoriler, araştırmalar, eğilimler (pp. 431–446). Ankara: Pegem Akademi.

Küçük, S., Kapakin, S., & Göktaş, Y. (2015). Medical faculty students views on anatomy learning via mobile augmented reality technology. *Journal of Higher Education and Science*, *5*(3), 316–323. doi:10.5961/jhes.2015.133

Küçük, S., Kapakin, S., & Göktaş, Y. (2016). Learning anatomy via mobile augmented reality: Effects on achievement and cognitive load. *Anatomical Sciences Education*. doi:10.1002/ase.1603 PMID:26950521

Küçük, S., Yılmaz, R. M., & Göktaş, Y. (2014). Augmented reality for learning english: Achievement, attitude and cognitive load levels of students. *Education and Science*, *39*(176), 393–404.

Kulik, J. (2003). *Effects of using instructional technology in elementary and secondary schools: What controlled evaluation studies say*. Arlington, VA: SRI International. Retrieved 14 May, 2016 from http://downloads.kennisnet.nl/mediawijzer/Onderzoeken/Kulik_ITinK-12_Main_Report.pdf

Lester, J. C., Converse, S. A., Kahler, S. E., Barlow, S. T., Stone, B. A., & Bhogal, R. S. (1997). The persona effect: affective impact of animated pedagogical agents.*Proceedings of conference on human factors in computing systems*, 359-366. doi:10.1145/258549.258797

Martín, S., Díaz, G., Cáceres, M., Gago, D., & Gibert, M. (2012). A Mobile Augmented Reality Gymkhana For Improving Technological Skills And History Learning: Outcomes And Some Determining Factors. In T. Bastiaens & G. Marks (Eds.), *Proceedings of E-Learn: World Conference on E-Learning in Corporate, Government, Healthcare, and Higher Education 2012* (pp. 260-265). Chesapeake, VA: Association for the Advancement of Computing in Education (AACE).

Matcha, W., & Rambli, D. R. A. (2013). Exploratory study on collaborative interaction through the use of augmented reality in science learning. *Procedia Computer Science, 25*, 144–153. doi:10.1016/j.procs.2013.11.018

Medicherla, P. S., Chang, G., & Morreale, P. (2010). *Visualization for increased understanding and learning using augmented reality.* Paper presented at 11th ACM SIGMM International Conference on Multimedia Information Retrieval (MIR '10), Philadelphia, PA. doi:10.1145/1743384.1743462

Milgram, P., & Kishino, F. (1994). A taxonomy of mixed reality visual displays. *IEICE Transactions on Information and Systems, 77*(12), 1321–1329.

New Media Consortium. (2005). *The Horizon Report. National Learning Infrastructure Initiative.* Stanford, CA: McGraw Hill.

Núñez, M., Quirós, R., Núñez, I., Carda, J. B., & Camahort, E. (2008, July). *Collaborative augmented reality for inorganic chemistry education.* Paper presented at 5th WSEAS / IASME International Conference on Engineering Education (EE'08), Heraklion, Greece.

O'Brien, H. L., & Toms, E. G. (2005). *Engagement as process in computer mediated environments.* Paper presented at ASIS&T, Charlotte, NC.

Oh, S., & Woo, W. (2008). *ARGarden: Augmented edutainment system with a learning companion* (Vol. 5080). Transactions on Edutainment I Lecture Notes in Computer Science.

Özarslan, Y. (2013). *The effect of augmented reality enhanced learning materials on learners' achievement and learners' satisfaction* (Unpublished Doctoral Dissertation). Anadolu University, Eskişehir, Turkey.

Papagiannakis, G., Singh, G., & Magnenat-Thalmann, N. (2008). A survey of mobile and wireless technologies for augmented reality systems. *Computer Animation and Virtual Worlds, 19*(1), 3–22. doi:10.1002/cav.221

Perez-Lopez, D., & Contero, M. (2013). Delivering educational multimedia contents through an augmented reality application: A case study on its impact on knowledge acquisition and retention. *Turkish Online Journal of Educational Technology, 12*(4), 19–28.

Rambli, D. R. A., Matcha, W., & Sulaiman, S. (2013). Fun learning with AR alphabet book for preschool children. *Procedia Computer Science, 25*, 211–219. doi:10.1016/j.procs.2013.11.026

Rosenbaum, E., Klopfer, E., & Perry, J. (2006). On location learning: Authentic applied science with networked augmented realities. *Journal of Science Education and Technology, 16*(1), 31–45. doi:10.1007/s10956-006-9036-0

Schmalstieg, D., & Wagner, D. (2007). *Experiences with handheld augmented reality.* Paper presented at 6th IEEE and ACM International Symposium on Mixed and Augmented Reality.

Serio, A. D., Ibanez, M. B., & Kloos, C. D. (2013). Impact of an augmented reality system on students motivation for a visual art course. *Computers & Education*, *68*, 586–596. doi:10.1016/j.compedu.2012.03.002

Shelton, B. E., & Hedley, N. R. (2002). *Using augmented reality for teaching earth-sun relationships to undergraduate geography students.* Paper presented at The First IEEE International Augmented Reality Toolkit Workshop, Darmstadt, Germany. doi:10.1109/ART.2002.1106948

Shelton, B. E., & Stevens, R. (2004). *Using coordination classes to interpret conceptual change in astronomical thinking.* Paper presented at 6th International Conference for the Learning Sciences, Santa Monica, CA.

Shirazi, A., & Behzadan, A. (2014). Design and assessment of a mobile augmented reality-based information delivery tool for construction and civil engineering curriculum. *Journal of Professional Issues in Engineering Education and Practice*, *04014012*. doi:10.1061/(ASCE)EI.1943-5541.0000229

Sırakaya, M. (2015). *Effects of augmented reality applications on students' achievement, misconceptions and course engagement* (Unpublished doctoral dissertation). Gazi University, Ankara, Turkey.

Sırakaya, M., & Seferoğlu, S. (2016). Öğrenme ortamlarında yeni bir araç: Bir eğitlence uygulaması olarak artırılmış gerçeklik. In A. İşman, H. F. Odabaşı, & B. Akkoyunlu (Eds.), *Eğitim teknolojileri okumaları 2016* (pp. 417–438). Ankara: Salmat Basım Yayıncılık.

Somyürek, S. (2014). Gaining the attention of generation Z in learning process: Augmented reality. *Eğitim Teknolojisi Kuram ve Uygulama*, *4*(1), 63–80.

Specht, M., Ternire, S., & Greller, W. (2011). Dimensions of mobile augmented reality for learning: A first inventory. *Journal of the Research Center for Educational Technology*, *7*(1), 117–127.

Squire, K., & Klopfer, E. (2007). Augmented reality simulations on handheld computers. *Journal of the Learning Sciences*, *16*(3), 371–413. doi:10.1080/10508400701413435

Squire, K. D., & Jan, M. (2007). Mad city mystery: Developing scientific argumentation skills with a place-based augmented reality game on handheld computers. *Journal of Science Education and Technology*, *16*(1), 5–29. doi:10.1007/s10956-006-9037-z

Sumadio, D. D., & Rambli, D. R. A. (2010). Preliminary evaluation on user acceptance of the augmented reality use for education.*Proceedings of Second International Conference on Computer Engineering and Applications*, 461–465. doi:10.1109/ICCEA.2010.239

Tarng, W., & Ou, K.-L. (2012). *A study of campus butterfly ecology learning system based on augmented reality and mobile learning.* Paper presented at 2012 IEEE Seventh International Conference on Wireless, Mobile and Ubiquitous Technology in Education. doi:10.1109/WMUTE.2012.17

Taşkıran, A., Koral, E., & Bozkurt, A. (2015). *Artırılmış gerçeklik uygulamasının yabancı dil eğitiminde kullanılması*. Paper presented at Akademik Bilişim 15, Eskişehir, Turkey.

Technopedia. (2016). *Mobile device*. Retrieved 20 May, 2016 from https://www.techopedia.com/definition/23586/mobile-device

Tian, K., Endo, M., Urata, M., Mouri, K., & Yasuda, T. (2014). Multi-viewpoint smartphone AR-based learning system for astronomical observation. *International Journal of Computer Th eory and Engineering, 6*(5), 396–400.

Tomi, A. B., & Rambli, D. R. A. (2013). An interactive mobile augmented reality magical playbook: Learning number with the thirsty crow. *Procedia Computer Science, 25*, 123–130. doi:10.1016/j.procs.2013.11.015

Tosik-Gün, E. (2014). *Effects of augmented reality applications on students' spatial ability* (Unpublished Master Thesis). Gazi University, Ankara, Turkey.

University of Calgary. (2016). *Augmented reality in education*. Retrieved 10 July, 2106 from http://679augmentedreality.wikispaces.com/History+of+AR

van Krevelen, D. W. F., & Poelman, R. (2010). A survey of augmented reality technologies, applications and limitations. *International Journal of Virtual Reality, 9*(2), 1.

Vilkoniene, M. (2009). Influence of augmented reality technology upon pupils' knowledge about human digestive system: The results of the experiment. *Online Submission, 6*(1), 36–43.

Wagner, D., Schmalstieg, D., & Billinghurst, M. (2006). Handheld AR for collaborative edutainment. *Advances in Artificial Reality and TeleExistence,* 85–96.

Wang, X. (2012). Augmented reality: A new way of augmented learning. *E Learn Magazine*. Retrieved 10 April, 2016 from http://elearnmag.acm.org/featured.cfm?aid=2380717

Warschauer, M. (2011). *Eventually tablets will facilitate more personalized and interactive learning*. Retrieved 10 May, 2016 from http://edutechdebate.org/tablet-computers-in-education/eventually-tablets-will-facilitate-more-personalized-and-interactive-learning/

Wenglinsky, H. (2006). Technology and achievement: The bottom line. *Learning in the Digital Age, 63*(4), 29–32.

Wojciechowski, R., & Cellary, W. (2013). Evaluation of learners attitude toward learning in ARIES augmented reality environments. *Computers & Education, 68*, 570–585. doi:10.1016/j.compedu.2013.02.014

Wu, H.-K., Lee, S. W.-Y., Chang, H.-Y., & Liang, J.-C. (2013). Current status, opportunities and challenges of augmented reality in education. *Computers & Education, 62*, 41–49. doi:10.1016/j.compedu.2012.10.024

Yen, J. C., Tsai, C. H., & Wang, J. Y. (2012). *The effects of augmented reality on students' moon phases concept learning and their conceptual changes of misconception*. Paper presented at 2012 International Conference on Business and Information, Sapporo, Japan.

Yen, J. C., Tsai, C. H., & Wu, M. (2013). Augmented reality in the higher education: Students science concept learning and academic achievement in astronomy. *Procedia: Social and Behavioral Sciences, 103*, 165–173. doi:10.1016/j.sbspro.2013.10.322

Yılmaz, R. M. (2014). *Effects of three dimensional storytelling developed with augmented reality technology on narrative skill and creativity* (Unpublished doctoral dissertation). Atatürk University, Erzurum, Turkey.

Yuen, S., Yaoyuneyong, G., & Johnson, E. (2011). Augmented reality: An overview and five directions for AR in education. *Journal of Educational Technology Development and Exchange*, *4*(1), 119–140.

Yusoff, Z., & Dahlan, H. M. (2013). *Mobile based learning: An integrated framework to support learning engagement through Augmented Reality environment.* Paper presented at 2013 International Conference on Research and Innovation in Information Systems (ICRIIS). doi:10.1109/ICRIIS.2013.6716718

Zarzuela, M. M., Pernas, F. J. D. P., Martínez, L. B., Ortega, D. G., & Rodríguez, M. A. (2013). Mobile serious game using augmented reality for supporting childrens learning about animals. *Procedia Computer Science*, *25*, 375–381. doi:10.1016/j.procs.2013.11.046

Zhou, Z., Cheok, A. D., & Pan, J. (2004). 3D story cube: An interactive tangible user interface for storytelling with 3D graphics and audio. *Personal and Ubiquitous Computing*, *8*(5), 374–376. doi:10.1007/s00779-004-0300-0

KEY TERMS AND DEFINITIONS

Augmented Learning: Application of knowledge and skills learned in a seamless way by combining learning environments with the real world we live in.

Augmented Reality: The integration of digital information with the user's environment in real time.

Educational Purposes: The purposes that guide teaching-learning and evaluating activities and direct teaching in education.

Mobile Augmented Reality: Enrichment of the information in the world we live in for a purpose by being combined with images and /or information obtained through a mobile device.

Mobile Device: A small computing device, small enough to hold and operate in the hand and having an operating system capable of running mobile apps.

Virtual Reality: An artificial, computer-generated simulation or recreation of a real life environment or situation.

Chapter 6
Existing Standards and Programs for Use in Mobile Augmented Reality

Gülay Ekren
Sinop University, Turkey

Nilgun Ozdamar Keskın
Anadolu University, Turkey

ABSTRACT

Augmented reality (AR) brings flexibility and openness to learning. It also strengthens connections between learners in multiple contexts by adding a new dimension to mobile learning. As well, AR facilitates the learning process of the learners by bringing different disciplines together with education. This chapter focuses on technologies, tools, programs, operating systems, a number of standards as well as some parts of web infrastructure standards for use in mobile AR applications and then discusses the current state, issues, and direction of the development and the use of these apps. The purpose of this chapter is to provide information on that can be used by developers of mobile AR applications for learning and to assist educators and instructional designers in developing mobile augmented reality apps for learning by using AR development tools that provide them to create custom mobile AR applications without programming skills.

INTRODUCTION

Mobile devices are affecting people's the way of work, spending of time, knowledge acquisition and learning/teaching methods besides making calls. Besides, using mobile devices in education is increasing depending on technologies that feeding them (Martin, Diaz, Sancristobal, Gil, Castro, & Peire, 2011; Özdamar Keskin & Kuzu, 2015). Augmented Reality (AR), which is considered to be one of these technologies, has not realized the expected impact in education before 2010. But after 2010, it is observed trends or developments from augmented reality towards mobile augmented reality associated with the additional capabilities of mobile devices such as camera, GPS (Global Positioning System), image rec-

DOI: 10.4018/978-1-5225-2110-5.ch006

ognition; compass (Johnson, Levine, Smith, & Stone, 2010). The basic hardware requirements of first AR applications independently from mobile devices are such as a display, tracking, input devices and computers (Jeon, Kim, & Lee, 2010). Nowadays, there has been significant progress in miniaturization and performance improvement of mobile computing platforms for integrating mobile augmented reality such as notebook computers, personal digital assistants, tablets, ultra-mobile PCs, mobile phones (or smartphones), and AR glasses (Huang, Hui, Peylo, & Chatzopoulos, 2013). Latest smartphones are equipped with brilliant cameras, accelerometers, GPS devices, compasses, barometers, step counters and many more sensors to capture contextual data. Thus, the smartphones and tablets are mainly seen suitable for AR platforms in recent years. On the other hand, mobile augmented reality is a relatively young technology, so the business models and usage patterns are still evolving (Makumburage, 2015). Augmented Reality reached the peak of its hype in 2010, and it has substantial long-term potential and will reach mainstream adoption within five to ten years. "Gartner Hype Cycle positioning of AR as of July 2012" is shown in Figure 1 (Padzensky, 2014).

AR is mainly related to the concept of virtual reality. However, virtual reality attempts to create an artificial world that a person can experience and explore interactively. AR makes way for an interactive experience but aims to supplement the real world, rather than creating an entirely artificial environment around the user (Hollerer, 2004). AR also includes activities providing augmentation or expansion of the physical environment that can be seen by reflecting real surfaces via mobile devices. AR is sometimes defined to produce any activity by adding a new dimension to the physical space, but with a web-enabled mobile device, layers can be hidden and suspended in nominal dimension and then may appear in space.

Figure 1. Gartner hype cycle positioning of augmented reality as of July 2012

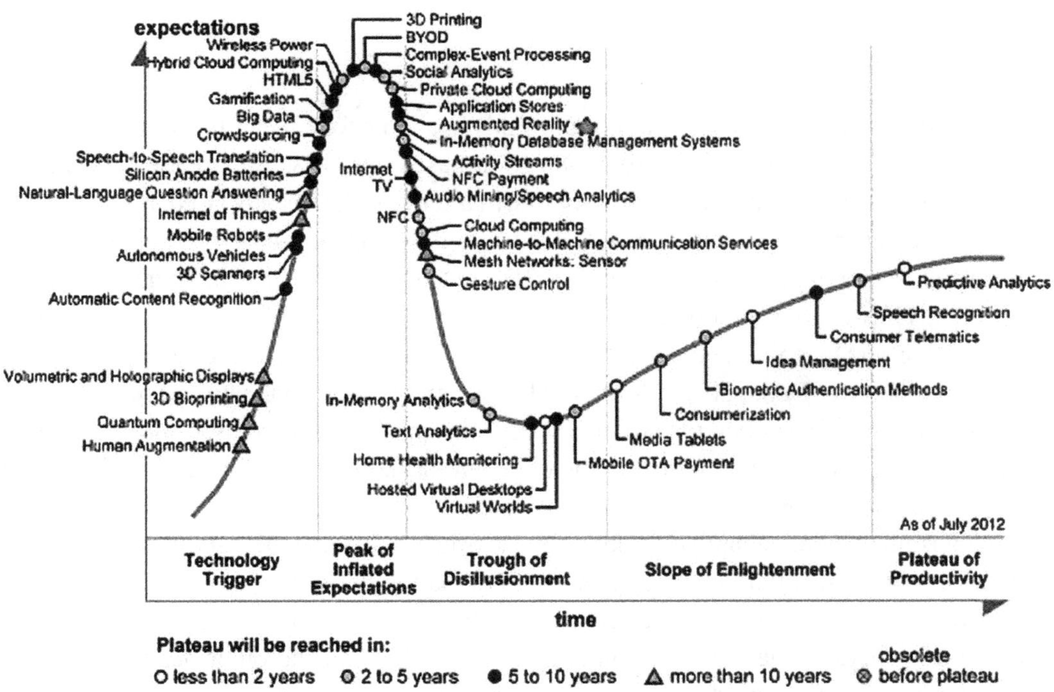

AR also brings flexibility and openness to learning. It strengthens connections between learners in multiple contexts. As well, it facilitates learning process by bringing different disciplines together with education (Munnerley, Bacon, Wilson, Steele, Hedberg, & Fitzgerald, 2012). Accordingly, mobile app designers and developers started to think more deeply about mobile users' context and situations. AR provides mobile learning which takes place in various contexts through contents and social interactions using personal electronic devices. (Crompton, 2013). Additionally, mobile learning is defined as a learning process which takes place via mobile technologies and enables accessing educational contents without a geographical restriction. It also dynamically making advantages from advanced services, providing to engage others and increasing the productivity and work performance of users by instantly responding to their individual needs (Özdamar Keskin & Kuzu, 2015). Along these lines, the mobile AR is starting to drive innovation within education as well as gaming, medical, automotive, and manufacturing markets (Vandebona, 2015).

AR could provide advanced learning experiences that are grounded in two interdependent theoretical frameworks; situated learning and constructivism (Dunleavy & Dede, 2014). Situated learning based on a specific context and the quality of the learning that is the result of interactions among the people, places, objects, processes, and culture (Brown, Collins, & Duguid, 1989). It also builds upon and extends other learning theories such as social learning and social development (Bandura, 1977; Vygotsky, 1978). Constructivism builds new knowledge and understandings based on what learners already know and learning is shaped by their developmental level, their prior experiences, their interactions, and their sociocultural background and context (Bruner, 1966; Vygotsky, 1978). As a cognitive tool or pedagogical approach, AR aligns well with situated and constructivist learning theory as it posits the learner within a physical and social context (Dunleavy & Dede, 2014). It also gives learning a new dimension with interactivity and makes classrooms more engaging.

Mobile AR Apps

There are two forms of mobile AR applications currently available for educators, or instructional designers: (1) location-based (can be named also geolocation-based, geo-based, markerless, location-aware) and (2) image-based (can be named also vision-based, marker-based, artefact-based, nonlocation-based). Both of them have enormous potential for distance learners. A location-based AR application presents digital media such as text, graphics, audio, video, 3D models to learners by moving through a physical environment with GPS-enabled mobile devices. On the contrary, an image-based AR application presents digital media to learners by pointing their webcam to an object that has QR (Quick Response) Code or has the 2D target (Dunleavy & Dede, 2014). Although location-based AR applications are designing mostly for large-scale outdoor AR applications, image based AR applications are used mostly as small-scale interior applications. Location-based AR applications are utilizing geographical data, named Point of Interests (POIs), by using positioning sensors such as GPS (Ahn, Ko, & Yoon, 2014) and an electronic compass to determine where the user is looking for (de Smit, 2010). Location-based AR allows users to edit virtual assets (text, 3D, and video, graphic) to POIs on the physical map. Thus, the users of these type applications can access to the required content in any area by exploring POIs. POIs can be arranged at any time at anywhere in the world (Munnerley et al., 2012). On the other hand, image-based AR applications work by tracking markers or patterns that placed on particular objects. Camera has been using for tracking marker and it captures AR code such as barcode, QR code, and predefined image to interact with the user. In such a way, information such as animations, videos, 3D pictures, games, etc.

can be initiated. (Munnerley et al., 2012). Distinctly from location-based AR, image-based AR generally uses image recognition by taking the part of any image from a live streaming camera that will be augmented (de Smit, 2010).

Standards Using in Mobile AR Apps

Many mobile AR applications directly or indirectly use a set of standards to strengthen the capabilities of the hardware. Besides, previous and emerging applications vary from device to device. Application Programming Interfaces (API) reduces the regulation of software requirements according to the needs of hardware platforms. Existing standards that can be used Geo Information System (GIS) based or geolocation based mobile AR applications are shown in Table 1 (Perey, Engelke, & Reed, 2011):

There are also two types of approaches for standards using in mobile AR apps; declarative approach and imperative approach. Imperative approaches are related to something like computer code that contains features such as flow control, storage, variables, and states. The design of such a structure can be made by the desired shape as long as it complied with the working system in the background. For example, JavaScript, which is a form of implementation of the ECMAScript language standard, is defined as an imperative approach, and is a part of all browsers. Declarative approaches are more restrictive than imperative approaches and are concerned with what will be calculated. Designs in declarative approaches must follow a certain behavior scheme or structure, but the creation and use of them are relatively easy. For example; XML, JSON (JavaScript Object Notation), HTML5, CSS3 (Cascading Style Sheets) standards can be defined as declarative approaches (Engelke, Becker, Wuest, Keil, & Kuijper, 2013).

Mobile AR applications based on Web technology use the HTTP protocol for client-server communication. "Layar" as a mobile AR application, is helping users to explore/scan their location in any subject nearest to them. AR browsers such as Wikitude and Layar are used to provide POIs or locations via web. While desktop AR applications require Flash and similar programs, these type native applications on mobile devices generally require programs such as JSON and XML (de Smit, 2010). The application frameworks have established to provide the portability of the content across different hardware platforms. HTML5 is a standard that converts browsers into an application platform by providing resources such as memory, threads, and sensors. In this way, web content gains portability. A modern browser is required main components such as JavaScript, DOM, and CSS3 in order to support an AR application. All mobile devices have a web browser that supports HTML5 standards. Since JavaScript directly provides to access DOM (Document Object Model) in any web browser, so it allows creating, investigating or removing any declarative items (Perey et al., 2011).

Junaio and ARchitect are HTML5-based AR content platforms. Junaio provides AREL (Augmented Reality Experience Language) in XML for content definition as well as a Javascript API for application logic. Besides, HTML is used to establish the relationship between the application logic in JavaScript and XML content definitions. Furthermore, Junaio provides latitude, longitude and altitude markers for marker-based positioning. ARchitect, as a JavaScript AR Software Development Kit (SDK) application, provides location-based AR support as well as visual-based AR support. Content structure of ARchitect has not trusted HTML content, so its application logic has not used the DOM APIs. With this approach, it is difficult to use AR content with standard HTML content or to implicate the interaction with DOM mechanism (Ahn, Ko, & Yoo, 2014).

WebGL standard is an interface extension that has integrated to web browser engines. WebGL is defined as a direct interface for Graphical Processing Unit (GPU) by using the OpenGL ES standard

Table 1. Existing standards for geolocation based mobile AR applications

Standards	Organization	Description
Geography Markup Language (GML)	OGC*, ISO	It is an XML** based encoding standard for geographic information
CityGML	OGC	It is an open data model and XML-based format for the storage and exchange of virtual 3D city models. It is an application schema for the GML3.
Keyhole Markup Language (KML)	OGC	It is an XML language focused on geographic visualization such as the presentation of graphical data on the globe, the control of the user's navigation in the sense of where to go and where to look, annotation of maps and images.
Sensor Model Language (SensorML)	OGC	It allows the developer to define models and XML Schema for describing any process, including measurement by a sensor system, as well as post-measurement processing.
Sensor Observation Service (SOS)	OGC	It is a data service that provides access to sensors and sensor systems that are consistent with all sensor systems including remote, in-situ, fixed and mobile sensors.
Web Map Service (WMS)	OGC, ISO	provides a simple HTTP interface for requesting geo-registered map images from one or more distributed geospatial databases
Style Layer Descriptor (SLD)	OGC	It provides constructs for the symbolization and coloring of geographic features and coverages.
OpenGL	Khronos	It is an application programming interface (API) for drawing 2D and 3D graphics. It is designed to be implemented mostly or entirely in hardware.
OpenGL ES	Khronos	It is a subset of the OpenGL computer graphics rendering API for rendering 2D and 3D computer graphics designed for embedded systems like smartphones, computer tablets, video game consoles and PDAs.
Open SL ES	Khronos	Open Sound Library for Embedded Systems, a free, cross-platform, hardware-accelerated, C-language audio API for 2D and 3D audio.
COLLADA	Khronos	It defines an open standard XML schema for exchanging digital assets among various graphics software applications
Vulkan	Khronos	Named as "Next Generation OpenGL Initiative" (glNext), it is a redesign effort to unify OpenGL and OpenGL ES into one common API.
Scalable Vector Graphics (SVG)	W3C	It is used to define vector-based graphics in XML format for the Web
HyperText Markup Language (HTML)	W3C	It is the core language of the World Wide Web
ECMAScript	ISO	It is a scripting language that forms the basis of JavaScript
X3D	Web3D, ISO	It is an open standards file format and run-time architecture to represent and communicate 3D scenes and objects using XML
GeoRSS	GeoRSS	It is an emerging standard for encoding location as part of a Web feed

*The OpenGIS Consortium
** eXtensible Markup Language

(Engelke, Becker, Wuest, Keil, & Kuijper, 2013). WebGL provides web developers to create any 3D content in HTML stack without requiring any plug-in. However, web content developers use declarative approaches for 3D visuals. For example, X3D standard allows images or graphics directly identified in conjunction with a standard such as MPEG-4 or OGC CityGML in a declarative way. Similarly, A2ML (Audio Markup Language) is used for the definition of sounds (Perey et al., 2011).

X3D is a three-dimensional content standard which describes 3D scenes, objects, animations, interactions and their behavior. This technology can directly be a part of DOM and can be accessed through JavaScript, so additional software has not needed for displaying any 3D content (Engelke et al., 2013). On the other hand, X3DOM is an open-source framework and runtime for 3D graphics on the web. X3DOM tries to support the use of declarative XML-based X3D scenes in the Web3D. This effort has strategic importance for Web3D and indeed for all 3D graphics since X3D is an interchangeable format for a wide variety of models (http://www.x3dom.org/, http://web3d.org/wiki/index.php/X3D_and_HTML5).

Programming Platforms for Mobile AR Apps

Mobile devices are gaining popularity with their features such as camera, music, GPS, accelerometer, etc. mostly available with all major platforms or operating systems (OS) such as Android, Blackberry, iOS, Symbian, Windows Mobile/Phone. According to the International Data Corporation (IDC), Android has been dominating the world market share over the last years (Figure 2). As it can be seen in Figure 2, in recent years, Android, iOS and Microsoft Windows Phone are the three major players of the market (Statista, 2016).

Above is the popularity graph which represents last seven years' trends. It can be seen clearly that Android is beating up all other operating systems, even the iOS. Symbian, Blackberry (RIM) and Windows Phone/Mobile (Windows) are also observing a diminishing slope.

Symbian OS is created by a company called Psion as an object-oriented operating system called EPOC, which was designed basically to fulfill the requirements of mobile computing devices. It was first introduced in the late 1980s. In 1998, EPOC is renamed to Symbian OS (Kamboj and Gupta, 2012). It is officially the property of Nokia, Panasonic, Psion, Samsung, Siemens and Sony Ericsson. It means that any other company will have to take permission from them before using this operating system. The Symbian Foundation is established in 2009. In 2010, the demand rate for Symbian has been continuously decreasing due to the changes in global economic and market conditions (http://licensing.symbian.org/). In 2011, Nokia announced that it will cut its support from Symbian as an operating system in its devices and will use the Windows Phone OS. After then, Nokia has lost a large part of developers in its platform.

Figure 2. Worldwide smartphone operating systems market share

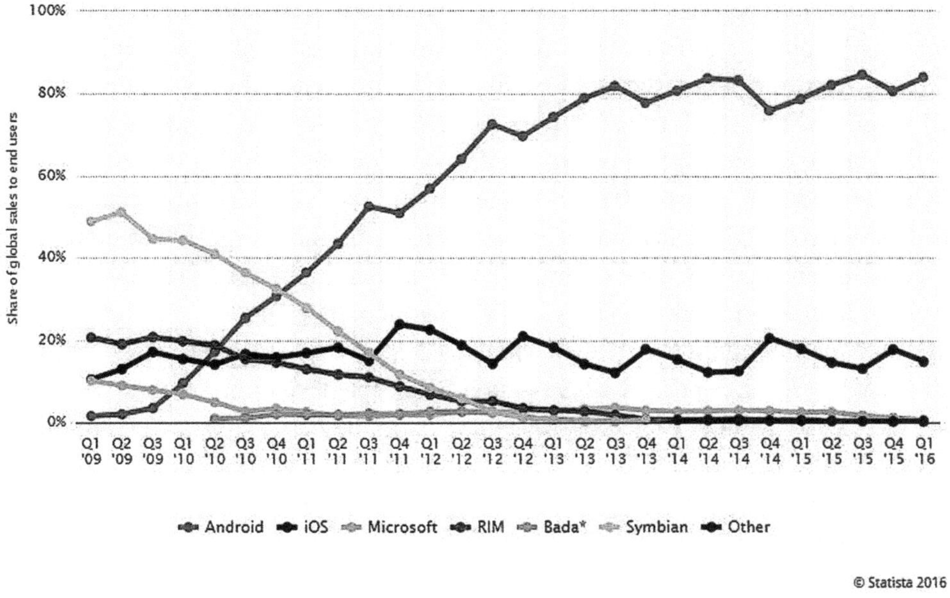

Blackberry OS was first released in 1999 and developed by RIM (Research in Motion) for its Blackberry line of smartphones. Blackberry is much different from other operating systems in terms of interface style, smartphone design, and keyboard. RIM has not allowed any other manufacturer to lay hands on its operating system like Apple (Guru, 2015). However, RIM provides tools for developing applications and themes for BlackBerry (Kamboj and Gupta, 2012).

iOS was introduced in 2007 when the first iPhone was developed by Apple. Apple has not allowed any other manufacturer to lay hands on its operating system. Unlike Android, Apple has more concentrated on the performance along with appearance. As the basic appearance of iOS is almost the same as it was in 2007. So far iOS has been used in all iPhones, iPod and iPad (Guru, 2015).

The first Android OS was released by Google in 2008 as the name of Astro. Then, its upgraded versions 'Bender' and 'Cupcake' were also released. It is an open source software. Google adopted the trend of naming Android versions as any dessert or a sweet such as Donut, Eclair, Froyo, Gingerbread, Honeycomb, Ice Cream Sandwich, Jelly Bean, Lollipop and Marshmallow. Google Play Store is an official app market that contains millions of different apps for Android devices. Samsung, HTC, Motorola and many other top manufacturers are currently using Android in their devices (Guru, 2015).

Bada was developed by Samsung Electronics and first released in 2010 with its software development kit to attract developers. The name 'bada' is derived from the Korean word for ocean or sea. It is designed for mid-range and high-end smartphones (Kamboj and Gupta, 2012).

Windows OS is used in computers all over the world and has also been used in mobile phones. Nokia and Microsoft joined hands to work together in 2013. Nokia Lumia series is completely windows based. Samsung and HTC also released some windows based phones. The computer like look has totally vanished from the windows phones with the release of Windows Phone 7 (Guru, 2015). The latest mobile OS released by Microsoft is known as Windows Phone 10.

Structures of Mobile Apps

The structure of mobile apps can be classified in three different ways; (1) mobile native apps, (2) mobile web apps, (3) mobile hybrid apps and cross-platforms mobile apps (Özdamar Keskin & Kılınç, 2015). Native apps tied to a mobile operating system, so it works on only its own platform. These apps also use programming language and framework of its suppliers, for example, the Apple iOS platform uses Objective C with Xcode, whereas the Google Android platform uses Java with Eclipse, and Microsoft Windows Mobile/Phone platform uses C# with Visual Studio (Serrano, Hernantes, & Gallardo, 2013).

On the other hand, mobile web apps are similar to any other web site which designed for mobile devices (Özdamar Keskin & Kılınc, 2015). Although mobile web apps are stored in servers, native apps are downloaded and installed on any mobile devices. However, the user interface controls of them such as buttons, selectors, and textboxes are optimized in a way that is similar to any native app. They can be accessed over the internet through web browsers of any mobile device.

Mobile hybrid apps are mobile web apps which packed into a native app. These apps are developed mostly with HTML5, JavaScript, and CSS3 which using for developing mobile web apps (Özdamar Keskin & Kılınç, 2015). Mobile web apps are converted to the native apps by compiling on all mobile platforms (iOS, Android, Windows, etc.). Besides, hybrid apps can fully access the specific hardware features of any mobile device such as compass, camera etc. that can be used in mobile apps partial. Some technical features of these apps are compared in Table 2.

Table 2. Comparison of native, hybrid and web apps (Serrano, Hernantes & Gallardo, 2013)

Technical Considerations	Mobile Native Apps	Mobile Hybrid Apps	Mobile Web Apps
Efforts of supporting platforms and versions	High	Medium	Low
Device capabilities access	Full	Full	Partial
User experience	Full	Full	Medium
Performance	Very high	Very high	High
Upgrade in the client	Needed	Needed	Not needed
Ease of publication/distribution	Medium	Medium	High
Approval cycle	Mandatory	In some cases	Not required
Monetization in app store	Available	Available	Not available

Cross-platform mobile apps have been developed to meet the needs of developers for writing source code once and running it on different platforms (Özdamar Keskin & Kılınç, 2015). By this way, developers do not need to know the APIs of each platform, but only the APIs provided by selected development tool. There are several cross-platform mobile development tools such as Rhodes, Phone Gap, Dragon Rad, MoSync, Cordova, JQuery Mobile, Adobe Air, Xamarin, and Sencha Touch (Palmieri et al., 2012). For instance, Adobe Phone Gap is a development tool that allows developers to build commercial or open-source hybrid apps with HTML, CSS, and JavaScript, and by using the functionalities of SDKs instead of using programming languages such as C, C++, Objective C, C# or Java. The experiences for multiple platforms can also be created with a single codebase (http://phonegap.com/).

The context of app affects users' expectations, in other words, the context for a single app may be different from one user to the next (Charland & Leroux, 2011). As stated by Smutni (2012), the context of the mobile app will drive a decision for choosing the best configuration. If a mobile app is mainly used to display and interact with online content or services, it is better to avoid native apps. However, if the mobile app is mainly used offline, a native app will offer a better user experience.

Developing a Mobile AR App

Major platforms (or operating systems) such as Android, iOS, Blackberry, Symbian, and Windows Mobile provide developing mobile apps. However, the main architecture and programming language supported by these platforms is different (Palmieri, Singh, & Cicchetti, 2012). They have been using some form of C programming language and Java as well as HTML, CSS, JavaScript for developing apps, but each platform has its own development environment and programming interface that can be seen in Table 3 (Charland & Leroux, 2011; Smith, 2013):

Differences among software development kits of these platforms make them complicated for mobile apps. Although there are different tools, APIs, structures, Integrated Development Environments (IDEs), and devices with different capabilities for each mobile platform, the only thing common is that all platforms ship with a mobile browser accessible programmatically from the native code. Each platform interacts with its JavaScript interface from native code, and also mobile browser allows calling native code from JavaScript (Charland & Leroux, 2011). Nevertheless, it is hard to build apps across different platforms for developers because their complex, proprietary structure.

Table 3. Mobile platforms required programming skills and development environments

Mobile Platforms	Required Skill Set	Development Environment
Android (Google)	Java or C++	Eclipse
iOS (Apple)	C, Objective-C	Xcode
Blackberry	Java	Eclipse
Symbian	C, C++, Python, HTML/CSS/JavaScript	Multiple choices
Windows Mobile (Microsoft)	C#,.NET	Visual Studio 2012
Windows 7 Phone (Microsoft)	C#,.NET	Visual Studio 2012
HP Palm WebOS	HTML/CSS/ JavaScript	Multiple choices
MeeGo	C, C++, HTML/CSS/ JavaScript	Multiple choices
Samsung Bada	C++	Multiple choices

Phone Gap interfaces the app with native APIs by using device-independent JavaScript APIs that provides rich functionality like accelerometer, barcode, Bluetooth, calendar, camera, capture, compass, network, contacts, file, geolocation, screen rotation, notification, and storage. The architecture of cross-platform development environment Phone Gap and interfacing among its components is shown in Figure 3 (Palmieri et al., 2012).

PhoneGap follows hybrid web model which supports interaction between web app and the device. Its apps are supported by platforms such as iOS, Android, Blackberry, Windows Mobile 6.5, Symbian, and Palm. Its apps can be built only in HTML, JavaScript, and CSS, so that removes the barriers of adoption for developers (Hartmann, Stead, & DeGani, 2011).

Figure 3. PhoneGap architecture and interfacing among components

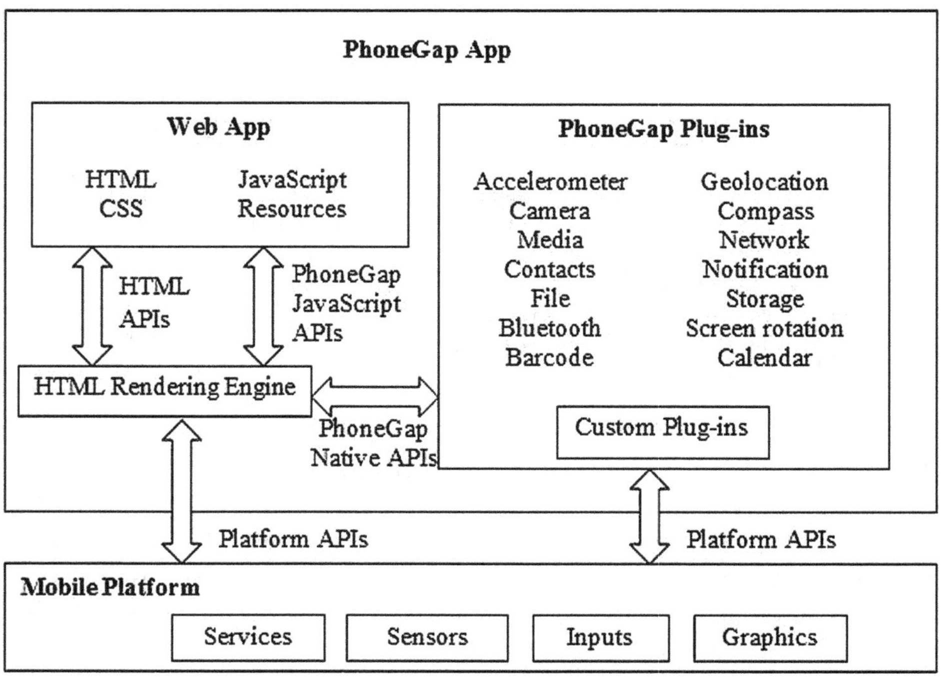

AR apps can be developed for any mobile platforms. For example, Android apps for AR can be built using the Android SDK as it provides the APIs for interacting with the camera. Android and iOS also supports 'OpenGL ES' specification, but Windows Mobile doesn't. However, Windows Mobile supports 'Direct 3D feature level 9_3'. So these operating systems are capable of running high-end AR apps with the help of the hardware. SDKs of these operating systems will provide APIs to accesses 2D and 3D graphic rendering, recognition and tracking facilities with hardware acceleration (Makumburage, 2015).

Programming Languages Use in Mobile Apps

HTML5 is a standard for developing browser-based mobile apps which are supported by different browsers such as Internet Explorer, Google Chrome, and Mozilla Firefox etc. However, it is not enough to create meaningful applications by using only HTML5, but also JavaScript. JavaScript provides using hardware features of mobile devices in cross-platform apps (Özdamar Keskin & Kılınç, 2015). JavaScript is a scripting language used for web browser applications. It is a useful language for adding additional features to web pages, but using only JavaScript is not seen as a good solution for developing mobile apps.

Java platform and also Enterprise JavaBeans (EJB) technology are using for developing mobile apps. Java is a commonly used object oriented programming language which was originally developed by Sun Microsystems, then continuing to be developed by Oracle. There are two different way to compile Java, one is in any browser window, and the other is in a virtual machine without requiring the browser. This provides flexibility to Java in such cases; reusing the language code or updating software. Besides, EJB technology is the server-side component architecture for Java Enterprise Edition (Java EE). This technology enables rapid and simplified development of distributed, transactional, secure and portable applications based on Java technology (www.oracle.com).

On the other hand, C is viable for cross-platform mobile apps. C is also providing the building blocks for the languages such as C#, Java, JavaScript, and Python. C++ is an extension of C language and is a general purpose object-oriented programming language. Therefore, it is possible to code C++ in a "C style" or "object-oriented style". It is considered to be an intermediate-level language, as it encapsulates both high and low-level language features (https://www.techopedia.com).

\C and C ++ principles are combined with C#, that is a general purpose language using to develop software for Microsoft and Windows platforms. C# is an object-oriented language that enables developers to build a variety of secure and robust applications that run on the.NET Framework such as Windows client applications, XML Web services, distributed components, client-server applications, database applications etc. Visual C# also provides an advanced code editor, convenient user interface designers, integrated debugger, and many other tools to make it easier to develop applications based on the C# language and the.NET Framework (https://msdn.microsoft.com).

Objective-C is a programming language used by the Apple operating system to create applications for the iPhone, iPad or iPod Touch. Objective-C is defined as a set of extensions to the C language. It's designed to give object-oriented programming capabilities of C, and to do so in a simple and straightforward way. Objective-C was layered on the top of C language. It means the extensions were added to C to create a new programming language that enabled objects to be created and manipulated. The language was based on a language called SmallTalk-80, one of the first object-oriented programming languages (Kochan, 2011).

When the iPhone was released in 2007, Apple allowed the developers to develop web-based apps which run under the iPhone Safari web browser and requires the user to connect the website that hosts the application in order to run it. Developers were not satisfied with web-based apps, so Apple announced that developers would be able to develop so-called native apps for iOS (Kochan, 2011).

Python is often used as a scripting language but is also used in a wide range of non-scripting contexts. It is a general-purpose, high-level programming language whose design philosophy emphasizes code readability. Python code can also be packaged into standalone executable programs and interpreted for many operating systems. Python's syntax allows programmers to express concepts in fewer lines of code than would be possible in languages such as C (Van Rossum, 2007).

Swift is a scripting language designed for composing application programs into parallel applications that can be executed on multicore processors, clusters, grids, clouds, and supercomputers. Swift focuses on the issues that arise from the concurrent execution, composition, and coordination of many independent computational tasks, apart from most other scripting languages. The swift script uses a C-like syntax to describe data, application components, invocations of application components, and the data flow among those invocations (Wilde, Hategan, Wozniak, Clifford, Katz, & Foster, 2011). It is a new programming language for iOS apps that builds on the best of C and Objective-C without the constraints of C compatibility. Objective-C developers feel familiar with Swift in that Swift adopts the readability of Objective-C's named parameters and the power of Objective-C's dynamic object model (https://developer.apple.com).

Software Development Tools for Mobile AR Apps

The features of mobile AR development platforms that have been aligned by educational objectives as follows (Dunleavy and Dede, 2014): (1*) browser-based editor* that allows designers to create custom AR experiences using an editing Web site interface, (2) *YouTube/Vimeo embed* that allows designers to embed videos into their AR experiences, (3) *roles* that allow designers to assign and differentiate between different participant roles for enabling individualized or team-based experiences, (4) *dynamic triggers* that allows for dynamic and cascading events within the AR experience, (5) *embedded assessment* that allows designers to embed assessments within their AR experience in multiple formats, (6) *data collection* in which app users will be able to capture and store data during the AR experience, (7) *device to device communication* in which app users will experience a single shared AR world with other users, (8) *QR code embed* allows designers to embed QR codes into an AR experience to act as markers or targets for triggering various media, (9) *image-based 3D model embed* allows designers to embed image-based or visual recognition AR to trigger interactive 3D models, (10) *social networking* allows designers to leverage social networking tools as a mechanic within the AR experience or as a way to share content or AR experiences.

There are relatively few stand-alone AR development tools that provide educators and instructional designers to create custom AR applications without programming skills. The main purposes of them are recognition, tracking and content rendering. The following AR development tools provide the majority of the previously outlined functions and not need any programming skills or server hosting (Dunleavy & Dede, 2014; Makumburage, 2015):

- **ARIS:** Is an open source platform for creating and playing mobile games, tours, and interactive stories. Using GPS and QR Codes, they experience a hybrid world of virtual interactive characters, items, and media placed in physical space (http://arisgames.org/).
- **BuildAR:** As allows creating media to tell stories about a particular place, or to display content around people wherever they are (location-based AR) and to display video, other images, interactive buttons or 3D objects when people scan relevant AR images (image-based AR) (http://buildar.com/).
- **FreshAiR:** Is a location-based storytelling and gaming app that reveals hidden stories about the areas around you as you drive, walk or play (http://playfreshair.com/).
- **Hoppola Augmentation:** Is a content platform for location-based augmented reality. It enables designers to create AR content once, and seamlessly to publish it in all three major AR browsers: Layar, Junaio and Wikitude (http://www.hoppala-agency.com/).
- **TaleBlazer:** Is a platform for making and playing location-based mobile AR games. It allows people to create new opportunities for dynamic storytelling, informal learning experiences, and community engagement (http://education.mit.edu/projects/taleblazer).
- **7Scenes:** Is a mobile storytelling platform that provides organizations with all the tools they need to create scenes – location based tours and games for mobile devices (http://7scenes.com/).
- **Catchoom CraftAR:** Offers tools to connect mobile apps with the CraftAR service in the cloud. Some of the key features of the CraftAR SDK are as camera capture management, cloud recognition, 3D object tracking, content rendering in AR view (http://catchoom.com/).
- **Aurasma:** Is available as a SDK or as a free app for iOS and Android mobile devices. Its image recognition technology uses the camera of a smartphone or a tablet to recognize real world images and then overlay rich media on them in the form of animations, videos, 3D models and web pages (https://www.aurasma.com/).
- **DroidAR:** Provides open source augmented reality SDKs for the Android platform. This framework can be used for many different scenarios such as footstep recognition, location based gaming etc. (https://bitstars.github.io/droidar/).

In addition to these, there are several toolkits available for creating AR apps on smartphones and wearable devices such as HandyAR, Vuforia, Layar, ARToolkit, LinceoVR, Wikitude, NyARToolkit, SketchUp (Ramakrishnamurthy, 2016). Some features of these development kits are as follows:

- **HandyAR:** Used a human hand as a tracking pattern to display augmented virtual objects on the hand. Handy AR has been using the user's hand for marker-free augmentation, but not for interaction with virtual objects (Lee & Höllerer, 2007; Lee, Green, & Billinghurst, 2008)
- **Vuforia AR SDK:** Supports Android and iOS smartphones and tablets as well as digital eyewear. Developers can build Vuforia apps in Eclipse, Xcode, and Unity cross-platform game engine. Vuforia-enabled apps can recognize a range of everyday objects such as books, magazines, toys, product packaging (http://www.vuforia.com/).
- **Layar:** Allows creating a personalized AR experience by enhancing flyers, postcards, packaging, catalogs, brochures, business cards and magazines or any other item are enabled with interactive content, including augmented video messages, active web and social links, photo slideshows, music clips etc. (https://www.layar.com/). Layar is one of the easiest ways to build AR apps without prior experience of AR.

- **ARToolkit:** Is an open source augmented reality SDK for iOS, Android, Linux, Windows and Mac OS. It includes OpenGL ES2.x support, integration with GPS and compass, and automatic camera calibration utilities. It can be integrated with a range of smart glasses, and allows calibration for new devices. Its video tracking libraries calculate the real camera position and orientation relative to physical markers in real time. This enables the easy development of a wide range of AR apps. (http://artoolkit.org/; Vandebona, 2015).
- **LinceoVR:** The authoring tool that enables to create apps both in virtual and augmented reality. Key features and benefits are such as sensor-based tracking, marker tracking, texture tracking, real-time rendering (environment reflections, 360^0 turnable on any axis, image or movie backgrounds, effects, materials, animations, sound, 3D data input etc.) (http://www.altairhyperworks.com/partner/LinceoVR).
- **Wikitude:** Is an augmented reality SDK that is a library for location based as well as vision-based AR, also has image recognition and tracking for Android, iOS, and BlackBerry10. No programming skills are needed for building AR apps (http://www.wikitude.com/).
- **NyARToolkit:** Is a vision-based Augmented reality library that based on ARToolKit. ARToolkit can be run on managed environment, such as Java or C #. In addition, the development environments of Unity or Processing package are published (http://nyatla.jp/nyartoolkit/wp/).
- **SketchUp:** Is a 3D modeling computer program for drawing architectural and engineering apps, interior designs or film, video and game designs. SketchUp™ 3D models can be visualized out of the digital workspace directly on users' desktop by connecting a simple webcam and by printing a suitable code. Users can create their own SketchUp™ Augmented Reality models and distribute them as independent, stand-alone files (http://www.armedia.it/sketchup_arplugin, http://www.sketchup.com/).

FUTURE RESEARCH DIRECTIONS

Mobile AR technology is seen useful for creating and developing learning activities for open and distance learners. This chapter provides a portrait of existing programs, standards or operating systems that can be used in mobile augmented reality. It also introduces a set of software development kits that are freely available on the net; with these tools, without needing to become a web developer expert, a trainer or a faculty can create a mobile augmented reality app or a media rich learning activity that meets the needs of open and distance learners. On future studies, the researchers can concern with the use of these programs, standards or mobile AR technologies in distance learning. They can be focused on the assessment and evaluation of the learning outcomes of the distance learners who are using mobile AR technology or mobile AR apps created by different programs, standards, and operating systems.

CONCLUSION

Mobile augmented reality apps are estimated as one of the most promising technology and has become an important manner to interact with the real world (Huang et al., 2013). Today, some applications of AR have been developed and are being used to enhance the education and training efficiency of students in higher education in different subjects such as astronomy, chemistry, biology, mathematics, geometry, and physics education. AR technology can also be applied to books by superimposing three-dimensional rendered models onto books, so these augmented books can be more interactive and realistic for learners. AR not only has the power to engage learners in a variety of interactive ways but also ease of providing education and training experiences to them. Therefore, education and training-oriented AR can improve the extent and quality of information in both school and business settings by making education and training environments more educational, productive, and contextual (Lee, 2012).

There are currently several tools for improving augmented reality apps available on mobile devices, but there is a need to apply them to open and distance learning. Improvements in mobile technology make it feasible to engage AR technology in any educational material or to create any educational content with AR. According to Di Serio, Ibanez and Kloos (2013), the use of augmented reality technology on learning environments had a positive effect on the motivation of learners. However, as stated by Kaufmann (2003), the potential of each AR feature needs careful reflection in order to be actually translated into educational efficacy. The matter is not questioning whether or not AR is useful to enhance learning, but understanding how to effectively exploit its potential.

The integration of mobile augmented tools in learning activities can be created an important effect on open and distance learners. In this chapter, a range of mobile AR applications are also presented that are feasible for open and distance learning environments. Besides, this chapter is aimed to give detailed information about the structure of mobile augmented reality apps in terms of a number of standards as well as some parts of web infrastructure standards, programming platforms and programming languages, software development tools. It provides information on that can be used by developers of mobile AR applications for learning. It is believed that the study will guide to those researchers and application developers who wants to create mobile AR application.

REFERENCES

Ahn, S., Ko, H., & Yoo, B. (2014). Webizing mobile augmented reality content. *New Review of Hypermedia and Multimedia*, *20*(1), 79–100. doi:10.1080/13614568.2013.857727

Bandura, A. (1977). *Social learning theory. Englewood Cliffs*, NJ: Prentice-Hall.

Brown, J. S., Collins, A., & Duguid, P. (1989). Situated cognition and the culture of learning. *Educational Researcher*, *18*(1), 32–42. doi:10.3102/0013189X018001032

Bruner, J. S. (1966). *Toward a theory of instruction*. Cambridge, MA: Belknap.

Catchoom. (2016). *Augmented Reality SDK*. Retrieved May 27, 2016 from http://catchoom.com/product/craftar/augmented-reality-sdk/

Charland, A., & Leroux, B. (2011). Mobile application development: Web vs. native. *Communications of the ACM*, *54*(5), 49–53. doi:10.1145/1941487.1941504

Crompton, H. (2013). Mobile learning: New approach, new theory. In Z. L. Berge & L. Y. Muilenburg (Eds.), *Handbook of Mobile Learning* (pp. 47–57). Florence, KY: Routledge.

Dalgarno, B., & Lee, M. J. (2010). What are the learning affordances of 3-D virtual environments? *British Journal of Educational Technology*, *41*(1), 10–32. doi:10.1111/j.1467-8535.2009.01038.x

Di Serio, Á., Ibáñez, M. B., & Kloos, C. D. (2013). The impact of an augmented reality system on students motivation for a visual art course. *Computers & Education*, *68*, 586–596. doi:10.1016/j.compedu.2012.03.002

Dunleavy, M., & Dede, C. (2014). Augmented reality teaching and learning. In *Handbook of research on educational communications and technology* (pp. 735–745). Springer New York. doi:10.1007/978-1-4614-3185-5_59

Engelke, T., Becker, M., Wuest, H., Keil, J., & Kuijper, A. (2013). MobileAR browser–a generic architecture for rapid AR-multi-level development. *Expert Systems with Applications*, *40*(7), 2704–2714. doi:10.1016/j.eswa.2012.11.003

Guru. (2015). *Top 10 Mobile Phones Operating Systems*. Retrieved May 27, 2016, from http://www.shoutmeloud.com/top-mobile-os-overview.html

Hartmann, G., Stead, G., & DeGani, A. (2011). *Cross-platform mobile development*. Cambridge: Mobile Learning Environment.

Hollerer, T. H. (2004). *User interfaces for mobile augmented reality systems* (Doctoral dissertation). Columbia University.

Huang, Z., Hui, P., Peylo, C., & Chatzopoulos, D. (2013). *Mobile augmented reality survey: A bottom-up approach*. arXiv preprint arXiv:1309.4413

Jeon, J., Kim, S., & Lee, S. (2010). Considerations of Generic Framework for AR on the Web. *Proceedings of W3C Workshop: Augmented Reality on the Web*.

Johnson, L., Levine, A., Smith, R., & Stone, S. (2010). The 2010 Horizon Report. *New Media Consortium*. Retrieved from http://www.nmc.org/pdf/2010-Horizon-Report.pdf

Kamboj, V., & Gupta, H. (2012). Mobile Operating Systems. *IJEIR*, *1*(2), 169–174.

Kaufmann, H. (2003). *Collaborative augmented reality in education. Institute of Software Technology and Interactive Systems*. Vienna University of Technology.

Kochan, S. G. (2011). Programming in Objective-C (3rd. ed.). Addison-Wesley Professional.

Lechner, M., & Tripp, M. (2010). ARML-an Augmented Reality Standard. *MobileARSummit*. Retrieved from http://www.perey.com/MobileARSummit/Mobilizy-ARML.pdf

Lee, K. (2012). Augmented reality in education and training. *TechTrends*, *56*(2), 13–21. doi:10.1007/s11528-012-0559-3

Lee, M., Green, R., & Billinghurst, M. (2008). 3D natural hand interaction for AR applications. *Image and Vision Computing New Zealand,23rd IEEE International Conference*, 1-6.

Lee, T., & Höllerer, T. (2007). Handy AR: Markerless inspection of augmented reality objects using fingertip tracking. *Wearable Computers,11th IEEE International Symposium*, 83-90.

Makumburage, T. A. (2015). *State of the Art in Mobile Augmented Reality Development Tools/APIs*. Retrieved May 30, 2016 from http://www.slideshare.net/maku13s1/stateoftheart-in-mobile-augmented-reality-development-toolsapis

Martin, S., Diaz, G., Sancristobal, E., Gil, R., Castro, M., & Peire, J. (2011). New technology trends in education: Seven years of forecasts and convergence. *Computers & Education*, *57*(3), 1893–1906. doi:10.1016/j.compedu.2011.04.003

Munnerley, D., Bacon, M., Wilson, A., Steele, J., Hedberg, J., & Fitzgerald, R. (2012). Confronting an augmented reality. *Research in Learning Technology, ALT-C 2012 Conference Proceedings*, 39-48.

Özdamar Keskin, N., & Kılınç, H. (2015). A comparison of development platforms for mobile learning applications and case studies. *Journal of Open Education Practices and Research*, *1*(3), 68–90.

Özdamar Keskin, N., & Kuzu, A. (2015). Development and testing of a m-learning system for the professional development of academics through design-based action research. *The International Review of Research in Open and Distributed Learning*, *16*(1), 193–220. doi:10.19173/irrodl.v16i1.1613

Padzensky, R. (2014). *06: Market Outlook*. Retrieved May 27, 2016 from http://augmera.com/?p=214

Palmieri, M., Singh, I., & Cicchetti, A. (2012). Comparison of cross-platform mobile development tools. *Intelligence in Next Generation Networks (ICIN),16th IEEE International Conference*, 179-186.

Perey, C., Engelke, T., & Reed, C. (2011). Current status of standards for augmented reality. In Recent Trends of Mobile Collaborative Augmented Reality Systems, (pp. 21-38). Springer. doi:10.1007/978-1-4419-9845-3_2

Ramakrishnamurthy, R. (2016). *How to Develop Augmented Reality Based Mobile Apps?* Retrieved May 27, 2016 from http://blog.contus.com/augmented-reality-apps-development/

Serrano, N., Hernantes, J., & Gallardo, G. (2013). Mobile web apps. *Software, IEEE*, *30*(5), 22–27. doi:10.1109/MS.2013.111

Smith, M. (2013). Developing Mobile Apps. *Journal of Management & Engineering Integration*, *6*(2), 79.

Smutni, P. (2012). Mobile development tools and cross-platform solutions.*Carpathian Control Conference (ICCC), 13th IEEE International Conference*, 653-656.

Statista. (2016). *Global market share held by the leading smartphone operating systems in sales to end users from 1st quarter 2009 to 1st quarter 2016*. Retrieved May 27, 2016 from http://www.statista.com/statistics/266136/global-market-share-held-by-smartphone-operating-systems/

Van Rossum, G. (2007). Python Programming Language.*USENIX Annual Technical Conference*, 41.

Vandebona, U. V. (2015). *State-of-the-Art in Mobile Augmented Reality Development Tools/APIs*. Retrieved May 30, 2016, from http://www.slideshare.net/UpekhaVandebona/literature-survey-report-13440722

Vygotsky, L. S. (1978). *Mind and society: The development of higher mental processes*. Cambridge, MA: Harvard University Press.

Wilde, M., Hategan, M., Wozniak, J. M., Clifford, B., Katz, D. S., & Foster, I. (2011). Swift: A language for distributed parallel scripting. *Parallel Computing*, *37*(9), 633–652. doi:10.1016/j.parco.2011.05.005

KEY TERMS AND DEFINITIONS

Application Programming Interface (API): A set of routine definitions, protocols, and tools for building apps.

AR Development Tools: AR development tool provide educators and instructional designers to create custom AR applications without programming skills, and purpose recognition, tracking and content rendering.

Augmented Reality (AR): AR integrates virtual information into a specific real world surrounding so that the physical environment becomes a context-aware user interface.

Cross-Platform Mobile App: There are different tools, building systems, APIs, and devices with different capabilities for each mobile platform. A cross-platform mobile app has a single codebase in one language and runs on several mobile platforms with little or no modification.

HTML5: A standard for developing browser-based mobile apps which are supported by different browsers such as Internet Explorer, Google Chrome, and Mozilla Firefox etc.

Mobile App: A computer program designed to run on mobile or wearable devices that are usually available through app stores of mobile platforms such as Google Play Store, Apple App Store, Windows Phone Store, and Blackberry App World etc.

Mobile Augmented Reality (MAR): MAR works virtually anywhere so it provides AR service by adding a layer of information to any environment whenever desired.

Mobile Hyprid Apps: Applications that are mobile web app which packed into a native app.

Mobile Learning: Learning that takes place in various contexts through contents and social interactions using personal electronic devices.

Mobile Native Apps: Applications that use language and framework of its suppliers, for example, the Apple iOS platform uses Objective C with Xcode.

Mobile Platform: A mobile platform is an operating system for smartphones, tablets, PDAs, or other mobile devices, such as Google Android, Apple iOS, Blackberry RIM, Windows Mobile etc.

Mobile Web Apps: Applications that are similar to any other web site which designed for mobile devices.

Open and Distance Learning (ODL): ODL is one of the most rapidly growing fields of education that represents approaches which focus on opening access to education and training provision without the constraints of time and place, and offers flexible learning opportunities to individuals and groups of learners.

PhoneGap: Hybrid web model which supports interaction between web app and the device and Its apps are supported by platforms such as iOS, Android, Blackberry, Windows Mobile 6.5, Symbian, and Palm.

Software Development Kits (SDKs): A set of programs that enables to develop apps for a specific platform. For example, the development of an Android app requires an SDK with Java or a Microsoft Windows app requires a.NET Framework SDK with.NET.

Virtual Reality (VR): VR attempts to create an artificial world that a person can experience or/and explore interactively over computer technologies.

Chapter 7
In Search for a "Good Fit" Between Augmented Reality and Mobile Learning Ecosystem

Miraç Banu Gundogan
Middle East Technical University, Turkey

ABSTRACT

Ecosystems are particular areas in nature where all living and nonliving components interact with each other and their environment. The term has also been used as a metaphor in scientific and social contexts. Learning ecosystem is one of these which studies the components and interactions of learning processes. Augmented reality is among the components of a (mobile) learning ecosystem. Potentials of integrating augmented reality in mobile learning are not denied, yet there are concerns that these might turn into short living fashion items if their long term consequences are not considered. Defining a mobile learning ecosystem, clarifying the position of augmented reality component within, describing its relations with other components and searching for a balance in these interactions would be an answer to these concerns. This chapter gives an answer by presenting mobile learning ecosystem and augmented reality definitions derived from a Delphi study carried out in 2016 in Turkey. The results and discussions present a "good fit" framework for a viable mobile learning ecosystem.

INTRODUCTION

Technology usage has always supported learning, moreover, the advances in Information and Communications Technology have made digital media part of our everyday lives, including education. In a survey conducted in 2008 for The Economist Intelligence Unit report, 63% of respondents have stated that technological innovation will have a major impact on teaching methodologies over the coming years and technology is marked as a core differentiator in attracting students and corporate partners regarding educational services. Scholars approach this situation cautiously; Ficheman and de Deus Lopes (2008) draw attention to the gap between learners and teachers in terms of technology usage. They state that learners mostly are digital natives familiar with the hyperlinked and interactive world

DOI: 10.4018/978-1-5225-2110-5.ch007

of digital technology and expect these in their learning processes, whereas, currently most teachers still being digital immigrants are struggling to adapt both to the technological advances and the demanding learners. Laziness is declared as another negative effect of wide technology usage in education by Ramey (2012) who states that technology users being so dependent on advanced tools become idle and less innovative and learners become more dependent on computing even to solve simple tasks. Bower, Howe, McCredie, Robinson, & Grover (2013) state that the pace of educational research is not keeping up with the advances in technology and as a result, not pedagogy but technology, accompanied by risks, is shaping the way people learn.

These risks need to be taken into account for mobile learning as well. In mobile learning, not only the tools and content, but the learner also is mobile. Any mobile learner using a mobile tool and has access to a network to communicate with others can reach a learning content. Although by definition, this technology is designed to facilitate, support, enhance and extend learning, literature presents important concerns. The Economist Intelligence Unit report (2008) presents assumptions which state that easy and ready access to mobile technologies may be disruptive in ways not intended such as a rise in student plagiarism, cheating and distractibility. Kearney, Schuck, Burden and Aubusson (2012) highlight the need for a pedagogical perspective in which authenticity, collaboration and personalization stand as critical concepts. They further state that the mobile environment needs to be designed cautiously since technology strongly influences learners' experience. Jackson (2013) highlights the difficulty in making technology decisions since hardware, software and operating systems are continuously changing and states that using massive expensive new technology may not end up improving the quality and effectiveness of learning.

Augmented reality, one of the supportive tools used in mobile learning also is facing concerns. Its contribution as a learning tool is not denied, yet, there are criticisms on its design and use. Although it provides a rich multimedia presentation which addresses different senses of learners and improves learning experiences by complementing the actual content, treating it as a tool or even a consumable within digital learning environments may have undesirable consequences. Höllerer (1997) highlights the risk of confusion which might take place when the user's view of the real world interferes with the virtual objects overlaid on the real world and Haag (2013) states that there are many examples of ineffective augmented reality applications which contain primitive forms of engagement, such as static graphics, and these cause distraction rather than engagement.

These opinions arise a challenging question; are we going to leave ways of learning into the hands of technology or should we search a conscious path encompassing the consideration of components and multidimensional interactions of any specific learning environment? One way to face this challenge is to refer to the ecosystem concept. Researchers with similar concerns have been studying these issues within ecosystem perspectives and presenting models where technology is analyzed as one of the many components interacting in a specific environment. Studies on digital learning ecosystems and usage of augmented reality in learning present valuable frameworks. The study presented in this chapter combines these frameworks with the results of a survey aiming to define a mobile learning ecosystem and discusses the position of augmented reality within this system.

The following sections present studies and definitions on digital and mobile learning ecosystems followed by guidelines on the design, implementation and evaluation of augmented reality applications in attaining efficient and sustainable learning experiences within these ecosystems.

DIGITAL AND MOBILE LEARNING ECOSYSTEMS

Sir Arthur Tansley (1935) was the first to voice the term ecosystem. He described it as a community of living organisms together with the nonliving components of their environment interacting as a system. Odum (1971) stated that this interaction occurs through nutrient cycles and energy flows, naming the living components as biotic and the nonliving as abiotic. According to Jax (1998), the term presents a unified perspective in studying a living community or grouping with its accompanying physical environment.

Living and nonliving components, a physical environment and interactions are core components of an ecosystem. Even though the term has its roots in Ecology, it has grown to describe human behaviors and human-generated processes in scientific and social contexts including education and learning. Among these, digital and mobile learning ecosystems are significant topics. Chang and Guetl (2007) state that the idea of ecosystem can be applied to any learning system and only by this way learning and knowledge enhancement and social and cultural development can be better understood. Sharples and Pea (2014) explain mobile learning as an environment in which multidimensional interactions between objectives, content, participants (backgrounds, motivations, and expectations) and tools are taking place. Udell and Woodill (2015) highlight the importance of the diversity of mobile tools and the multiplicity of content and operating platforms in defining mobile learning ecosystems. Winters (2013) focuses on the sustainability of mobile learning in terms of content development by declaring that mobile learning would not be an alternative to the formal system unless teachers currently lacking sufficient expertise with digital technologies become powerful and dominant and create adequate content mobile learning necessitates.

Mobile learning is enabled through digital technologies and tools, thus a mobile learning ecosystem needs to be studied in accordance with digital learning ecosystem models. Chang and Guetl (2007) list the outstanding characteristics of a digital learning ecosystem as (1) an interactive community equipped with multiple channels for interaction, (2) a strong information infrastructure, (3) smart information use and (4) a cross-disciplinary engagement enhancing productivity, prosperity and international competitiveness. In a study which holds learners' access to technology vital, Ficheman and de Deus Lopes (2008) describe nonliving units as hardware, software, network and database technologies and state that pedagogies are supportive in building effective interactions with the living factors of the ecosystem. The complex interactions between the participants and content in digital learning environments are grouped as the Teaching Niche and the Learning Niche by Reyna (2011) where the nonliving components are listed as tools that enable content access, the Internet connection, the learning interface or portal, and the content and teaching and learning are labeled as the sources of energy powering the ecosystem. In a recent study by Laanpere, Pata, Normak, & Põldoja (2014) participants and content are set as living modules and the learning environment as the nonliving part. Clark (2014) underlines the importance of an awareness on learning when and how to use the right tool for a task, further stating that an attractive digital content and a stimulating culture of inquiry and innovation where multiple forms of assessment would guide the design for differentiation and accessibility. The author states that an integrative approach regarding these components would foster a sustainable ecosystem which will take on a life of its own. Gundogan (2014) underlines usability, pleasure, satisfaction and sustainability issues and draws attention to necessity of studying the sustainability of digital ecosystems in terms of waste generation.

Within the framework of the above given definitions, the components of a mobile learning ecosystem can be defined as follows:

Teaching, learning, culture of inquiry and multiple forms of assessment are the external inputs, participants, institute ions and content are the living parts, and technology, the Internet, hardware, software and network form the nonliving parts.

Pedagogy, theory, usability and pleasure are the nutrient cycles within the ecosystem and the energy flows occur through multiple channels of interaction and necessitate a design approach which encompasses differentiation and accessibility. This ecosystem is presented in Figure 1.

Pedagogy and learning theories provide solutions for the risks of inconsiderate technology integration and alongside, sustainability stands out as an assistive framework. In environmental sciences and studies, sustainability highlights the risks created by the consumption of natural resources and the harm made by technology and sets out provisions for keeping resources undamaged and renewable so that they continue to serve future generations. Natural ecosystems continue living if and only if the environment is able to flourish and support itself without outside influence or assistance; everything is provided within and no waste is generated. Sustainability defines how systems would remain diverse and productive over time and is a precondition for the human wellbeing. Since humans are part of the living communities of digital learning ecosystems, sustainability measures need to be reflected to mobile learning ecosystem definitions.

Sridharan, Deng and Corbitt (2010) list critical success factors for sustainable digital learning ecosystems as (1) adoption of appropriate pedagogical strategies, (2) technologies supporting the chosen pedagogical strategies, (3) understanding the theory and power of those technologies, (4) proactive involvement of teachers in encouraging and fostering appropriate use of the chosen strategy and management and (5) reuse of resources. Laanpere et al. (2014) support the sustainability emphasis by putting desirable pedagogical beliefs, strategies and learning activity patterns in the core of a digital learning ecosystem.

Figure 1. Components of the mobile learning ecosystem

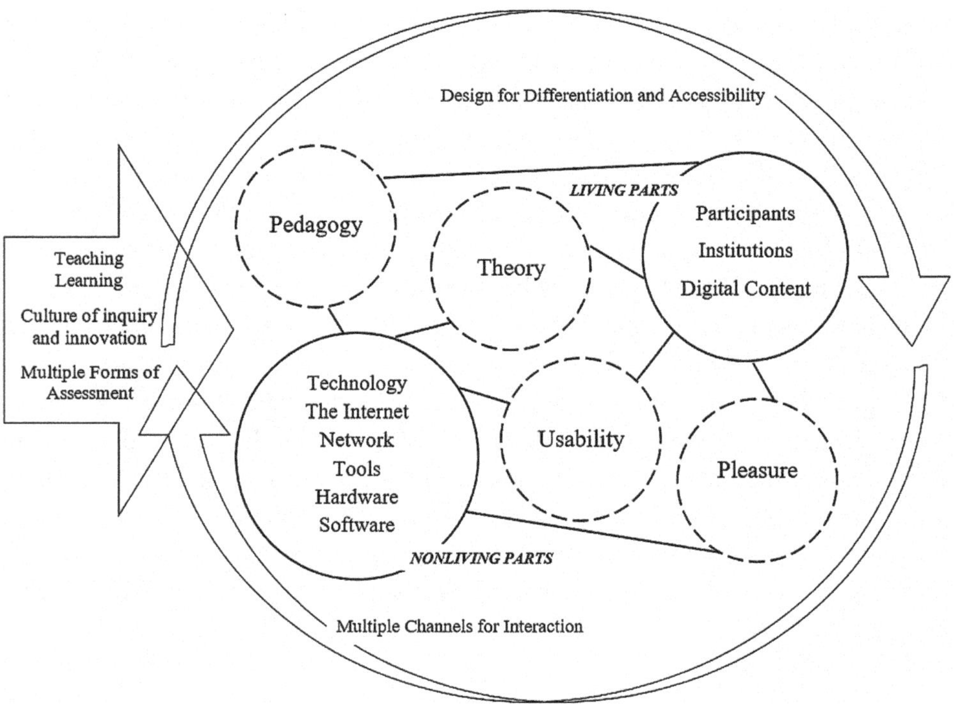

They present a model in which first a detailed needs analysis for teaching and learning purposes in a specific context is tailored, then the most appropriate method is defined, and finally the technological requirements to make it work is described.

Research shows that digital learning ecosystem models not only provide valuable perspectives for designing enhanced learning environments, but also underline provisions for keeping them sustainable to serve future generations. Especially, defining the interactivity component on a pedagogical rationale supports cautious technology decisions and in the case of augmented reality applications, would be assistive in overcoming the potential risks highlighted by scholars.

THE DESIGN, IMPLEMENTATION AND EVALUATION OF AUGMENTED REALITY IN MOBILE LEARNING

Current augmented reality applications enhance the actual content by providing rich multimedia presentations. Azuma (1997) describes augmented reality as a step further than virtual reality in which virtual objects overlay the real world, overcoming the unfamiliarity of the synthetic environment created by virtual reality technologies. With the use of augmented reality, different senses of learners are addressed, information is displayed in real time and real world contexts and an environment which otherwise is impossible to be sensed is created. As stated in Zheng (2015), the usage of 3D environments, augmented reality creates a real learning environment and provides high-quality learning experiences where the interaction of teachers and students are enriched and consequently, learning experiences are improved.

Augmented reality applications are also valued for increasing the interactions among the participants of mobile learning. FitzGerald, Adams, Ferguson, Gaved, Mor, & Thomas (2012) state that the mobility of the user, the physical place where learning can occur and formal learning connections to informal learning are fostered when augmented reality is combined with mobile learning. They further state that mobile augmented reality can be studied in terms of tools, modes of interaction, type of media, learner experience and learning outcomes.

The contribution of augmented reality applications as a learning tool is not denied, yet, there are criticisms on their design and use. Azuma (1997) lists accuracy problems, distortion and tracking errors caused by system delays as obstacles in building effective experiences. Haag (2013) declares concerns on the modes of interaction in the sense of placing the learner in an active or passive situation. The author states that, if the concept behind this new technology is not fully understood, then, when using it as a learning tool, a need for technical support might arouse and this may cause detraction from the learning experience. In an article listing the best augmented reality applications, Widder (2016) states that "choosing which apps are worth the user's smartphone or tablet's precious memory is tougher than using the apps themselves". These statements point out the concern on sustainability; it is important to question if augmented reality applications would survive within a mobile learning ecosystem, or even, in the long term, would the ecosystem's life be determined by these applications? This question also introduces a new challenge regarding the design and use of augmented reality applications in mobile learning ecosystems; the position of augmented reality component with respect to the mobile learner's context of interaction and expected learning experiences needs to be clarified. Bower et al. (2013) suggest that implementing strategies which encourage students to design augmented reality applications through the mobile learning process would be a method for bridging the gap between educational research and technological advances. This supports the findings of FitzGerald et al. (2012) where augmented reality has proven to be successful where collaboration and student inquiry were integrated as pedagogical inputs. Clark (2014) believes that facilitating a sustainable digital learning environment that endures over time could be realized by being conscious on when and how to use the right technological tool for a task. Since each mobile learner uses his or her own mobile tool, the context of interaction together with expected learning experiences are important. Within this framework, defining a sustainable mobile learning ecosystem with augmented reality component would provide answers for the challenges and concerns mentioned above.

Being aware of the importance of sustainability in a mobile learning ecosystem and the state of augmented reality component within this system, a Delphi study was carried out in 2016 with 27 senior computer education and instructional technology department students of a state university in Turkey. The study aimed to compile a mobile learning ecosystem definition based on the characteristics of natural ecosystems; the external inputs, producers, consumers, decomposers, organic, and inorganic compounds. The study integrated a brainstorming session aiming to clarify the position of augmented reality within this ecosystem. Details and findings are presented in the following section.

DELPHI STUDY DEFINING THE MOBILE LEARNING ECOSYSTEM AND THE AUGMENTED REALITY COMPONENT OF THIS ECOSYSTEM

The Delphi studies aim gathering data that is similar to focus groups. Individuals who have knowledge necessary to analyze a specific problem are carefully selected to form focus groups. Haughey describes it as a method of generating ideas and facilitating consensus among individuals who have special knowledge to share and the anonymity encourages openness and allows them to express their opinions freely.

Considering the concerns derived from literature which draw attention to the expectations of current learners, namely digital citizens, 27 senior computer education and instructional technology department students of a state university were invited to take part in this study. The study was planned and conducted in three rounds, all presented as digital surveys accessed on the Internet. Round 1 aimed to collect definitions on the external inputs, producers, consumers, decomposers, organic, and inorganic compounds of a mobile learning ecosystem based on the characteristics of natural ecosystems. Round 2 aimed to reach a consensus in terms of definitions by tabulating all answers given to each question of the first round 1 with a five-level Likert scale. Moreover, an open ended question was added and the participants were asked to describe the concept of 'waste' in a mobile learning ecosystem. A final round was designed as a brainstorming session in which participants were asked to clarify the position of augmented reality within this ecosystem. Questions of round 1 are given in Table 1.

The results of the first round are presented in Table 2 and 3. Table 2 presents the components of a mobile learning ecosystem based on the characteristics of natural ecosystems. Table 3 shows the main inputs and outputs described by participants.

In the second round of the study, all answers given to each question of Round 1 were listed together with a five-level Likert scale. The aim was to establish a collaborative framework regarding the inputs, outputs, producers, consumers, decomposers, organic and inorganic compounds of a mobile learning ecosystem. Table 4 shows consensus items for Mobile Learning Ecosystem components and Table 5 presents the main inputs and outputs for this ecosystem based on 70% evaluations as *agree* and *strongly agree*.

Table 1. Delphi study round 1 questions

1. Natural ecosystems are open systems. In an open system, where matter or energy can flow into and/or out of the system, living organisms continuously need external inputs in order to continue functioning. The most important external input needed is the solar energy. Without the continued input of solar energy, biological systems would quickly shut down. According to you, what would be the external inputs of a mobile learning ecosystem?
2. Primary producers are the major functional aspect of an ecosystem. All photosynthetic plants are primary producers. Energy from the sun is captured by the process of photosynthesis and is the first step in the production of energy for living things. According to you, what would be the primary producers of a mobile learning ecosystem?
3. In ecosystems, living organisms consume the energy produced by the primary producers. Organisms which obtain their energy by consuming plants or plant products are named as primary consumers (herbivores). According to you, what would be the primary consumers of a mobile learning ecosystem?
4. In ecosystems, organisms which obtain their energy by consuming the primary consumers are named as secondary consumers (carnivores). According to you, what would be the secondary consumers of a mobile learning ecosystem?
5. Dead tissue and waste products are produced at all levels in an open ecosystem. Decomposers collectively account for the use of all such waste. Named as bacteria and fungi, they decompose the waste both for their own consumption and for the re-use of other living things. According to you, what would be the decomposers of a mobile learning ecosystem?
6. An open ecosystem consists of both the living and the non-living environment. Inorganic compounds, largely consumed by living organisms, are an important part of the non-living environment. Inorganic compounds like water, carbon dioxide and salt are inanimate, and cannot be obtained from the living things. According to you, what would be the inorganic compounds of a mobile learning ecosystem?
7. The productivity of an ecosystem mostly relies on organic compounds. These compounds derive from dead animal or plant tissues and have a carbon basis. Carbohydrates, proteins and vitamins are organic compounds. According to you, what would be the organic compounds of a mobile learning ecosystem?
8. An ecosystem's capacity to survive, continue functioning and be sustainable, in which all organisms and compounds continue to live and flourish for many generations, depends on equilibrium between all input and outputs. According to you, what would be the inputs of a mobile learning ecosystem?
9. According to you, what would be the outputs of a mobile learning ecosystem?

Table 2. Components of a mobile learning ecosystem

Questions 1-7	Answers	# of Responses
The External Input	(Implementation Of New) Technologies	10
	Information / Content / Resources	9
	Learners / Learner Needs	7
	Tools	6
	The Internet	4
	Interaction / Collaboration	1
	Need To Learn	1
	Trusted Resources	1
	Aim	1
	Feedback	1

continued on next page

Table 2. Continued

Questions 1-7	Answers	# of Responses
Primary Producers	Developers (Content And Software)	6
	Instructional Technology	6
	Mobile Learning Tools	6
	Educators/ Instructors / Facilitators	5
	Learners / Learner Skills	4
	The Internet	4
	Software	2
	Support Mechanism	2
	Educational Content	2
	Mobile Learning Tool Producers	2
	Motivation	1
	Organization / Institutional Strategies	1
	Research(Er)	1
Primary Consumers	Learners	12
	Educators/ Instructors / Facilitators	5
	Companies / Businesses	4
	Developers (Content And Software)	4
	Modern Technologies And Learning Tools	1
Secondary Consumers	Learners	8
	Companies / The Business	8
	The Society	4
	Educators/ Instructors / Facilitators	3
	Developers (Content And Software)	2
	Communication Tools / The Internet	1
	Websites	1
Decomposers	Educators/ Instructors / Facilitators	4
	Renewable Technology	2
	Developers (Content And Software)	2
	Educational Institutions	2
	Learners	2
	Added Value Of Employing Technology Experts	1
	Content	1
	Advertisements	1
	Evaluation / Assessment	1
	New Developing Needs	1
	Servers / Network Components	1
	Software	1
	System Analysis	1
	Mobile Learning Tools	1
	The Internet	1
	Up To Date Content	1

continued on next page

Table 2. Continued

Questions 1-7	Answers	# of Responses
Inorganic Compounds	Technology Infrastructure (Servers-Network)	10
	Computer Applications & Software	6
	Mobile Learning Tools	5
	Content	4
	The Internet	2
	Institutions	1
	Learners' Positive Attitude, Concentration, Discipline, Self-Confidence	1
	Educational Institutions	1
	Presentations	1
	Support Systems	1
Organic Compounds	Experience	3
	Knowledge	3
	Developers (Content And Software)	3
	Content	3
	Learning Environment	3
	Motivation	3
	Resources	3
	Learner Needs	2
	Computer Applications & Software	2
	Learner Responsibilities	1
	Learners	1
	Educators/ Instructors / Facilitators	1

In the second round of the study, participants were asked to describe the concept of 'waste' in a mobile learning ecosystem by answering an open-ended question. Answers are listed below in alphabetical order;

- Addiction and overuse.
- Applications, programs, games, publishes pictures, interpretations, or videos, sounds we do not use anymore.
- Being careless, inefficient, ignorant, unaware about learner's needs, goals and characteristics.
- Biases and prejudices.
- Ideas that might be considered wrong and dropped out of the system other than working on them and finding out what their owners thought when they came up with those ideas.
- Inappropriate content.
- Inefficient content.
- Insecurity.
- Learning management.
- Massive investment.
- Mismanaged technological tools.
- No updates.

Table 3. Main Inputs and Outputs of a Mobile Learning Ecosystem

Questions 8-9	Answers	# of Responses
Main Inputs	(Up To Date) Content	8
	Learning Environment	7
	Learners	6
	Resources	6
	Educators/ Instructors / Facilitators	5
	Interaction	5
	Learner Needs & Characteristics	5
	Technology, Tools, Materials	5
	Information	3
	Feedback	3
	Collaboration	3
	Cost	2
	Computer Applications & Software	2
	Knowledge	2
	Aim	2
	Educational Technologies	1
	Time	1
	Usability	1
	Accessibility	1
Main Outputs	Knowledge	10
	Innovative Ideas/Products	6
	Learning Environment Design	3
	Content Delivery	2
	Experience	2
	Learners	2
	Success	2
	Qualified Workforce	2
	Administration	1
	Communication	1
	Interaction	1
	Accessibility	1
	Reusable Learning Objects	1

Table 4. Consensus items for a mobile learning ecosystem

Components of a Mobile Learning Ecosystem		%
External Inputs	Information / Content / Resources	100
	Interaction / Collaboration	89
	The Internet	82
	Tools	79
	(Implementation Of New) Technologies	78
	Learners / Learner Needs	78
	Need To Learn	70
Primary Producers	Developers (Content And Software)	100
	Instructional Technology	100
	Educators/ Instructors / Facilitators	95
	Research(er)	95
	Organization / Institutional Strategies	81
	Mobile Learning Tools	77
Primary Consumers	Learners	95
	Educators/ Instructors / Facilitators	91
	Companies / Businesses	77
Secondary Consumers	The Society	100
	Companies / The Business	87
	Learners	77
	Educators/ Instructors / Facilitators	74
	Communication Tools / The Internet	73
Decomposers	Evaluation / Assessment	91
	Educational Institutions	85
	Added Value Of Employing Technology Experts	76
Inorganic Compounds	Technology Infrastructure (Servers-Network)	90
	The Internet	86
	Content	81
	Support Systems	78
	Learners' Positive Attitude, Concentration, Discipline, Self-Confidence	76
	Computer Applications & Software	72
Organic Compounds	Knowledge	91
	Experience	80
	Motivation	77
	Educators/ Instructors / Facilitators	73
	Learning Environment	73
	Learner Needs	73

Table 5. Consensus items for a mobile learning ecosystem (main inputs and outputs)

Main Inputs	Technology, Tools, Materials	98
	Resources	95
	Aim	91
	Feedback	91
	(Up To Date) Content	86
	Educational Technologies	82
	Learner Needs & Characteristics	82
	Collaboration	77
	Learning Environment	76
	Interaction	73
	Learners	73
Main Outputs	Knowledge	100
	Experience	91
	Accessibility	91
	Qualified Workforce	82
	Innovative Ideas/Products	82
	Success	82
	Reusable Learning Objects	76

- Repeated contents.
- Social problems like less face-to-face communication.
- Time.
- Uncategorized knowledge.
- Unnecessary learning materials.
- Unnecessary information in lessons.
- Unnecessary use of storage areas.
- Unreachable bad interfaces.
- Using internet with no goal.

The final round of the study was designed as a brainstorming session in which participants were asked to clarify the position of augmented reality within a mobile learning ecosystem. The contributions can be grouped into 3 categories; familiarity, supportiveness and its future potential. These are presented in Table 6.

Table 6. The position of augmented reality within a mobile learning ecosystem

Categories	Definitions
Familiarity with Augmented Reality (AR)	• We are already using AR in our daily lives • Modifying or adding elements to a photo in our mobile phones is the simplest example of using AR • Most of the children have mobile tools like phones and tablets
Supportiveness of Augmented Reality (AR)	• Helps to visualize and understand abstract concepts • Avoids misunderstandings • Enhances motivation • Facilitates instruction, thus reduces the workload of the instructor • Provides a better learning environment • Provides an interactive learning environment • Visualizing or understanding concepts no more depend on the students ability to imagine • Speeds up the learning process • Makes learning memorable • Enables designing interactive projects that educate and entertain learners • Usable at all ages
Future Potential of Augmented Reality (AR)	• Certainly will develop • Will be an indispensable element in mobile learning • Will enable proficient homeschooling • Any place can become a learning environment

FINDINGS AND DISCUSSIONS

The Mobile Learning Ecosystem

In the first sections of this chapter, mobile learning ecosystem is discussed with reference to digital learning ecosystems. The reviewed literature has shown that sustainability of a mobile learning ecosystem and the situation of augmented reality component within this system is important. Within this framework, the Delphi study explained above was carried out. Round 1 aimed to collect definitions on the external inputs, producers, consumers, decomposers, organic, and inorganic compounds of a mobile learning ecosystem based on the characteristics of natural ecosystems. The results are presented in Figure 2.

Previously in this chapter, the external inputs of a mobile learning ecosystem were defined as teaching, learning, culture of inquiry and multiple forms of assessment within the framework of the literature review based on digital and mobile learning ecosystems, The mobile learning ecosystem defined in the Delphi study reveals more elements as external inputs, namely; technology, tools, materials, resources, aim, feedback, content, learners, collaboration, the learning environment and interaction. Studies on digital and mobile learning ecosystems present more holistic definitions whereas the participants of the Delphi study, the digital citizens, have listed components in detail. This is noteworthy since items like aim, feedback, collaboration and interaction tell what current learners expect from mobile learning environments, and moreover, these items need to be projected to sustainability. The living and nonliving parts being common in both cases shows that the results of the Delphi study are in consent with the literature.

Ecosystems continue living if and only if the environment is able to flourish and support itself without outside influence or assistance. This highlights the importance of a balance between the inputs and outputs of the system. Based on the findings of the study, the balance conditions are presented in Figure 3.

Figure 2. The mobile learning ecosystem defined in the Delphi Study

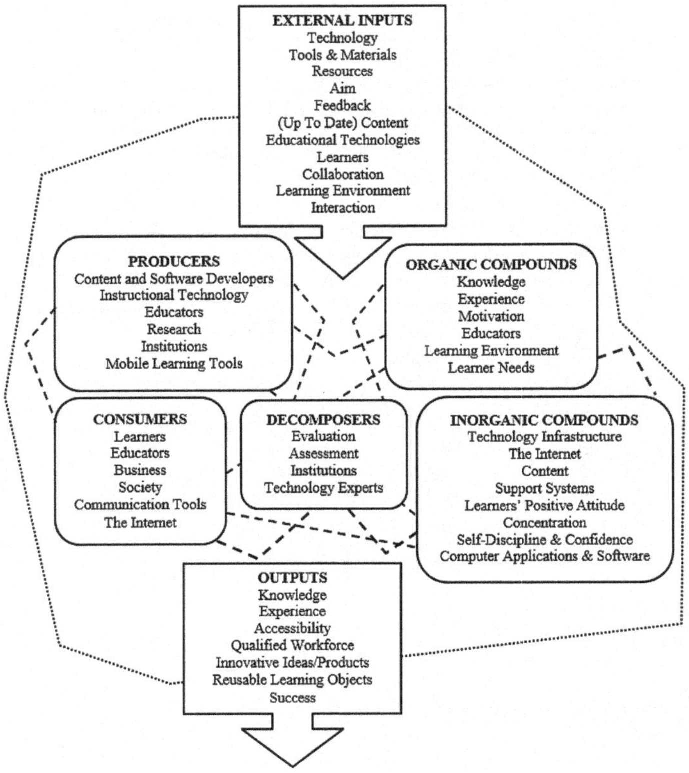

Figure 3. Balance conditions in a mobile learning ecosystem

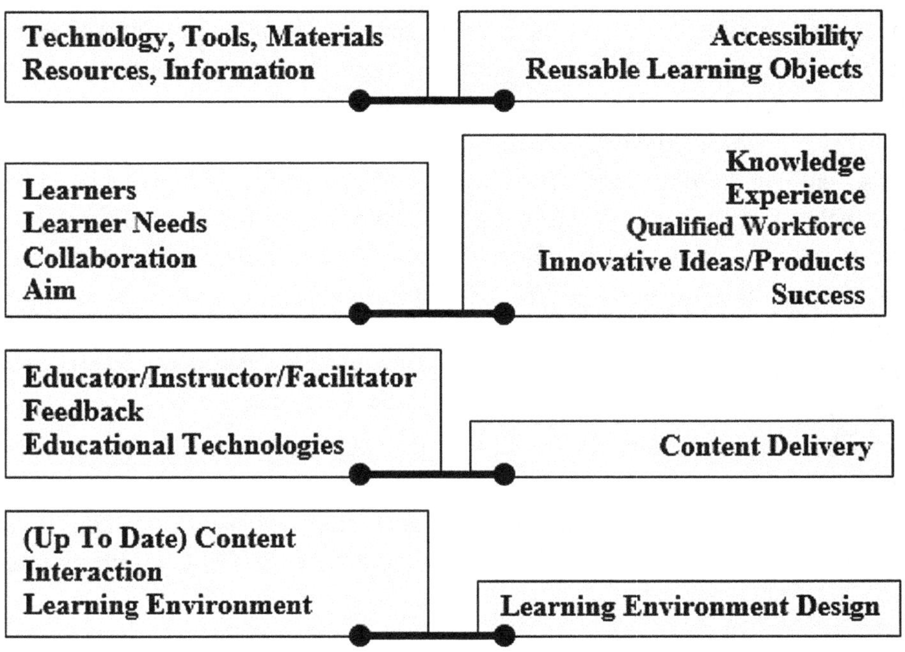

Augmented Reality Component of the Ecosystem

Participants of this study state that augmented reality is most supportive in visualizing and understanding abstract concepts and by providing interactivity, better learning environments are created. Since Learning Environment and Interaction are named as major inputs to a Mobile Learning Ecosystem, then, augmented reality is an input to this system.

Augmented reality is found supportive in avoiding misunderstandings, enhancing motivation, speeding up the learning process and making learning memorable. With reference to Figure 2, in terms of motivation, learner needs and experience, augmented reality serves as an organic compound, and in terms of computer applications and software, it is also an inorganic compound.

The results also reveal that by using augmented reality, visualizing or understanding concepts no more depend on the learners' own imagination abilities. Moreover, designing interactive projects are enabled for learners of all ages. These statements reinforce augmented reality's position as a producer, a component which provides energy for the ecosystem.

Even though augmented reality is among the inputs, producers, organic and inorganic compounds of a mobile learning ecosystem, analyzing the waste descriptions show that it also has the potential of being a waste in terms of:

- Addiction and overuse - in compliance with Ramey (2012),
- Massive investment- in compliance with FitzGerald et al. (2012) and Clark (2014),
- Inefficient content - in compliance with Haag (2013),
- Unnecessary learning materials and information- in compliance with Höllerer (1997),
- Misuse of storage areas - in compliance with Widder (2016), and
- Bad interfaces or other usability issues- in compliance with Azuma (1997) and Gundogan (2014).

Natural ecosystems rely on decomposers for eliminating waste and to sustain balance between the inputs and outputs of the system, waste must be prevented. In terms of augmented reality as an input, producer, organic and inorganic compound and a possible waste item, prevention and elimination of waste is discussed in the coming section.

A 'Good Fit' Between the Mobile Learning Ecosystem and the Augmented Reality Component

The balance conditions described in Figure 3 show that input items such as tools, technologies and information to be accessible and designed as reusable learning objects. Augmented reality as a major input in terms of tools and technologies should also be able to function in different instructional contexts and be easily accessible and designed as a reusable leaning object.

Augmented reality is considered as an item which is already a part of learners' daily lives; participants of this study have stated that learners are already familiar with modifying or adding elements even to a photo in their mobile phones. In terms of balancing input units such as learners, their needs and their aim, augmented reality can further be used to create collaborative projects which aim to support innovative ideas or products.

The importance of integrating the pedagogical frameworks, educator experiences and feedback to the content design delivery methods stands as another balance condition from which the design of augmented reality applications would benefit. With these reflections, augmented reality can become an indispensable element in mobile learning. Since any place can become a learning environment by using augmented reality, it is essential that the design of this environment should be an iterative process in which feedbacks from all participants are collected and analyzed to form new inputs.

Waste is produced at all levels in a natural ecosystem., but decomposers like bacteria and fungi collectively decompose the waste both for their own consumption and for the re-use of other living things. According to the findings of this study, Evaluation, Assessment, Educational Institutions and Technology Experts as decomposers are responsible of taking care for the possible waste condition of augmented reality. This leads to the recommendation that institutions which provide mobile learning opportunities must assess educational needs, evaluate the quality of their services and employ both pedagogical and technological experts. By this way, possible wastes such as massive investment, unnecessary learning materials and information, misuse of storage areas can be avoided or decomposed. Moreover, technical experts could be supportive in training the educators so that they become powerful and dominant in creating adequate mobile learning content. Figure 4 shows how augmented reality fits in this system both as a compound and a manageable waste item.

CONCLUSION

Technological innovations influence teaching and learning, digital media is already a part of everyday lives and learners, mostly digital natives, expect these in their learning processes as well. Mobile learning is one answer to these expectations and integrating augmented reality within this process comes both

Figure 4. Provisions of a good fit between augmented reality and mobile Learning ecosystem

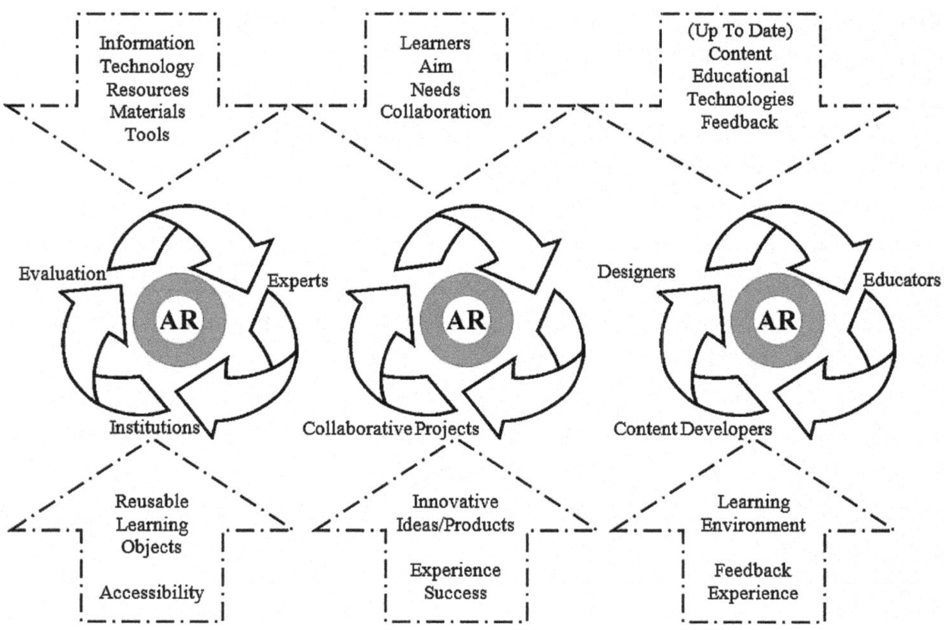

with promises and concerns. Either to leave ways of learning into the hands of technology or searching a good fit between components and multidimensional interactions of the learning environment is the challenge. The study presented in this chapter analyses this within the framework of an ecosystem approach by defining a mobile learning ecosystem, clarifying the position of augmented reality component within, describing its relations with other components and searching for a balance in these interactions. The results highlight the importance of pedagogy, evaluation, assessment, educator feedback and institutional strategies which support collaborative projects. Moreover, responsibilities of designers and developers in terms of creating and implementing augmented reality in reaching a "good fit" in a mobile learning ecosystem is underlined.

In natural ecosystems living organisms continuously need external inputs in order to continue functioning. The living organisms, the learners and educators of a mobile learning ecosystem also needs inputs and in this study, augmented reality is defined as one of them. Even if the potential of its usage in learning is not denied, concerns will always be present. Turning concerns into promises is a must for a viable mobile learning ecosystem.

REFERENCES

Azuma, R. T. (1997). A Survey of Augmented Reality. *Teleoperators and Virtual Environments*, *6*(4), 355–385. doi:10.1162/pres.1997.6.4.355

Bower, M., Howe, C., McCredie, N., Robinson, A., & Grover, D. (2013) Augmented reality in education – cases, places, and potentials. In *Proceedings of the IEEE 63rd Annual Conference International Council for Educational Media* (ICEM). doi:10.1109/CICEM.2013.6820176

Chang, V., & Guetl, C. (2007). E-learning ecosystems (ELES). In *Proceedings of the Inaugural Conference on Digital Ecosystems and Technologies* (pp. 420- 425). IEEE. doi:10.1109/DEST.2007.372010

Clark, T. (2014). *The Components of a Digital Age Learning Ecosystem*. Retrieved October 29, 2016, from https://byotnetwork.com/2014/07/06/the-components-of-a-digital-age-learning-ecosystem/

Ficheman, I. K., & de Deus Lopes, R. (2008). Digital learning ecosystems: authoring, collaboration, immersion and mobility. In *Proceedings of the 7th International Conference on Interaction Design and Children* (pp. 9-12). ACM. doi:10.1145/1463689.1463705

FitzGerald, E., Adams, A., Ferguson, R., Gaved, M., Mor, Y., & Thomas, R. (2012). Augmented Reality and Mobile Learning: The State of the Art. In *Proceedings of the 11th International Conference on Mobile and Contextual Learning* (pp. 62-69). Retrieved October 29, 2016, from http://ceur-ws.org/Vol-955/

Gündoğan, M. B. (2014). E" is for "Ecology": An Ecosystem Approach to Sustainable E-Learning. In T. V. Yuzer & G. Eby (Eds.), *Handbook of Research on Emerging Priorities and Trends in Distance Education: Communication, Pedagogy, and Technology* (pp. 103–119). Hershey, PA: Information Science Reference. doi:10.4018/978-1-4666-5162-3.ch008

Haag, J. (2013). *Using Augmented Reality for Contextual Mobile Learning*. Retrieved October 29, 2016, from http://www.learningsolutionsmag.com/articles/1310/using-augmented-reality-for-contextual-mobile-learning

Haughey, D. (n.d.) *Delphi Technique: a Step-By-Step Guide*. Retrieved October 29, 2016, from https://www.projectsmart.co.uk/delphi-technique-a-step-by-step-guide.php

Höllerer, T., Feiner, S., Terauchi, T., Rashid, G., & Hallaway, D. (1999). Exploring MARS: Developing Indoor and Outdoor User Interfaces to a Mobile Augmented Reality System. *Computers & Graphics*, *23*(6), 779–785. doi:10.1016/S0097-8493(99)00103-X

Jackson, A. (2013). *Mobile learning: The shortest lived technology fad ever?* Retrieved October 29, 2016, from http://www.trainingzone.co.uk/develop/business/mobile-learning-the-shortest-lived-technology-fad-ever

Jax, K. (1998). Holocoen and Ecosystem: On the Origin and Historical Consequences of Two Concepts. *Journal of the History of Biology*, *3*(1), 113–142. doi:10.1023/A:1004261607170

Kearney, M., Schuck, S., Burden, K., & Aubusson, P. (2012). Viewing Mobile Learning From a Pedagogical Perspective. *Research in Learning Technology*, *20*(0). doi:10.3402/rlt.v20i0.14406

Laanpere, M., Pata, K., Normak, P., & Põldoja, H. (2014). Pedagogy-driven Design of Digital Learning Ecosystems. *Computer Science and Information Systems*, *11*(1), 419–442. doi:10.2298/CSIS121204015L

Odum, E. P. (1971). *Fundamentals of Ecology*. New York: Saunders.

Ramey, K. (2012). *Technological Advancements and Their Effects on Humanity*. Retrieved October 29, 2016, from http://www.useoftechnology.com/technological-advancements-effects-humanity/

Reyna, J. (2011) Digital Teaching and Learning Ecosystem (DTLE): A Theoretical Approach for Online Learning Environments. In *Proceedings ASCILITE 2011* (pp. 1083 – 1088). Hobart: The University of Tasmania. Retrieved October 29, 2016, from http://www.ascilite.org/conferences/hobart11/downloads/ProceedingsV3.pdf

Sharples, M., & Pea, R. (2014). Mobile Learning. In K. Sawyer (Ed.), *The Cambridge Handbook of the Learning Sciences* (2nd ed.; pp. 501–521). New York, NY: Cambriudge University Press. doi:10.1017/CBO9781139519526.030

Sridharan, B., Deng, H., & Corbitt, B. (2010). Critical success factors in e-learning ecosystems: A qualitative study. *Journal of Systems and Information Technology*, *12*(4), 263–288. doi:10.1108/13287261011095798

Tansley, A. G. (1935). The Use and Abuse of Vegetational Concepts and Terms. *Ecology*, *16*(3), 284–307. doi:10.2307/1930070

The Economist Intelligence Unit. (2008). *The Future of Higher Education: How Technology Will Shape Learning*. Retrieved October 29, 2016, from https://www.nmc.org/pdf/Future-of-Higher-Ed-(NMC).pdf

Udell, C., & Woodill, G. (2015) *The Mobile Learning Ecosystem*. Retrieved October 29, 2016, from https://www.td.org/Publications/Newsletters/Links/2015/12/The-Mobile-Learning-Ecosystem

Widder, B. (2016). *Best Augmented Reality Apps*. Retrieved October 29, 2016, from http://www.digitaltrends.com/android/best-augmented-reality-apps/

Winters, N. (2013). Mobile learning in the majority world: A critique of the GSMA position. In S. Price, C. Jewitt, & B. Brown (Eds.), *SAGE Handbook of Digital Technology Research* (pp. 402–412). London: SAGE Publications Ltd. doi:10.4135/9781446282229.n27

Zheng, S. (2015). Research on Mobile Learning Based on Augmented Reality. *Open Journal of Social Sciences*, *3*(12), 179–182. doi:10.4236/jss.2015.312019

Chapter 8
An Augmented-Reality-Based Intelligent Mobile Application for Open Computer Education

Utku Köse
Usak University, Turkey

ABSTRACT

Objective of this chapter is to introduce an Augmented Reality based intelligent mobile application (M-Learning application) to support courses of Computer Education. In the study, it was aimed to provide an alternative way of improving M-Learning experiences by employing both Augmented Reality and Artificial Intelligence based approaches in a common environment. Briefly, the application is able to use an intelligent mechanism evaluating students' several dynamic learning parameters to match them with the most appropriate course materials provided over the system. So, each student can encounter with appropriate course materials matching with their states over the application system. The related course materials include both AR based ones and standard ones as uploaded by teachers. An evaluation based flow has been run in the study by using the developed application through courses of Computer Education and the obtained results have shown that the application is effective and successful enough at improving students' learning experiences and achieving a good Open Computer Education.

INTRODUCTION

Because of needs appeared as a result of transformation of the society, there has been a remarkable interest in developing computer and communication oriented technologies to overcome the issue of gathering the desired information, revising it and even sharing it. As a result, many different technological approaches, methods, and techniques have appeared in order to adapt individuals to the new society and the new 'brave world'. It is clear that computer and communication oriented technologies have a great role on the whole changes, improvements, and developments so far and will continue to shape the future.

When the use of computer and communication oriented technologies in the modern life is considered, developments within both hardware and software sides of the computer technology should be discussed

DOI: 10.4018/978-1-5225-2110-5.ch008

and enough consideration should be given to the revolutionary communication technology called as Internet. Today, Internet and the Web as its logical side for the users have a remarkable role on providing more digital and practical ways of works in almost all fields. Among all different fields, education has a remarkable value because it has a great potential to benefit from the latest technologies and determining future states of the technologies to make education better and better.

Today's education field is associated with highly use of both computer and communication oriented technologies. Especially since the first introduction of distance education, taking part in educational process whenever and wherever individuals want has been a popular trend to be adapted to the rapidly changing life and fast information flow thanks to advantages of the digital world. In time, the approach of distance education has caused more advanced technological techniques like E-Learning. As a result of increasing use of especially mobile technologies, today's E-Learning applications are based on mostly Mobile Learning (M-Learning) technique. Thanks to also rapid developments experienced in mobile software and hardware technologies, it is now possible for every individual to use their own mobile phones in order to join to M-Learning based applications provided by educational institutions.

As like many other educational technologies, M-Learning also has been improved in time and still there is a remarkable effort in the associated literature to achieve better M-Learning experiences. At this point, it has been a popular approach to support M-Learning with different scientific methods to improve outputs of typical M-Learning processes. In the associated literature, it can be observed that researchers are in an effort of making M-Learning applications more effective and efficient. In this context, use of 'intelligent' components to improve using features and functions of M-Learning tools is one of the most popular research interest that researchers focus on. Because of multidisciplinary nature of Artificial Intelligence, it has been widely used to achieve intelligent software systems. So, M-Learning has become an important research field in which Artificial Intelligence based approaches, methods, and techniques are widely employed to improve everything in an educational manner. On the other hand, it has also been an important research interest to enable individuals to keep their interaction with the real-world alive while they are using a computer based educational tool. So, technological approaches like Augmented Reality, which allows mixing both virtual and real world, have been started to be used widely in different kinds of mobile applications. Here, M-Learning supported with Augmented Reality has gained popularity in a short time and the associated literature has received many research works focused on this solution way. If it can be thought further about improving such applications of Augmented Reality supported M-Learning, use of Artificial Intelligence is still one of the best choice leading to developments of educational tools of the future.

Objective of this chapter is to introduce an Augmented Reality based intelligent mobile application (in other words, M-Learning application) to support courses of Computer Education. As it was just indicated at the end of the last paragraph, it is a great way of improving M-Learning experiences by employing both Augmented Reality and Artificial Intelligence based approaches in a common environment. So, the study explained here has included design and development of such mobile application. Furthermore, the study has run also an evaluation based flow by using the developed application through courses of Computer Education. Briefly, the mobile application introduced in this chapter has a remarkable potential to improve learning experiences by combining the latest scientific technologies and approaches together and allows readers to have enough idea which step that the associated literature has just arrived to. At this point, the application is able to use an intelligent mechanism, which evaluates students' several dynamic learning parameters to match them with the most appropriate course materials provided over the system. So, each student can encounter with appropriate course materials matching with their states over the application system. The related course materials include both AR based ones and standard ones as uploaded by teachers.

Remaining content of the chapter is organized as follows: The next section is about the essential approaches forming the related application introduced in this chapter. At this point, the section provides some brief information regarding to Augmented Reality, M-Learning, Artificial Intelligence, and Open Education. After this section, the third section explains the designed and developed application. The section provides information about design and development processes of the application and introduces essential using features and functions of it. Next, the fourth section explains the evaluation works done in order to understand about effectiveness of the developed application in teaching Computer Education subjects and achieving that thanks to Augmented Reality, M-Learning, and Artificial Intelligence oriented approaches. Finally, the fifth section briefly focuses on future research plans and the last section discusses about conclusions.

ESSENTIAL APPROACHES ON THE BACKGROUND

On the background of the developed application in this study, there are four approaches that were taken into consideration by the author. These are Augmented Reality, M-Learning, Artificial Intelligence, and Open Education respectively. In order to understand better about what have been done within this study, it is a good idea to explain these approaches briefly.

Augmented Reality

As today's one of the most popular software technology approaches, Augmented Reality (AR) is a real-world based environment in which both 3D virtual objects and 3D real objects can take active part together (Azuma, 1997; Falconi, 2010; Zhou, Duh, & Billinghurst, 2008). Thanks to AR, it is possible for software developers to provide a unique reality experience for users, by achieving a mixed environment in which everything that is allowed by virtual reality can be done by also without leaving the real world. Because of this function, AR has been used widely in many different areas since its first introduction (Figure 1).

By using interactive sides of both virtual and real worlds, AR provides many different advantages that allow us to design and develop solutions for improving real-world based works. By taking some ideas expressed in the literature (Carmigniani et al., 2011; Kose, 2015; Van Krevelen & Poelman, 2010), it is possible to discuss about main advantages of AR as follows:

- AR has the ability of providing high interactivity by combining both virtual and real environment.
- AR can open our minds to benefit from a mixed reality and understand both virtual and real-world from this perspective.
- Thanks to the AR, it is possible to develop more effective simulation based applications by adding virtual objects / events to instant real environments.
- AR allows developers to use every type of multimedia objects (i.e., animations, sounds, videos) to improve user experience and keep it high always.
- Because of its flexible structure to run every kind of application supported with interactive elements, it is possible to run AR in almost every field of the modern life.

Figure 1. Some examples of Augmented Reality based applications

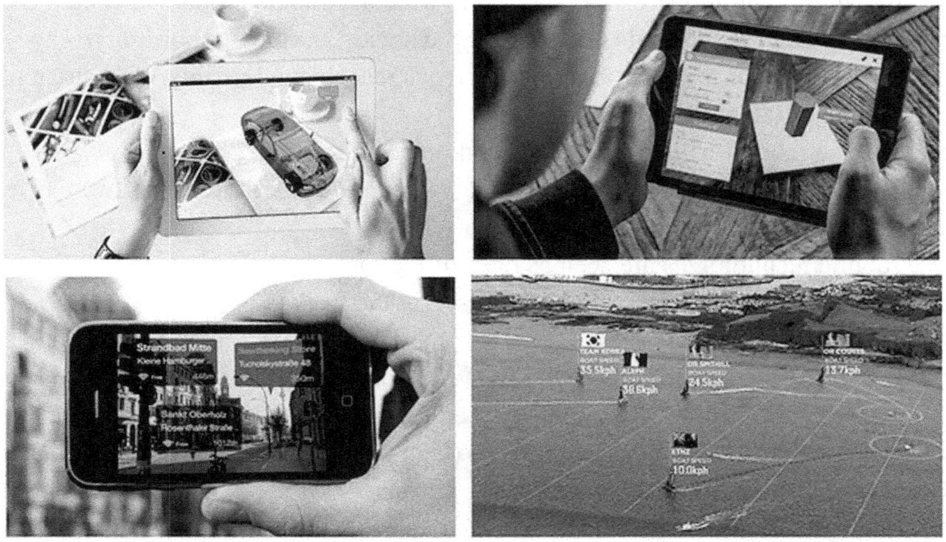

- AR is an easy-to-understand and easy-to-use technological approach, which allows developing practical applications that cannot be achieved by using other alternative software oriented techniques.

Education is one of the most remarkable fields in which AR is widely employed. In the literature, it is indicated that AR can improve students' motivation for learning, affect their many senses towards effective learning, enhance real-world based learning experiences and provide both strong, contextual, onsite learning experiences along with also serendipitous exploration and discovery over the real world (Cascales, Laguna, Pérez-López, Perona, & Contero, 2013; Chang, Morreale, & Medicherla, 2010; Johnson, Levine, Smith, & Stone, 2010). All of these functions of AR in educational manner have led the author to employ it within the M-Learning application introduced in this study.

M-Learning

Rapid developments in communication technologies have caused mobile technologies to improve more in time and take active part in humankind's life day by day. It can be clearly seen that especially use of advanced software technologies has enabled people to use their mobile devices for many different kinds of activities. As a result of using E-Learning in mobile devices, a new concept of learning called as M-Learning has appeared. As a combination of E-Learning and mobile computing oriented approaches (Trifonova, 2003), M-Learning is currently often used for educational purposes.

By considering also the associated literature, it is possible to main advantages of M-Learning as follows (El-Hussein & Cronje, 2010; Georgiev, Georgieva, & Smrikarov, 2004; Quinn, 2001; Preece, 2000; Sharples, 2002; Soloway et al., 2001):

- M-Learning provides the 'education at anywhere – anytime' function of Distance Education / E-Learning.

- According to traditional education, M-Learning provides a low-cost teaching-learning opportunity.
- Thanks to M-Learning, it is possible to reach to wide age range of people needing education.
- M-Learning provides a personal (self-learning), distance learning opportunity.
- Thanks to M-Learning, it is possible to experience teaching – learning with just mobile devices like smart phones, tablet PCs, PDAs…etc.
- M-Learning is a more flexible, interactive, and practical way of E-Learning.
- By using M-Learning, it is possible to provide a wide range of effective software oriented educational materials and even multimedia sources.
- Because mobile technologies are seen as ones of the most remarkable technologies of the future, M-Learning has a strong potential to be still used in the future.

As the AR is a key element for the research followed in this study, it has been applied in the context of an M-Learning based approach. It can be already seen from the associated literature, the most common employment of AR is related to mobile technologies (Doswell, 2006; Doswell, Blake, & Butcher-Green, 2006; FitzGerald et al., 2013; Liestol, 2011; Specht, Ternier, & Greller, 2011). So, combination of both these essential approaches has been an important research rule followed by the author.

Artificial Intelligence

As an important sub-field of Computer Science, Artificial Intelligence (AI) is known as the science of the future. Briefly, AI is the science of making intelligent machines to simulate human behaviors and natural dynamics to ensure effective and efficient solutions for real-world based problems (McCarthy, 2007; Nabiyev, 2005). As a result of developments in computer technology, AI has improved in time and today, it has been a widely used research field with its strong logical and mathematical solution approaches, methods, and techniques. Because of flexible, easy-to-apply algorithmic structures of AI based solutions, the related approaches, methods, and techniques can be applied to almost all fields of the modern life.

By considering the associated literature, major advantages of AI on solving real-world based problems can be expressed as follows (Flasiński, 2016; Ginsberg, 2012; Nabiyev, 2005; Nilsson, 2014; Russell, Norvig, Canny, Malik, & Edwards, 2003; Strong, 2016):

- AI has all necessary logical and mathematical abilities to simulate human behaviors and any dynamics in the nature that can be used as some kind of intelligent problem solution approaches.
- AI based systems can be designed and developed easily.
- AI can benefit from computer based technologies and calculating capabilities to achieve more effective, efficient, and faster ways of problem solving.
- Because logical and mathematical aspects of AI are flexible enough to be adapted to different kind of problems, AI has a great potential of used multidisciplinary.
- AI has many relations with different technologies like computers, electronics, soft computing, so it is always affected by developments and improvements within the related technologies.
- AI based methods and techniques can be used with alternative multidisciplinary approaches to form stronger, hybrid solution models.

It can be seen from especially the literature of education, AI is often used within today's software oriented teaching-learning tools to provide more effective ways of educational experiences for both teachers and students (Cuéllar & Pegalajar, 2014; Jain, Gurupur, Schroeder, & Faulkenberry, 2014; Jovanovic & Jovanovic, 2015; Nesbit, Adesope, Liu, & Ma, 2014; Walia, Singhal, & Sharma, 2015). In this study, Artificial Neural Networks (ANN), which is a well-known AI technique, has been used to form an intelligent evaluation mechanism for improving educational functions of the designed and developed mobile application.

Open Education

As a result of the transformation to the informatics society, individuals are required to have skills on reaching to instant knowledge rapidly, sharing the obtained knowledge with other individuals, and adapting it to new forms that can be used for solving problems of the modern life. This is generally because of technological improvements and also effects of the changed – developed technologies to the living standards of the society. This paradox has caused changes in also all fields of the modern life including education. In addition to improvements in educational technologies, the related transformations – changes have also caused new educational philosophies to appear. Open Education is known as one of these philosophies by "encompassing resources, tools, and practices, which employ a framework of open sharing to improve educational access and effectiveness worldwide" (OpenEducationWeek.Org, 2016).

Actually, the most effective way of achieving Open Education is to use the latest educational technologies associated with especially computer based systems to provide educational materials and sources to all over the world. Under this approach, the following rules – mechanisms of the Open Education should be run (OpenEducationWeek.Org, 2016; OpenSource.com, 2016; SparcOpen.Org, 2016):

- Everyone all over the world should receive education.
- The most effective computer and communication technologies (i.e., Internet) should be employed to provide effective education and knowledge.
- Everyone all over the world should be enabled to way "produce, share, and build on knowledge".
- The related teaching – learning materials should be open for everyone and free to use.
- In the sense of location and time, there should be no barriers against the education.

Nowadays, the most popular application of Open Education is Massive Online Open Courses (MOOCs). Web platforms based on MOOCs are widely visited by users to receive immediate learning sessions. Readers are referred to (Bonk, Lee, Reeves, & Reynolds, 2015; Daniel, 2012; Gaebel, 2014; Liyanagunawardena, Adams, & Williams, 2013; Siemens, 2013; Yuan, Powell, & Cetis, 2013) in order to learn more about MOOCs.

The related AR based intelligent M-Learning application introduced in this study has been employed through the Open Education approach to teach – learn courses of Computer Education. In this way, it is aimed to enable everyone to gain necessary knowledge and abilities regarding to the subjects of computer and its supportive technologies.

An Augmented Reality Based Intelligent Mobile Application

The AR based intelligent M-Learning application, which was designed and developed in this study is a typical application, which can be installed and used directly in Google Android based mobile devices. Because of the Open Education approaches followed by the application, the.apk file can be downloaded by everyone interested in taking online courses of Computer Education. Thanks to using features and functions combining effective approaches like AR and AI, the application aims to improve learning experiences and provide a successful educational tool that can be used easily by everyone.

In order learn more about the application, this section is devoted to some brief explanations about design – development process of the application, its essential using features and functions, and finally AI based mechanism employed in it.

Design and Development Process

Design and development of the M-Learning application has consisted of using some software platforms to obtain a software system, which combines typical Android based application functions, AR based using features, and also AI infrastructure to run an intelligent mechanism to improve learning experiences. At this point, ideas from a group of education experts have also been used to get better form of application. Some brief information regarding to that process can be given as follows (Figure 2):

- At the start of February 2016, pre-analyze of the idea of developing such application has been done. In this context, experiences done with previous M-Learning and AR based research works have also been evaluated. Readers interested some of these works are referred to (Kose, 2009; Kose, Koc, & Yucesoy, 2013; Tufekci, Ekinci, & Kose, 2013). At the end of a one-month time period, a draft form of AR based intelligent M-learning application has been obtained.
- By using the draft form of the application, an additional pre-analyze session for one-month time period has been completed with a group of education experts (who are from Gazi University, Afyon Kocatepe University, and Aegean University of Turkey, respectively) in order to determine

Figure 2. A brief look at to the design and development process of the application

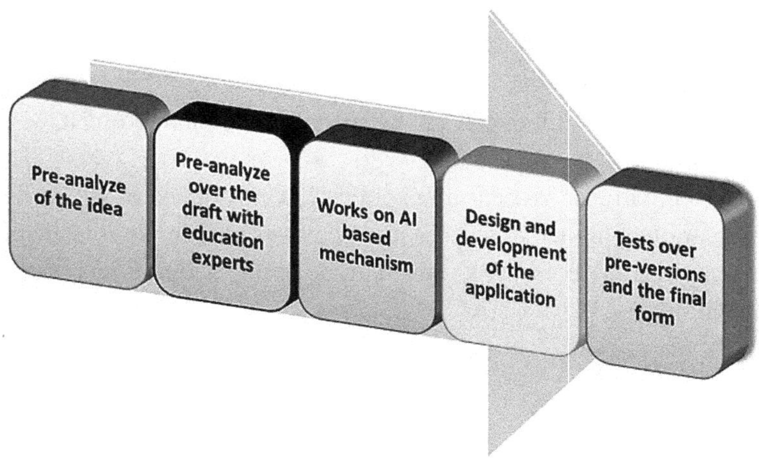

essential educational approaches that will be employed by the application and main using features – functions meeting with these approaches.

- After the related two months, details of the AI based mechanism that will be included in the application have been figured out in a two-month work. At this point, formerly obtained ideas from the education experts and the author's educational experiences have been combined in order to build a total of five different ideas of intelligent functions. By considering details of these ideas in the context of their simplicity to code and effectiveness to improve learning experiences, the intelligent function, which is able to match an active student with the next course material that will be viewed to him / her, has been chosen to be included. Details regarding to this intelligent mechanism have been provided under next paragraphs / sub-sections.
- For next three and half months, the application has been coded under the Eclipse platform for developing Google Android based applications. Meanwhile, interfaces that will be used over the application have been designed by using two widely used graphic design programs, which are Adobe Illustrator and Adobe Fireworks. On the other hand, animated components and also most of AR course materials have been designed by using Adobe Flash environment. It should be indicated that the whole materials regarding to the courses of Computer Education, which is considered in this study, have been designed by using the related Adobe programs. On the background, the database structure of the application has been designed by using Android SQLite Database.
- After the related design and development processes, the application has its final form at around October 2016, after some tests done over two pre-versions.

Essential Using Features and Functions

M-Learning application developed in this study employ easy-to-use interfaces with simple using features and functions. Briefly, there are three types of users enrolled in the application system as 'admin', 'teacher', and 'student'. After logging into the system, each user type encounter with the related options that are specific for their tasks over the system. The whole system has been designed and developed to be used over especially touch-screen mobile devices. So, interfaces include low number of simple control to achieve easy-to-use, fast using experiences for all user types, by focusing on only educational mission and usability in this manner.

In the context of essential using features and functions, a typical educational flow over the application is done briefly as follows:

- Teachers can use their interfaces to manage their own courses. In this sense, they can add – remove, manage course subjects and materials. In details, it is possible for teachers to define details of course subjects, upload materials (including AR based opens) to them and tracking students over the system. On the other hand, creating a course and all the related management options over the courses and the application system (i.e., defining students and teachers, adjusting some system parameters, defining system using – storage limits) belong to admins.
- Regarding to the intelligent mechanism of the application system, teachers can determine which student parameters – states will be considered by the intelligent mechanism. Over the system, there are some typical dynamic learning parameters like students' homework study grades, number of views of a specific course (or course subject), time spend over a course (or course subject)

to use for course material matching approach. It is possible for teachers to use them after they upload a material to the application system (More details about the intelligent mechanism are provided under the next sub-section).

- In addition to tasks over courses, course materials, and students enrolled in the courses, it is possible for teachers to also define some homework studies, quizzes, and examinations. While defining them, the application system automatically adds 'grade parameter' to the students associated with them (So, these learning parameters can be used later for the intelligent matching mechanism over course materials).
- By taking the tasks performed by teachers, it is possible to understand essential using features and functions provided for students over the system. Briefly, students can view the courses in which they are included by the teachers. While viewing these courses, the related materials are provided to them by the intelligent mechanism of the application system. On the other hand, students can also do homework studies, quizzes and examinations provided over the application.
- As general, all teacher and student interfaces provide some help options to direct users about application use. But it is thought that the application has easy-to-use interfaces that anyone can adapt to it in a short time and focus on the exact educational process.

Figure 3 presents some views from use of the developed AR based intelligent M-Learning application.

Regarding to the AR function of the application, there are some label types that can be used by teachers to activate AR materials. These labels can be provided over the application so students can print these labels to use them or just save them as image files to view on another screen to be focused by a mobile device. The application provides some pre-defined labels to be used by teachers. But it is also possible for teachers to upload their own appropriate image files to match them with AR materials. In addition to the function of activating AR materials, it is possible for teachers to also connect standard course materials (i.e., text content, videos, image files, document files) with the labels. Pre-defined labels included in the application system can be adjusted by only admins. Some label types used over the application are shown in Figure 4.

Figure 3. Some views from use of the developed application

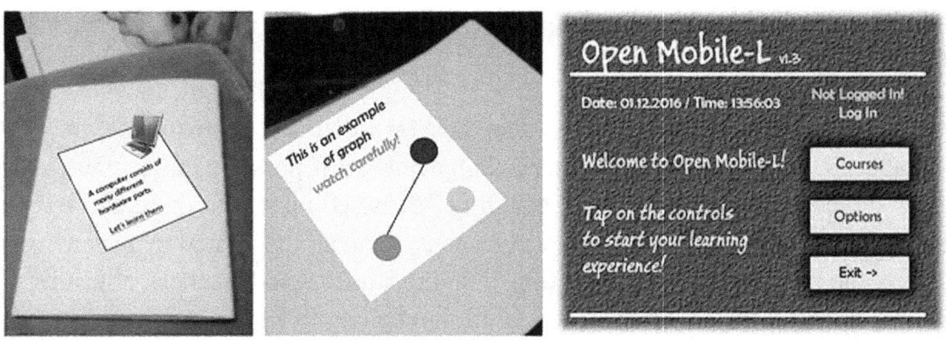

Figure 4. Some label types used for activating AR based materials provided over the application

Intelligent Mechanism to Match Student and Course Material

Intelligent mechanism of the M-Learning application has been designed over a widely-used AI technique: Artificial Neural Networks (ANN). A typical ANN model provides some kind of parallel and distributed data processing systems, which are able to learn from the environments and improve the solution process via learning to be used for classification, prediction, optimization, pattern recognition, control, and associative memory (Carpenter, 1989; Cochocki and Unbehauen, 1993; Hassoun, 1995; Miller, Werbos, & Sutton, 1995; Ripley, 1994). In order not to make the chapter content more complex, readers interested in having more information about ANN are referred to (Fyfe, 2005; Hassoun, 1995; Mehrotra, Mohan, & Ranka, 1997; Yegnanarayana, 2009).

Intelligent mechanism developed here thanks to the ANN model achieves an accurate matching function between students and the course materials. In detail, when a student is in a course interface, each material that will be viewed to this student is automatically determined by the ANN model by considering the student's current learning parameters to get an output value of 'choosing rate', which is defined for each uploaded material by the teacher(s). It is important that the ANN model is trained before by feeding it with some data sets that can be set by admins of the application system. More technical details regarding to training of an ANN model can be obtained from the references indicated at the end of the previous paragraph.

As it can be understood, the ANN model is able to choose the best course material meeting with the active student's needs and improve his / her learning experiences through the most appropriate set of course materials provided over the application. A summary regarding to the intelligent mechanism is provided in Figure 5.

Figure 5. A summary of the intelligent mechanism employed in the application

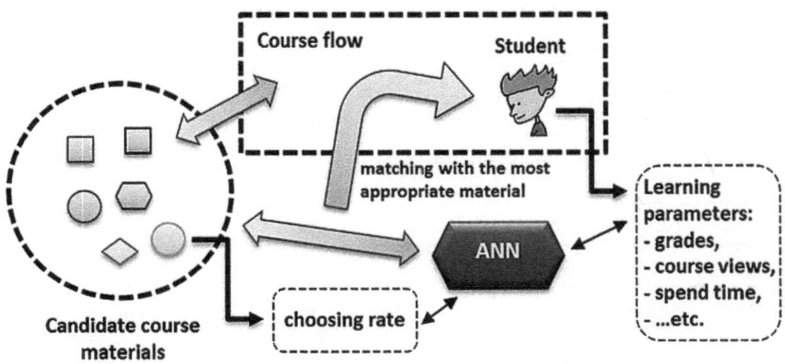

EVALUATION

In order to understand more about success and effectiveness of the M-Learning application, two different evaluation works have been completed after a using period. Before focusing on the evaluation works done, it is also important to provide brief information regarding to using period of the application:

- Although the application system allows creating different type of courses, Computer Education has been the main focused field in this study.
- The application has been used for three weeks as free by a total of 453 users (students – learners) from different places of Turkey.
- Over the system, five different Computer Education oriented courses, which are called respectively as 'Essentials of Computer Programming', 'Operating Systems', 'Computer Networks', 'Computer Hardware', and 'Graphic Design and Animation with Computers' have been provided.
- Each provided course consists of a total of 15 subjects including many activities supported by AR and alternative components. In order to improve the related learning processes, some quizzes and homework studies have also been provided for the users over the application.
- It is clear that each user's – student's interaction with the courses are controlled by the intelligent mechanism and learning processes are directed by providing appropriate course materials according to their dynamic learning parameters over the M-Learning application system.
- After the using period, the related users – students are requested to take final examinations and fill the evaluation survey, which are explained as follows:

Final Examinations and Findings

At the end of the using period, the users (they will be called as students from now on) have requested to complete final examinations for each course provided over the M-Learning application. At this point, each final examination has included a total of 40 questions including multi-choice and classical ones (Final grade is determined by summing point of each true answer as determined according to importance of each question). Students, who had the final grade as at least 60 or higher than it, have accepted as passed the related course.

Table 1 provides findings regarding to the final examinations done by the objective students over the application.

Table 1. Findings regarding to the final examinations done over the application

Course	Number of Students Passed	Mean Final Grade
Essentials of Computer Programming	344	70,56
Operating Systems	401	73,93
Computer Networks	363	74,54
Computer Hardware	412	86,12
Graphic Design and Animation with Computers	389	77,17

As it can be seen from the Table 1, students took active part over the M-Learning application have provided successful performance within final examinations of the related courses, after the three week of learning period.

Evaluation Survey and Findings

Next to the final examinations, students have also wanted to fill an evaluation survey, which includes a total of 10 statements regarding to the M-Learning application and the experienced learning process over the courses of Computer Education. In this context, feedbacks from the students have gathered by using a Likert-type scale on five-level responses.

Evaluation survey statements and the obtained findings are provided briefly in Table 2.

Regarding to the findings provided in Table 2, it can be expressed that the students are satisfied with the used M-Learning application and the experienced learning process for courses of Computer Education. As general, positive feedback came from the students have been key factor to understand effectiveness and efficiency that can be achieved thanks to combination of the related components considered in this study.

In addition to the responses given for the evaluation survey, the students have also wanted to express their opinions if they desire. Some remarkable opinions received from them are listed as follows:

- *This has been a surprising opportunity to learn something about computers.*
- *I think number of such free learning systems should be increased.*
- *I can now have more enjoyable time with my phone. Because it allows me to learn!*
- *I wish I could take part in some courses regarding to History.*
- *I wish there could be some opportunities to contact with other users over the system.*
- *I think the course process that I have just experienced is the future of education.*
- *I have excellent computer programming skills and I would like to take part in developing such entertaining education approach.*

Table 2. Evaluation survey statements and the obtained findings

St. No	Statement	Responses*				
		1	2	3	4	5
1	*"I have enjoyed using the mobile application."*	2	13	26	51	361
2	*"I have enjoyed the learning process supported with AR."*	6	9	21	45	372
3	*"The mobile application was easy-to-use and highly interactive."*	0	2	13	37	401
4	*"I have encountered with the course materials I need to learn."*	1	4	9	28	411
5	*"I think the provided AR approach was effective at teaching."*	0	10	46	42	355
6	*"I have improved my knowledge on essentials of computers."*	0	3	23	22	405
7	*"I do not want to take part in such educational process again."*	413	28	8	2	2
8	*"This educational process provides a good open education."*	1	5	11	39	397
9	*"This mobile application should include some other courses, too."*	0	12	20	40	381
10	*"The application is not a good educational tool."*	408	27	10	8	0

*** Likert Scale: 1: 'I totally disagree'; 2: 'I disagree'; 3: 'No opinion'; 4: 'I agree'; 5: 'I totally agree'**

- *I know AR but I had not experienced it in my learning processes. It is fantastic and should be used more in education!*
- *Employing such educational process in all open education institutes would be great.*

FUTURE RESEARCH DIRECTIONS

Positive feedback received from the objective students have encouraged the author to think about some further research works. In this context, future works regarding to the developed mobile application consists of evaluating it and the related teaching-learning approach more from different perspectives and also improving it with additional components. The related future works can be explained briefly as follows:

- The mobile application will be employed in different departments (including both social and natural sciences) to see if its employment in different disciplines can affect objective students' academic achievements.
- Regarding to Open Education oriented objectives, feedback received from the objective audience will be get by using also wide surveys evaluating the application and the performed learning processes.
- Developments for newer versions of the mobile application will be continued. In this context, there will be updates regarding to making the application more personal (i.e., enabling users to adjust interfaces, placement of application controls) and interactive (i.e., including social media oriented components).
- As it was indicated before, the application introduced in this study is currently for Google Android operating system. So, it is planned to develop iOS, and Windows Phone versions of the application system in the future.
- Artificial Intelligence infrastructure of the mobile application will be improved by adding some additional techniques to adjust parameters regarding to teaching-learning processes. Except from the current ANN model, the application will be supported with Genetic Algorithm, and also some Swarm Intelligence based algorithms to perform some optimization based solutions on adjusting parameters regarding to students, teachers, and maybe the application system. The related student and teacher parameters to be adjusted with the related Artificial Intelligence techniques will be determined with additional works to get feedback about this issue from objective learners.
- In future versions of the mobile application, there will be newer features – functions enabling teachers to adjust more about AR infrastructure of the application (i.e., providing AR to the chosen students after some kind of competitions, making the AR adaptive to hardware features of the mobile device used).

CONCLUSION

In this chapter, an AR based intelligent mobile application for Open Computer Education has been introduced. Along the M-Learning approach followed thanks to this application, it is possible to improve students' learning experiences by providing especially AR based course materials. Here, the application is able to use an intelligent mechanism, which evaluates students' several dynamic learning parameters to match them with the most appropriate course materials provided over the system. Briefly, this intelligent mechanism is run via an ANN model and in this way, each student can encounter with appropriate course materials matching with their states over the application system. Eventually, effectiveness and efficiency of students' learning experiences are improved thanks to this AR based intelligent M-Learning application.

A remarkable side of the research approach followed in this study has been its Open Education philosophy. In this sense, the M-Learning application has been employed for the courses of Computer Education over the approach of Open Education. The application has been run for three weeks in order to provide online learning opportunities to the interested audience. Additionally, in order to understand more about success and effectiveness of the application – study, final examinations and an evaluation survey work have been done at the end, with the contribution by a total of 453 students – learners, who were randomly chosen. Obtained findings from the final examination and the filled survey have shown that the developed AR based intelligent M-Learning application is successful enough to enable people for taking part in effective and efficient, highly interactive learning processes and caused positive ideas about application of such AR and AI based mobile application for educational purposes. The results obtained in this study have also encouraged the author to focus on some future works as explained in detail under the previous section.

REFERENCES

Azuma, R. (1997). A survey of augmented reality. *Presence (Cambridge, Mass.)*, *4*(4), 355–385. doi:10.1162/pres.1997.6.4.355

Bonk, C. J., Lee, M. M., Reeves, T. C., & Reynolds, T. H. (Eds.). (2015). *MOOCs and open education around the world*. Routledge.

Carmigniani, J., Furht, B., Anisetti, M., Ceravolo, P., Damiani, E., & Ivkovic, M. (2011). Augmented reality technologies, systems and applications. *Multimedia Tools and Applications*, *51*(1), 341–377. doi:10.1007/s11042-010-0660-6

Carpenter, G. A. (1989). Neural network models for pattern recognition and associative memory. *Neural Networks*, *2*(4), 243–257. doi:10.1016/0893-6080(89)90035-X

Cascales, A., Laguna, I., Pérez-López, D., Perona, P., & Contero, M. (2013). An experience on natural sciences augmented reality contents for preschoolers. In R. Shumaker (Ed.), Virtual, Augmented and Mixed Reality. Systems and Applications (LNCS Vol. 8022, pp. 103-112). Berlin: Springer-Verlag. doi:10.1007/978-3-642-39420-1_12

Chang, G., Morreale, P., & Medicherla, P. (2010). Applications of augmented reality systems in education. In *Society for Information Technology & Teacher Education International Conference 2010* (pp. 1380-1385). Chesapeake, VA: AACE.

Cochocki, A., & Unbehauen, R. (1993). *Neural networks for optimization and signal processing*. John Wiley & Sons, Inc.

Cuéllar, M. P., & Pegalajar, M. C. (2014). Design and implementation of intelligent systems with LEGO Mindstorms for undergraduate computer engineers. *Computer Applications in Engineering Education, 22*(1), 153–166. doi:10.1002/cae.20541

Daniel, J. (2012). Making sense of MOOCs: Musings in a maze of myth, paradox and possibility. *Journal of Interactive Media in Education, 2012*(3), 18. doi:10.5334/2012-18

Doswell, J. T. (2006). Context-aware mobile augmented reality architecture for lifelong learning. In *Sixth IEEE International Conference on Advanced Learning Technologies (ICALT'06)* (pp. 372-374). IEEE. doi:10.1109/ICALT.2006.1652448

Doswell, J. T., Blake, M. B., & Butcher-Green, J. (2006). Mobile augmented reality system architecture for ubiquitous e-learning. In *2006 Fourth IEEE International Workshop on Wireless, Mobile and Ubiquitous Technology in Education (WMTE'06)* (pp. 121-123). IEEE. doi:10.1109/WMTE.2006.261358

El-Hussein, M. O. M., & Cronje, J. C. (2010). Defining Mobile Learning in the Higher Education Landscape. *Journal of Educational Technology & Society, 13*(3), 12–21.

Falconi, R. F. (2010). *Usability and game design: Improving the MITAR Game Editor* (Master of Eng. Thesis). Massachusetts Institute of Technology, Cambridge, MA.

FitzGerald, E., Ferguson, R., Adams, A., Gaved, M., Mor, Y., & Thomas, R. (2013). Augmented reality and mobile learning: The state of the art. *International Journal of Mobile and Blended Learning, 5*(4), 43–58. doi:10.4018/ijmbl.2013100103

Flasiński, M. (2016). *Introduction to artificial intelligence*. Springer. doi:10.1007/978-3-319-40022-8

Fyfe, C. (2005). Artificial neural networks. In *Do Smart Adaptive Systems Exist?* (pp. 57–79). Springer Berlin Heidelberg. doi:10.1007/3-540-32374-0_4

Gaebel, M. (2014). *MOOCs: Massive open online courses*. EUA.

Georgiev, T., Georgieva, E., & Smrikarov, A. (2004). M-learning-a New Stage of E-Learning. In *International Conference on Computer Systems and Technologies-CompSysTech* (*Vol. 4*, No. 28, pp. 1-4). Academic Press.

Ginsberg, M. (2012). *Essentials of artificial intelligence*. Newnes.

Hassoun, M. H. (1995). *Fundamentals of artificial neural networks*. MIT Press.

Jain, G. P., Gurupur, V. P., Schroeder, J. L., & Faulkenberry, E. D. (2014). Artificial intelligence-based student learning evaluation: A concept map-based approach for analyzing a students understanding of a topic. *IEEE Transactions on Learning Technologies, 7*(3), 267–279. doi:10.1109/TLT.2014.2330297

Johnson, L., Levine, A., Smith, R., & Stone, S. (2010). *Simple augmented reality. The 2010 Horizon Report* (pp. 21–24). Austin, TX: The New Media Consortium.

Jovanovic, D., & Jovanovic, S. (2015). An adaptive e-learning system for Java programming course, based on Dokeos LE. *Computer Applications in Engineering Education, 23*(3), 337–343. doi:10.1002/cae.21603

Kose, U. (2009). *An Advanced M-Learning System for Teaching Artificial Intelligence Techniques, International Conference of Education, Research and Innovation 2009* (ICERI 2009). Madrid, Spain: Academic Press.

Kose, U. (2015). Augmented Reality Based E-Learning Applications. In M. Khosrow-Pour (Ed.), *Encyclopedia of Information Science and Technology* (3rd ed.; pp. 7507–7518). Hershey, PA: IGI Global; doi:10.4018/978-1-4666-5888-2.ch739

Kose, U., Koc, D., & Yucesoy, S. A. (2013). An augmented reality based mobile software to support learning experiences in computer science courses. *Procedia Computer Science*, *25*, 370–374. doi:10.1016/j.procs.2013.11.045

Liestol, G. (2011). Learning through situated simulations: Exploring mobile augmented reality. *Research Bulletin (Sun Chiwawitthaya Thang Thale Phuket)*, *1*, 2011.

Liyanagunawardena, T. R., Adams, A. A., & Williams, S. A. (2013). MOOCs: A systematic study of the published literature 20082012. *The International Review of Research in Open and Distributed Learning*, *14*(3), 202–227. doi:10.19173/irrodl.v14i3.1455

McCarthy, J. (2007). *What is Artificial Intelligence?* Online: http://www-formal.stanford.edu/jmc/whatisai/whatisai.html

Mehrotra, K., Mohan, C. K., & Ranka, S. (1997). *Elements of artificial neural networks*. MIT Press.

Miller, W. T., Werbos, P. J., & Sutton, R. S. (1995). *Neural networks for control*. MIT Press. doi:10.1007/978-3-642-57760-4

Nabiyev, V. V. (2005). *Artificial Intelligence (Problems – Methods – Algorithm)* [Yapay Zekâ (Problemler – Yöntemler – Algoritma)]. Ankara, Turkey: Seçkin Press.

Nesbit, J. C., Adesope, O. O., Liu, Q., & Ma, W. (2014). How Effective are Intelligent Tutoring Systems in Computer Science Education? In *Advanced Learning Technologies (ICALT), 2014 IEEE 14th Int. Conference on* (pp. 99-103). IEEE.

Nilsson, N. J. (2014). *Principles of artificial intelligence*. Morgan Kaufmann.

OpenEducationWeek.Org. (2016). *What is Open Education?* Online: https://www.openeducationweek.org/page/what-is-open-education

OpenSource.Com. (2016). *What is Open Education?* Online: https://opensource.com/resources/what-open-education

Preece, J. (2000). *Online communities: Designing usability and supporting socialbilty*. John Wiley & Sons, Inc.

Quinn, C. (2001). Get ready for m-learning. *Training & Development*, *20*(2), 20–21.

Ripley, B. D. (1994). Neural networks and related methods for classification. *Journal of the Royal Statistical Society. Series B. Methodological*, 409–456.

Russell, S. J., Norvig, P., Canny, J. F., Malik, J. M., & Edwards, D. D. (2003). *Artificial intelligence: a modern approach* (Vol. 2). Upper Saddle River, NJ: Prentice hall.

Sharples, M. (2002). Disruptive devices: Mobile technology for conversational learning. *International Journal of Continuing Engineering Education and Lifelong Learning*, *12*(5-6), 504–520. doi:10.1504/IJCEELL.2002.002148

Siemens, G. (2013). Massive open online courses: Innovation in education. *Open Educational Resources: Innovation, Research and Practice, 5.*

Soloway, E., Norris, C., Blumenfeld, P., Fishman, B., Krajcik, J., & Marx, R. (2001). Log on education: Handheld devices are ready-at-hand. *Communications of the ACM*, *44*(6), 15–20. doi:10.1145/376134.376140

SparcOpen.Org. (2016). *Open Education.* Online: http://sparcopen.org/open-education/

Specht, M., Ternier, S., & Greller, W. (2011). Mobile augmented reality for learning: A case study. *Journal of the Research Center for Educational Technology*, *7*(1), 117–127.

Strong, A. I. (2016). *Applications of Artificial Intelligence & Associated Technologies. Science.* Academic Press.

Trifonova, A. (2003). *Mobile Learning – Review of the Literature.* University of Trento Technical Report.

Tufekci, A., Ekinci, H., & Kose, U. (2013). Development of an internet-based exam system for mobile environments and evaluation of its usability. *Mevlana Int. J. of Education*, *3*(4), 57–74. doi:10.13054/mije.13.59.3.4

Van Krevelen, D. W. F., & Poelman, R. (2010). A survey of augmented reality technologies, applications and limitations. *International Journal of Virtual Reality*, *9*(2), 1.

Walia, A., Singhal, N., & Sharma, A. K. (2015). A Novel E-learning Approach to Add More Cognition to Semantic Web. In *Computational Intelligence & Communication Technology (CICT), 2015 IEEE Int. Conference on* (pp. 13-17). IEEE.

Yegnanarayana, B. (2009). *Artificial neural networks.* PHI Learning Pvt. Ltd.

Yuan, L., Powell, S., & Cetis, J. (2013). *MOOCs and open education: Implications for higher education.* JISC Cetis Publications.

Zhou, F., Duh, H. B. L., & Billinghurst, M. (2008). Trends in augmented reality tracking, interaction and display: A review of ten years of ISMAR. In *IEEE and ACM ISMAR 2008* (pp. 193–202). Cambridge, UK: IEEE and ACM.

ADDITIONAL READING

Anabuki, M., Kakuta, H., Yamamoto, H., & Tamura, H. (2000, April). Welbo: An embodied conversational agent living in mixed reality space. In CHI'00 Extended Abstracts on Human Factors in Computing Systems (pp. 10-11). ACM.

Andersson, R., Anker, M., Dunford, A., Lundqvist, J., & Weiss, A. (2016). HAMLIN: An augmented reality solution to visualize abstract concepts for science education. In Proceedings of SIDeR'16–student interaction design research conference.

Bertelsen, O. W., & Nielsen, C. (2000). Augmented reality as a design tool for mobile interfaces. In Proceedings of the 3rd conference on Designing interactive systems: processes, practices, methods, and techniques (pp. 185-192). ACM. doi:10.1145/347642.347714

Brusilovsky, P. (1999). Adaptive and intelligent technologies for web-based eduction. KI, 13(4), 19-25.

Brusilovsky, P., & Peylo, C. (2003). Adaptive and intelligent web-based educational systems. *International Journal of Artificial Intelligence in Education*, *13*(2), 159–172.

Cavus, N., & Ibrahim, D. (2016). Learning English using childrens stories in mobile devices. *British Journal of Educational Technology*, n/a. doi:10.1111/bjet.12427

Cetin, S. (2016). General trends in master and doctorate thesis in the area of computer education and ınstructional technologies in Turkey. *International Journal of Innovative Research in Education, 2*(2), 55-67.

Chatzidimitris, T., Gavalas, D., & Michael, D. (2016, April). SoundPacman: Audio augmented reality in location-based games. In Electrotechnical Conference (MELECON), 2016 18th Mediterranean (pp. 1-6). IEEE.

Chen, D. M., Tsai, S. S., Vedantham, R., Grzeszczuk, R., & Girod, B. (2009). Streaming mobile augmented reality on mobile phones. In Mixed and Augmented Reality, 2009. ISMAR 2009. 8th IEEE International Symposium on (pp. 181-182). IEEE. doi:10.1109/ISMAR.2009.5336472

Clow, D. (2013). MOOCs and the funnel of participation. In *Proceedings of the Third International Conference on Learning Analytics and Knowledge* (pp. 185-189). ACM. doi:10.1145/2460296.2460332

Czerkawski, B. O. (Ed.). (2010). *Free and Open Source Software for E-Learning: Issues, Successes and Challenges: Issues, Successes and Challenges*. IGI Global.

Downes, S. (2005). E-learning 2.0. Elearn magazine, 2005(10), 1.

Ekengren, B. (2009). Mobile augmented reality. Skolan för datavetenskap och kommunikation, Kungliga Tekniska högskolan.

Fedosov, A., Elhart, I., Niforatos, E., North, A., & Langheinrich, M. (2016). SkiAR: Wearable Augmented Reality System for Sharing Personalized Content on Ski Resort Maps. In *Proceedings of the 7th Augmented Human International Conference 2016* (p. 46). ACM. doi:10.1145/2875194.2875234

Feiner, S., Macintyre, B., & Seligmann, D. (1993). Knowledge-based augmented reality. *Communications of the ACM*, *36*(7), 53–62. doi:10.1145/159544.159587

Henrysson, A., & Ollila, M. (2004). UMAR: Ubiquitous mobile augmented reality. In *Proceedings of the 3rd international conference on Mobile and ubiquitous multimedia* (pp. 41-45). ACM. doi:10.1145/1052380.1052387

Hollerer, T., & Feiner, S. (2004). *Mobile augmented reality. Telegeoinformatics: Location-Based Computing and Services* (p. 21). London, UK: Taylor and Francis Books Ltd.

Hollerer, T., Feiner, S., Hallaway, D., Bell, B., Lanzagorta, M., Brown, D., & Rosenblum, L. et al. (2001). User interface management techniques for collaborative mobile augmented reality. *Computers & Graphics*, *25*(5), 799–810. doi:10.1016/S0097-8493(01)00122-4

Hollerer, T., Feiner, S., Terauchi, T., Rashid, G., & Hallaway, D. (1999). Exploring MARS: Developing indoor and outdoor user interfaces to a mobile augmented reality system. *Computers & Graphics*, *23*(6), 779–785. doi:10.1016/S0097-8493(99)00103-X

Huang, T. C., Chen, C. C., & Chou, Y. W. (2016). Animating eco-education: To see, feel, and discover in an augmented reality-based experiential learning environment. *Computers & Education*, *96*, 72–82. doi:10.1016/j.compedu.2016.02.008

Ibanez, M. B., Di-Serio, Á., Villaran-Molina, D., & Delgado-Kloos, C. (2016). Support for Augmented Reality Simulation Systems: The Effects of Scaffolding on Learning Outcomes and Behavior Patterns. *IEEE Transactions on Learning Technologies*, *9*(1), 46–56. doi:10.1109/TLT.2015.2445761

Iiyoshi, T., & Kumar, M. V. (2008). *Opening up education: The collective advancement of education through open technology, open content, and open knowledge*. Mit Press.

Jennings, N. R. (2000). On agent-based software engineering. *Artificial Intelligence*, *117*(2), 277–296. doi:10.1016/S0004-3702(99)00107-1

Katier, B. (2011). Mobile Augmented Reality (Doctoral dissertation, Master Thesis, University of Amesterdam).

Keegan, D. (2004). *Foundations of Distance Education*. New York: Routledge.

Kose, U. (2009). Teach-MS: Developing a Management System for M-Learning, The 5th International Conference on Open and Distance Learning (ICODL 2009), Athens, Greece, (D), 69–77.

Kose, U. (2010). *Web 2.0 Technologies in E-learning. Free and Open Source Software for E-learning: Issues, Successes and Challenges* (pp. 1–23). Hershey: IGI Global.

Kounavis, C. D., Kasimati, A. E., & Zamani, E. D. (2012). Enhancing the tourism experience through mobile augmented reality: Challenges and prospects. International Journal of Engineering Business Management, 4.

Liebowitz, J., & Frank, M. (Eds.). (2016). *Knowledge management and e-learning*. CRC press.

Littlejohn, A., & Hood, N. (2016). How educators build knowledge and expand their practice: The case of open education resources. *British Journal of Educational Technology*, n/a. doi:10.1111/bjet.12438

Liu, J. C., Gu, X., Sun, Y., & Wang, C. X. (2016). Learning and Knowledge Analytics in Open Education: A Brief Overview. *TechTrends*, *60*(1), 85–86. doi:10.1007/s11528-015-0005-4

MartínNúñez, J. L., Caro, E. T., & González, J. R. H. (2016). From Higher Education to Open Education: Challenges in the Transformation of an Online Traditional Course. *IEEE Transactions on Education*.

O'Shea, P. M., & Elliott, J. B. (2016, June). Augmented Reality in Education: An Exploration and Analysis of Currently Available Educational Apps. In *International Conference on Immersive Learning* (pp. 147-159). Springer International Publishing. doi:10.1007/978-3-319-41769-1_12

Park, N. (2016). Development of Computer Education Program Using LOGO Programming and Fractals Learning for Enhancing Creativity: Focus on Creative Problem-Solving. *International Journal of u-and e-Service. Science and Technology*, *9*(2), 121–126.

Ponce, J., Ornelas, F., Álvarez, F., & Toscano, B. (2016). Use of Augmented Reality a New Vision on the Massive Open Online Courses. User-Centered Design Strategies for Massive Open Online Courses (MOOCs), 254.

Pryss, R., Geiger, P., Schickler, M., Schobel, J., & Reichert, M. (2016). Advanced Algorithms for Location-Based Smart Mobile Augmented Reality Applications. *Procedia Computer Science*, *94*, 97–104. doi:10.1016/j.procs.2016.08.017

Rekimoto, J., & Ayatsuka, Y. (2000). CyberCode: designing augmented reality environments with visual tags. In *Proceedings of DARE 2000 on Designing augmented reality environments* (pp. 1-10). ACM. doi:10.1145/354666.354667

Rodriguez, C. O. (2012). MOOCs and the AI-Stanford like courses: Two successful and distinct course formats for massive open online courses. *European Journal of Open, Distance and E-Learning, 15*(2).

Rosenberg, M. J. (2001). *E-learning: Strategies for delivering knowledge in the digital age* (Vol. 3). New York: McGraw-Hill.

Sabah, N. M. (2016). Exploring students awareness and perceptions: Influencing factors and individual differences driving m-learning adoption. *Computers in Human Behavior*, *65*, 522–533. doi:10.1016/j.chb.2016.09.009

Seely Brown, J. (2008). Open education, the long tail, and learning 2.0. *EDUCAUSE Review*, *43*(1), 16–20.

Sharp, V. F. (2001). *Computer education for teachers: Integrating technology into classroom teaching*. McGraw-Hill, Inc.

Specht, M. (2014). Design of Contextualised Mobile Learning Applications.Increasing Access, 61.

Stead, G. (2014). Open Formats for Mobile Learning. Increasing Access, 99.

Wagner, E. D. (2005). Enabling mobile learning. *EDUCAUSE Review*, *40*(3), 41–42.

Woolf, B. P. (2010). *Building intelligent interactive tutors: Student-centered strategies for revolutionizing e-learning*. Morgan Kaufmann.

Zhi-jun, Y. I. N. (2006). Reflection of Computer-Education Reform in University [J]. Theory and Practice of Education, 12, 021.

Zhou, F., Duh, H. B. L., & Billinghurst, M. (2008). Trends in augmented reality tracking, interaction and display: A review of ten years of ISMAR. In *Proceedings of the 7th IEEE/ACM International Symposium on Mixed and Augmented Reality* (pp. 193-202). IEEE Computer Society.

KEY TERMS AND DEFINITIONS

Artificial Intelligence: (1) A term that is used to describe the feature, function or characteristic of computer systems or machines that try to simulate human-thinking behavior or human intelligence. (2) A field of Computer Science, which is based on research studies or developments on providing intelligent systems simulating the human-thinking behavior or human intelligence.

Augmented Reality: Some kind of reality in which virtual world based elements – objects are added to the real world based environment.

Computer Education: A process of education including the subjects regarding to essentials of computer systems and its sub-elements like hardware and software technologies.

E-Learning: Learning activities, which are supported and performed via electronic media – sources, by enabling individuals to experience educational processes on anytime and anywhere.

Intelligent M-Learning: M-Learning supported with Artificial Intelligence based approaches, methods, or techniques.

Mixed Reality: The reality including elements – objects from both real world and virtual world based approaches.

M-Learning: A type of E-Learning done via mobile technologies including both software and hardware oriented factors.

Mobile Learning Environments: Software oriented platforms (especially over Web) that is used for M-Learning purposes.

Open Education: A process of education, which is accessible for all individuals removing place and time limitations.

Chapter 9
Learning in a Virtual Environment:
Implementation and Evaluation of a VR Math-Game

Christof Sternig
Graz University of Technology, Austria

Michael Spitzer
Graz University of Technology, Austria

Martin Ebner
Graz University of Technology, Austria

ABSTRACT

With the introduction of Google Cardboard, a combination of mobile devices, Virtual Reality (VR) and making was created. This "marriage" opened a wide range of possible, cheap Virtual Reality applications, which can be created and used by everyone. In this chapter, the potential of combining making, gaming and education is demonstrated by evaluating an implemented math-game prototype in a school by pupils aged 12-13. The aim of the virtual reality game is to solve math exercises with increasing difficulty. The pupils were motivated and excited by immerging into the virtual world of the game to solve exercises and advance in the game. The results of the evaluation were very positive and showed the high motivational potential of combining making and game-based learning and its usage in schools as educational instrument.

DOI: 10.4018/978-1-5225-2110-5.ch009

INTRODUCTION

Since the dawn of virtual reality (VR) headsets there have been major improvements in the head-mounted display technology in recent years. With the evolution of VR headsets and the announcements of products as Oculus Rift, HTC Vive, Samsung Gear VR etc., more and more manufacturers will release their headsets to the market in the near future to make Virtual Reality available to the public. Mobile devices are complex computers which are gradually replacing Personal Computers (PCs) in all-day life (Gartner Press Release, 2016). With their high availability, the variety of embedded sensors, cameras and the availability of high computational power, mobile phones especially lie in the focus of researchers and developers. A new movement emerged in the last few years, namely, the Maker Movement (Schön, Ebner, & Kumar, 2014). Small computers as the Raspberry Pi or the Arduino offer new possibilities to invent and experiment with ideas and technologies. Makers meet at annual maker fairs to showcase inventions and experiments. When Google introduced Google Cardboard, a virtual reality headset can be made of a cardboard combined with a mobile phone. Therefore, making was combined with VR and mobile technology. The marriage of making, VR and a mobile device offers a great possibility to use concepts of digital game-based learning to create an immersive learning experience. Using VR techniques for learning, education or advanced training is an interesting research field with high potential.

The following research questions were examined:

RQ1: How could a VR device be integrated into a mathematical learning scenario?
RQ2: Which lessons could be learned while evaluating the learning scenario in a secondary school class?
RQ3: How could the learning experience be improved?

For this work, a simple virtual reality game has been implemented to evaluate the need for such educational games, supporting and motivating children in exercising math and the educational and motivational effects of combining making, gaming and learning. Finally, this chapter gives some recommendations how to integrate and augment school lessons with making and game-based learning.

BACKGROUND

In the following three sections, some background information is given about virtual reality (VR), game-based learning and the maker movement. The sections outline the key definitions used throughout this chapter.

Virtual Reality

Virtual Reality is a computer-generated illusion of the real world. The perfect Virtual Reality manipulates the human senses to be indistinguishable from the real world. Although, it is a utopian desire to create such a perfect virtual world (Stanković, 2015, p. 4). Common sense defines VR by head-mounted displays as Oculus Rift and/or data gloves, but this is not an adequate definition (Burdea, & Coiffet, 2003, p. 1). VR can also be accomplished by projectors in combination with Personal Computers (PCs) called the CAVE (Cruz-Neira, Sandin, & DeFanti, 1993). All sorts of computer games are also capable of creating a VR experience. The differences between VR and 3D movies are the possibilities to interact with the created world, change the state of the world and get a feedback (Stanković, 2015, p. 9). By having the possibility of interaction one can immerge into the virtual world. According to Burdea and Coiffet (2003, p. 3), interaction and immersion are two of three key features of VR. The third feature is imagination. Virtual Reality is often used to simulate real world processes. To restrict the parameters of the simulation to fully map a real world process, without breaking the simulation, is a difficult task which has to be solved by VR-developers and their imagination. Another important definition with respect to Virtual Reality is the Virtual Environment (VE). "VEs provide the illusion of presence in a place different from one's current physical surrounding." (Stanković, 2015, p. 10). Stanković (2015) differs between four different types of VEs.

- **Single User, Text Only (2D):** i.e. text adventures.
- **Single User, Realistic 3D:** AAA games (video games of high quality and produced with high development budgets are called "triple-A"-games) on modern console generations or the PC.
- **Multi User, Text Only (2D):** Social networks like Facebook.
- **Multi User, Realistic 3D:** Massively Multiplayer Online (MMO) games or general purpose VEs.

The rise of VR started again after Palmer Luckey started a successful Kickstarter campaign on August 1, 2012, for his VR headset Oculus Rift. Luckey showed the world that it is possible to create a high-end VR headset priced under $300. Facebook acquired the company in 2014 for 2 billion dollars and showed the world VR is, again, the next big thing (Kuusisto, 2015). Virtual Reality already was the "next big thing" back in the 1990s but Virtual Reality devices failed to enter the market (Stanković, 2015). The desire of buying a head-mounted display (HMD) to enter virtual worlds rose when affordable Virtual Reality headsets and applications where available. Although the available devices are using the newest technologies to create those virtual worlds, the used concepts behind all modern HMDs go back to 1968, when Sutherland (1969) introduced the first head-mounted display. It featured two cathode ray tube (CRT) displays in front of every eye and a tracking device to track movements of the head.

In 2014 Google introduced Google Cardboard at its Google I/O conference. It combines a mobile phone and a cardboard construction to create an HMD. With the high availability of mobile phones in the public nowadays, immerging into a virtual world is possible for everyone simply by owning a mobile phone. The prototype implemented and described in this work is developed in Java for Android devices and uses Google Cardboard to create the needed headset. This artefact enables pupils to immerge into the created virtual world to learn math. Considering the previous definitions, the game implements a single user, 3D Virtual Environment.

Video Games in Education

The Video game industry has steadily grown over the years. With the evolution of games and game mechanics, realistic graphics and sounds, more and more people are playing video games. The Entertainment Software Association (ESA) releases a yearly report, showing essential facts about the computer and video game industry. The current report points out that the average American gamer is 35 years old (Entertainment Software Association [ESA], 2015), in 2005 the average gamer's age was 30 (ESA, 2005). 51% of American households own a dedicated game console and 42% are playing video games on a regularly basis (three hours or more a week) (ESA, 2015). With this statistic increasing on a yearly basis, using video games for educational purposes gets an interesting industrial factor. Kapp (2012) reviews researches and studies about the effectiveness of games and game elements used in learning contexts and he concludes, that learners can benefit from such technologies when implemented and presented in the right way.

The following sections explain some key terms as digital game-based learning and gamification and their use in learning applications.

(Digital) Game-Based Learning

The term digital game-based learning (DGBL) describes the process of learning, while playing a computer game (Prensky, 2007). It can be seen as a specialization of the umbrella term game-based learning (GBL) which describes the process of learning by playing a game. In terms of GBL any game, i.e. board games or card games, match the definition. This work focuses on DGBL, therefore the following definitions concentrate on learning by playing video games.

Since the creation of the first video games in the 70s and 80s, the idea of learning by playing computer games was present. With the evolution of the game market, from a small industry for young men interested in technology, to a billion dollar industry, more and more studies about learning in combination with video games emerged. From the early 2000s the two terms 'Serious Games' and 'Digital Game-Based Learning' replaced the term 'Edutainment', short for education and entertainment (Breuer, 2011).

The term serious game, defined by Abt in 1971, describes all games with an educational purpose without the intention to be played only for amusement (Abt, 1987). Digital game based learning should feel like playing a computer game without recognizing the learning effect or discovering the content to be taught by the game (Prensky, 2007). It is "any marriage of educational content and computer games" (Prensky, 2007, p. 145). Breuer (2011) points out, that DGBL and serious games as defined before, have common elements as shown in the following enumeration, but serious games go beyond, when for example being used as distraction for painful therapies. Although studies and researches differentiate between DGBL and serious games, the aim of both concepts is to use games for an educating purpose going beyond mere entertainment. Breuer (2011) lists six common features of DGBL and serious games:

- **Interactivity:** Learning by doing and experimenting.
- **Multimedia-Based:** Visualizing/preparing content and feedback by using 3D models, audio etc.
- **Involvement:** The game should be fully engaging to keep the player from distractions.
- **Challenge:** Increasing difficulty but beginner friendly, the game should always challenge the individual skills to keep players motivated.
- **Reward:** Rewards and feedback of progression should push self-efficacy and motivation.
- **Social Experience:** Providing communication channels to connect players.

Considered as sets, DGBL is a proper subset of serious games and game-based learning and further of entertainment education. Serious games, e-learning as whole and entertainment education have common properties but differentiate in their base definitions. Figure 1 shows the different sets and their connections (Breuer, 2011).

Figure 1. The different learning approaches displayed as sets and their connections (Breuer, 2011)

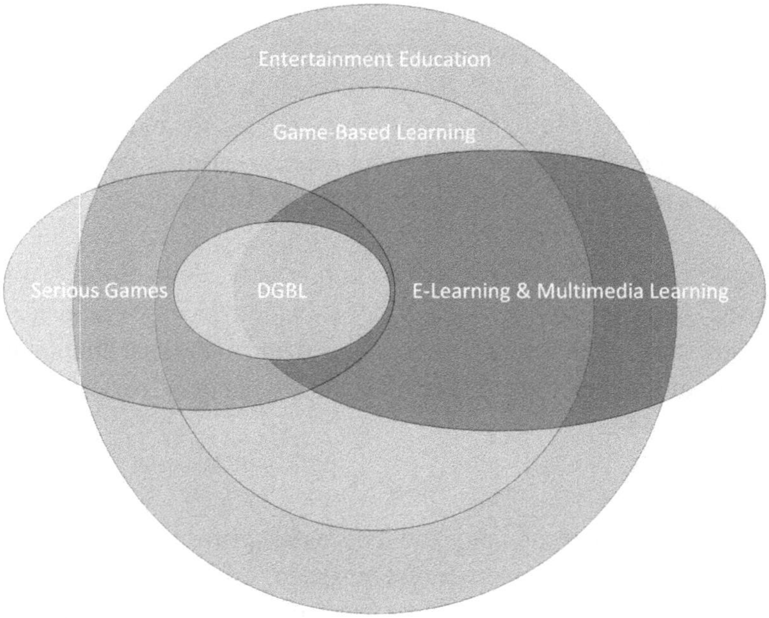

Gamification

According to Deterding, Dixon Khaled and Nacke (2011) gamification has been a contested term since its first appearance in 2008. There have been several definitions of gamification. With the lack of academic attempts to define gamification, Deterding mentions two ideas. Current usages of the word are fluctuating between: "The first is the increasing adoption, institutionalization and ubiquity of (video) games in everyday life" (Deterding et al., 2011, pp. 9-10). The second idea is the usage of game design and game elements in non-game contexts or products, adding enjoyment which in turn is leading to engagement, motivation and duration. Deterding et al. (2011) define gamification as "the use of game design elements in non-game contexts" (p. 10) and thus favors the second idea stated before.

Kapp (2012) defines gamification as the usage of "game-based mechanics, aesthetics and game thinking to engage people, motivate action, promote learning, and solve problems" (p. 17). This definition of gamification will be used when the term is used within this work. It fits the implemented prototype and its methods to convey the learning content efficiently. The next section will discuss different types of gamification and sample applications to get a better understanding how gamification can be used and implemented.

Applications

Implementing learning applications using game elements, gamifying existing systems and contents or designing a digital game-based learning application from scratch is a challenging task. Kapp, Blair and Mesh (2014) define the term Interactive Learning Event (ILE) to combine games, gamification and simulation in one term. To create a successful ILE, Kapp et al. (2014) list four important points to keep in mind:

- Game-based (fun) elements and instructional, non-entertaining elements should co-exist and grow together to keep the right balance.
- Interactivity ensures player engagement which results in learning more and keeping the knowledge for a longer time.
- Create a good story. The better the story and the storytelling, the more the player is engaged. Story elements and learning goals should be linked for a better learning experience.
- Proper testing needs to be done in order to evaluate the ILE and its elements.

For the simple implemented prototype described in this work, three of the four previous mentioned principles (game-based elements, interactivity and the evaluation) have been considered. Creating a good story and implementing story elements was not considered due to the first attempt but should be implemented in later versions of the prototype to keep players motivated and to achieve a better engagement.

Making and Education

With today's availability of cheap technology and tools as 3D printers, Personal Computers (PCs), tiny computers as the Raspberry Pi or the Internet of Things (IoT), inventing, creating and making can be done by everyone. Magazines as 'Make', FabLabs, conferences and fairs made 'Making' popular and more and more people are joining this movement. The maker movement stands for creating and developing new things using the aforementioned technologies. This can happen in fabrication labs (FabLabs), Makerspaces or on the desktop at home (Schön, Ebner, & Kumar, 2014). This section outlines the history of the maker movement and shows how making and education can be combined to create a new learning experience, not only for children, but students and adults.

According to Libow and Stager (2013), Seymour Papert can be seen as the father of today's maker movement. He wanted to revolutionize education by using modern technologies as computers not only for problem solving, as they are and have been used, but even more for creating some action. In the paper "Twenty Things to Do with a Computer" these aforementioned actions as composing music, programming, etc. are described and shown (Solomon & Papert, 1971). But Papert not only showed how to use technology for better educational methods, he was a maker himself. He invented the programming language Logo to teach children the concepts of programming. He also worked on the first programmable robotic construction kits for LEGO (Libow & Stager, 2013). He coined the term 'Constructionism' which means that learners gain knowledge by using tools to realize their ideas (Papert, 1986).

With the invention of computers, the desire of using them to teach pupils was evident. But the first computers were big machines and programming was done by experts. Adults, apart from experts, did not know how to program these complex computers, so how could it be taught to children? With the previously mentioned invention of Logo by Seymour Papert, things changed and schools used Logo to teach programing and mathematics to pupils. The same applies to the maker movement. As stated before, FabLabs teach how to use technology to create new tools, technology or products. Policy recognized the importance of making. Curricula of schools were expanded to teach these new working skills, but eventually, in 1999 this skill-based approach was abandoned due to the rapid evolution of technology. The aim is to teach technology and how it works rather than teaching skills how to work with specific items (Blikstein & Krannich, 2013).

Nevertheless, Halverson and Sheridan (2014) point out that "learning in making is, emphatically, not interchangeable with schooling" (p. 498). With the invention of FabLabs and adaptations for schools, the learning focus is placed on "principles of engineering, robotics, and design" (p. 499). Researches show how to rebuild a classroom to turn it into a makerspace, to enable schoolkids experimenting with tools and technology. There has not been found a consensus yet, how making should fit into existing curricula or if it should replace those curricula. How tests and grading should look like as making could not only combine the learning of different skills as mathematics, physics etc., but also leadership (Thompson, 2014). Makers fear that institutionalizing making through school programs "will quash the emergence, creativity, innovation and entrepreneurial spirit that are hallmarks of the "maker revolution"" (Halverson & Sheridan, 2014, p. 500). According to Halverson and Sheridan (2014) "the great promise of the maker movement in education is to democratize access to the discourses of power that accompany becoming a producer of artifacts, especially when those artifacts use twenty-first-century technologies." (p. 500). Learning through making has the potential to reach the institutional and policy goals for learning in science, technology, engineering and mathematics (Halverson & Sheridan, 2014), so-called STEM education.

Closely related to making is the concept of Project-Based Learning (PBL). It was introduced by John Dewey in the 1890s. He noticed the importance to integrate real-world problems and situations with the contents taught in school. Project-Based Learning evolved and developed over the years and there exist many definitions and different interpretations of the term (Habók & Nagy, 2016). Markham, Lamer and Ravitz (2003) define PBL as "a systematic teaching method that engages students in learning knowledge and skills through an extended inquiry process structured around complex, authentic questions and carefully designed products and tasks" (p. 4). PBL has become more popular but faces challenges as how to integrate it into school lessons and how teachers and students should handle this form of schooling. Studies showed that Project-Based learning results in higher engagement of students or pupils and also teaching how to plan and organize a project and communicate with others about aspects of the project (Hwang & Kim, 2006). Although, Project-Based Learning combines many positive aspects of teaching and learning, it is not suitable for learning basic skills. These basic skills or knowledge as math, writing, reading, etc. are required beforehand to accomplish the PBL project (Markham et al., 2003).

This work uses a kind of an inverse approach to PBL. The described math game was implemented and evaluated in school. After the evaluation, a tinkering project was realized with the evaluated children to create their own cardboard headsets to play the prototype on their own. After the evaluation the pupils were highly motivated and everyone managed to realize the complex tinkering project. Figure 2 shows a final cardboard headset, work in progress and a pupil who is crafting his cardboard headset.

MAIN FOCUS OF THE CHAPTER

Nowadays mobile devices are gradually replacing PCs or Laptops. The increasing computational power and their available embedded sensors open up many possibilities for new applications. Especially smartphones lie in the focus of developers due to their high availability in public. Children are growing up with these technologies and are therefore used to mobile devices. Children know how to interact with them and how to use them for surfing the internet, messaging, gaming, etc. (Grimus & Ebner, 2014). When Google introduced Google Cardboard at its Google I/O conference in 2014 they managed to combine the Maker Movement with mobile phones, to create a cheap virtual reality headset. Figure 3 shows the cardboard headset used during the evaluation of the prototype.

Figure 2. The image shows three pictures of a pupil crafting his cardboard headset

Figure 3. An image of the cardboard headset used to evaluate the prototype

As stated in the introduction, the aim of this work was to evaluate the need for educational games used to teach content to young pupils. Learning basic math operations as multiplication and addition need a lot of repetition to internalize them. The idea was to create an educational math game to test a game-based approach for learning and repeating these operations. This app was developed based on several years of experience in designing mathematical learning applications (Ebner, 2015; Ebner, Schön, Taraghi, & Steyrer, 2014). To excite the pupils and keep them motivated to play the game, Google Cardboard was used to create a mobile, virtual reality game. By testing the game, the pupils should also be motivated to create their own headsets to play the game, and, therefore, learn at home.

There is no other mathematical learning VR game available for google cardboard yet (Jan, 2016) hence his concept was tested from scratch. The next section describes the game mechanics, game items and how to play the game.

Game Mechanics

The game mechanics section is divided in three main parts. The navigation mechanics of the game are described in the first part of this section. The different game items and the level structure are outlined in the second part. The third part describes the gameplay, how to solve exercises, finish the game and the possible "Game Over"-states.

Navigation

An easy navigation and good controls are key elements of successful games. Sophisticated control mechanics can decrease the overall gaming experience of the player. Due to the lack of input devices, a very simple navigation scheme was implemented within this prototype. When a new level is loaded, the player's avatar is placed in the center of the level platform. The camera used to render the scene sits atop the avatar as illustrated in Figure 4. The avatar automatically moves along the viewing direction with constant speed. By moving around the own axis, the player changes the viewing direction and thus, the moving direction of the avatar in the virtual world.

Figure 4. A simple cone model representing the player's avatar. The camera, visualized as the frustum of a pyramid, is placed on top of this cone

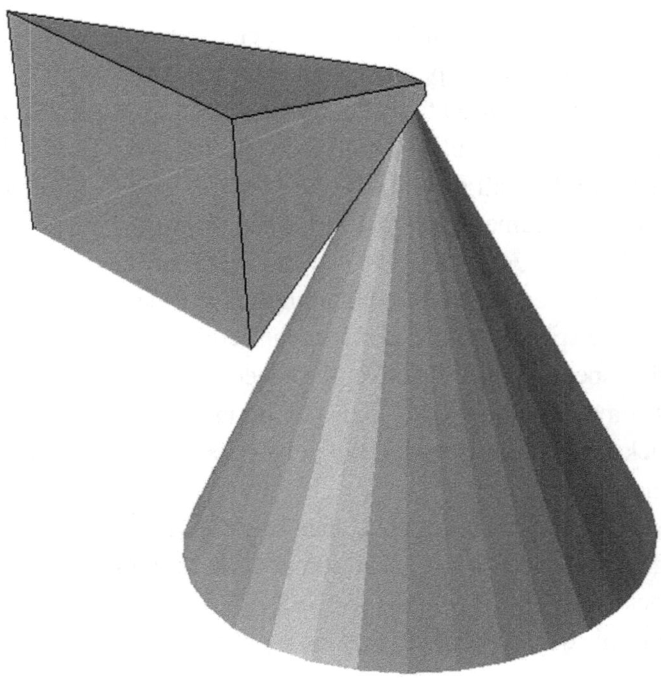

Due to the moving avatar, suggesting the player's brain a motion although the player himself is not moving, causing a discrepancy. This discrepancy, called visually induced motion sickness, "typically occurs in the absence of real physical motion, thus, motion sickness in simulators, cinemas, video-games, or virtual reality" (Keshavarz & Hecht, 2014, p. 521) causes symptoms like nausea and dizziness. There have been researches about motion sickness prevention i.e. using pleasant music as countermeasure against visually induced motion sickness (Keshavarz & Hecht, 2014). The induced symptoms could lead to difficulties in keeping the balance while standing and playing the game. To keep the prototype's implementation simple, no visually induced motion thickness prevention methods have been implemented. The simplest prevention of losing the balance is to sit down on a revolving chair and do the navigation by rotating around the chair's axis.

The evaluation of the game showed that neither motion sickness nor symptoms like nausea or dizziness were a problem for the children testing the game. The previous simple prevention method of using a revolving chair was not needed by any of the pupils.

Game Items and Level Structure

Every level of the game consists of three key elements; the avatar, the floor and collectible items. All game objects are represented by simple geometric forms to keep the rendering simple and fast and therefore, prevent frame rate issues.

The avatar of the player, a simple cone illustrated in Figure 4, represents the player in the virtual world. The player can navigate the avatar by changing the view direction as described before, to collect the collectible items and solve exercises. The avatar is placed onto a platform, the floor, when the game level is loaded. The floor is another key element and is represented by a scaled cube. It bounds the area and therefore the virtual world, in which the player can navigate his avatar. Collectible items are divided into two types, the digits and pickups, respectively.

Digits are represented by their corresponding 3D objects and are collected to solve the current exercise. The pickups are green, yellow and red colored cubes with a symbol on each side. The color visualizes the possible effect of the pickup, green stands for a positive, yellow for a neutral and red for an undesired effect. The symbols are used to distinguish between pickups of the same color resulting in different effects. Pickups and digits are placed onto the floor when the level is loaded.

The following enumeration lists all pickups available in the game with images and descriptions of their effects; Figure 5 shows the game representation of the pickups.

- **Time Bonus:** Adds thirty seconds to the remaining game time.
- **Speed Up:** Doubles the speed of the avatar for ten seconds. If a second Speed-Up-pickup is collected, while the first speed-up is still active, the speed is doubled again.
- **Jump:** The avatar jumps up into the air. This can be used to get a better view onto the platform or to jump over pickups and digits which the player does not want to collect. Navigation is still possible while airborne.
- **Invert Controls:** The controls are inverted, changing the viewing direction to the right results in the avatar moving left and vice versa. Additionally, the moving direction of the avatar changes from forward to backward. The effect lasts five seconds. This is rather short but as testing showed, inversion of the controls quickly leads to dizziness or motion sickness.

Figure 5. The four different pickups collectible in the game. In the top left corner the Time-Bonus-pickup, top right the Invert-Controls-pickup, bottom left the Speed-Up-pickup and bottom right the Jump-pickup

With the description of the navigation mechanics, the key game items and the structure of the levels finished, the gameplay is described in detail in the next section.

Gameplay

This section explains in detail how to solve exercises and collect items to accomplish a level. When the level is loaded, the avatar is placed in the center of the platform and the collectible items are distributed randomly over the gaming platform. More precisely, ten digits, nine possibly wrong results and the correct results, and ten pickups are created. The generation of the pickups is completely random; there are no restrictions or rules the pickup creation follows. It is possible that seven time bonuses and three speed-up pickups are created, but no control-inversion or jump-pickup exists.

The generation of the exercises depends on the chosen settings from the main menu. Possible settings are only additions, only multiplications or a mix of these two operations. The generation of the exercises follows the rules of the regular expression shown in listing 1. Before the level starts, five exercises are generated. The player has to solve them sequentially to advance to the next level.

Listing 1: The regular expression used to generate the exercises for each level

```
[1-9]? [0-9] ((+[1-9] ? [0-9]) | (*[1-9])
```

When the level is loaded successfully and all items have been generated, the avatar starts moving and the clearing time for the level starts decreasing. The clearing times vary from level to level and are chosen tighter for higher levels. The remaining time can be increased by solving exercises or collecting time-pickups. Collectible items can be "picked" up by bumping the avatar into the desired item. When the avatar collides with the object, a sound effect is played to indicate the player that the item was successfully collected. Three different scenarios are possible when collecting items: a pickup is collected, the player picked the wrong digit or the player solved the current exercise by choosing the correct digit. The following enumeration explains every scenario in detail:

1. **A Pickup is Collected:** The simplest of all scenarios, the pickup is applied. The cube representing the pickup is destroyed. The possible effects of pickups are explained in the previous section.
2. **The wrong Result is Collected:** A sound effect and a message displayed on the screen signaling an incorrect calculation. The player's remaining time is decreased by five seconds, the overall score decreases by ten points. The collected digit is destroyed and a new one is generated randomly, the current exercise does not change.
3. **The Correct Result is Collected:** As with the wrong result, a sound effect and a message indicate that the player solved the current calculation. The remaining time is increased by five seconds and additionally, 50 points are added to the player's score. The next calculation is displayed on the screen and a digit representing the correct result is generated and placed randomly on the game platform.

While looping through the above scenarios, the player may solve five exercises within the time limit of the level. If so, the level is completed, the remaining time is added to the player's high score and the next level, with increased difficulty resulting in a tighter time limit and more challenging calculations, is loaded. The game ends if one of the following scenarios occurs.

1. **Time Limit is Exceeded:** The remaining time reaches zero and the player was not able to solve five calculations.
2. **Falling Off the Platform:** If the avatar is navigated beyond the borders of the game floor, the avatar falls into the void and the game ends immediately, even though time is still remaining.
3. **The Player Completes All Challenges:** If the player solves 25 exercises within the time limits of the five levels, the player has completed the game and wins.

Regardless of which ending scenario occurred, the player's current score can be saved and viewed in the high score list. Even though a non-winning scenario is reached, the points, won by previous mastered levels or correct calculations, can be saved. Table 1 shows the time limits and number ranges for the generated calculations. Figure 6 shows a screenshot of the game with descriptions of the User

Table 1. The level properties for the five levels of the prototype

Number	Name	Time Limit	Addition Range	Multiplication Tables
1	Plaza	60	0 – 9	3
2	City	55	0 – 19	5
3	Jungle	50	0 – 30	6
4	Desert	45	0 – 50	8
5	Mountain	40	0 – 99	7

Figure 6. The game screen from the player's point of view

Interface (UI) and the visible game items. Figure 7 shows the virtual world from different angles from the user's point of view.

In Figure 6, the green circles indicate the digits; the yellow circles are showing pickup items. The bar in the center of the screen represents the remaining time; the current calculation is displayed in the upper half. The green disc at the bottom displays a small fraction of the avatar

The implemented math game covers five of the six principles of DGBL defined by Breuer (2011) in the background section of this chapter. The player can interact with the virtual game world (Interactivity

Figure 7. The game screen from the player's point of view from different angles

and Multimedia). The virtual reality effect creates high engagement and the feeling of presence in the world keeps the player focused on the current exercise (Involvement). The difficulty is increasing with every level by generating harder exercises and a tighter time limit (Challenge). Different themed levels and a simple scoring system should keep players motivated to play on (Reward). According to Kapp et al. (2014) the implemented prototype can be classified as a testing game, where previous learned mathematical skills are tested via solving exercises. A learning process is created by repeatedly playing the game and getting better and faster in solving the exercises.

This part of the chapter described the key game elements and the gameplay in detail. The next section examines the evaluation of the prototype by pupils in a secondary school.

Prototype Evaluation

This section describes the evaluation process and its results in detail. It is divided into two main parts. The first part explains the evaluation process and setup, the second part shows the results of the evaluation and discusses issues and observations made during the evaluation.

Evaluation Setup

The evaluation took place on the 18th of December 2015 in the "Neuen Mittelschule (NMS) Fröbel" (secondary school) and lasted about two hours. The class, taught by Mrs. Silvana Aureli, counts 15 pupils, one child was missing on the evaluation day. The proportion of boys and girls was almost equal; the children were aged 12 to 13. After a short introduction to the game, the evaluation and their tasks, the pupils were split into groups of three and one group of two. The evaluation took place in a separate room with only the current testing group present. Every group ran through a four-step-process to test and evaluate the game.

1. **Introduction:** The evaluating group of pupils got a brief introduction about the game and its mechanics. Neither of the children in the group had played the game before nor had experience with VR-headsets or VR games. Therefore, the children got instructions on how to use the cardboard VR headset used for the evaluation, how to navigate through the virtual world and what the aim of the game is.
2. **First Play-Through:** After introducing the game and the VR headset, the first child of the group started playing the game for the first time. Every child had the chance to play until it failed or completed the game. Each child of the testing group played the game once and after that first play-through, possible questions were answered, some advice was given to avoid common mistakes made in the first iteration of the test.
3. **Second Play-Through:** The second play-through did not differ from the first; the children played the game again, considering the tips given after the first play-through.
4. **Evaluation:** The evaluation of the game consisted of five statements about the game. The current evaluation group had to rate the statements by marking them with one of four differently tempered smileys shown in Figure 8. One statement had to be rated with one smiley by the whole group. This method was used to gain some extra information about the feelings of the children during their discussion on how to rate the current statement. This approach has already been used in previous studies (Spitzer & Ebner, 2015).

The evaluation process lasted, as stated before, approximately two hours. Each of the five evaluation groups ran through the above described evaluation process and rated the five evaluation statements. Two months after the evaluation took place, the pupils created their own VR-headsets in six school lessons. The construction manual needed to tinker the headsets and the bi-convex lenses to create the stereoscopic vision of the 3D world created by the prototype, were offered by the Technical University of Graz. According to the teachers supervising the tinkering lessons, all pupils enjoyed creating their own personalized headsets.

The results of the evaluation and the ratings of the statements are presented in the upcoming section.

Evaluation Results

As described in point 4 of the evaluation process, each evaluation group rated five statements by marking them with smileys. The statements were read out loud with additional explanations about each statement's intentions, to be sure every child understands the meaning of every statement, being able to rate them properly. To get a measureable result, every type of smiley is assigned a score from -2 to +2. The scores are added up group by group with the sum being the final score of a statement.

The meanings and scores of the different smileys can be found in Figure 9. The following sections describe each statement, the group-specific answers and a brief interpretation of each rating. In parentheses below each statement, the original German statement, as used during the evaluation, is shown. Observations made during the play-through-steps of the evaluation process and the discussions, on how to rate the statements, respectively, are described and interpreted afterwards.

Figure 8. The image shows the five statements used for evaluating the prototype and the smileys used to rate each statement

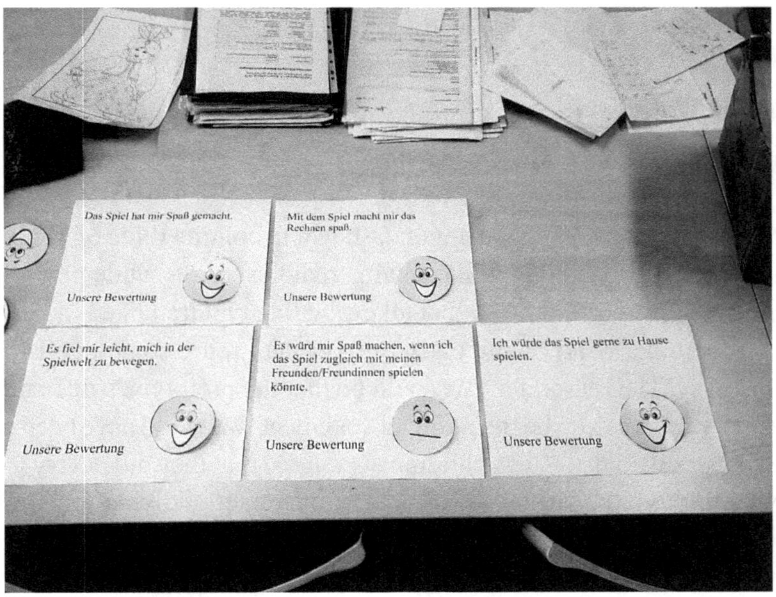

Figure 9. The meaning of the smileys used to rate the statements and the points of each smiley used to interpret the results

Smileys and their Meaning		**Points**
	The evaluating group totally agrees with the statement	2
	The group was not fully convinced but positive about the statement	1
	The children were sceptical about the statement, but did not completely disagree	-1
	The group totally disagrees with the statement	-2

Statement 1

The game was fun.

(Das Spiel hat mir Spaß gemacht.)

The second column in Table 2 shows the final results for statement one. The groups all agreed to the first statement and rated it with the highest possible score. The children liked to immerge into the virtual world and navigate through it. It was an overwhelming experience for the children to be part of a virtual world. The fact that no child of the evaluating groups had experience with Virtual Reality certainly had a big influence on the rating.

Statement 2

Calculating was fun using the game.

(Mit dem Spiel hat mir das Rechnen Spaß gemacht.)

Statement two had the same rating as statement one, shown in column three of Table 2. The intention of this statement was to check, if the children like solving math exercises while playing a computer game. On the one hand the high rating could mean that all evaluating children liked the game-based approach of learning math, on the other hand, while discussing in the group, some pupils pointed out that the calculations were too easy. The reason for this could be that the prototype's difficulty was designed for pupils attending primary school. Maybe the rating of statement two could have been worse by generating calculations with a higher difficulty. A few children also stated that they did not try to solve the exercises but they liked moving through the virtual world. Again, the effect of diving into a virtual world via the VR headset could have influenced the rating of this statement.

Statement 3

It was easy to navigate through the virtual world.

(Es fiel mir leicht, mich in der Spielwelt zu bewegen.)

A good navigation mechanic is an important requirement to create a good user experience. The intention of statement three was to evaluate the implemented control mechanics of the game. With a final score of seven points, all groups rated the game's navigation positive, although three groups did not rate the statement with the full score. Playing a computer game always implies the use of an input device as a game controller, a keyboard or fingers touching the screen of a mobile device to control and navigate the game's avatar. Since the evaluating pupils were not used to Virtual Reality games and VR-Headsets, respectively, navigating by moving the whole body took some time until the children got used to it. All evaluating children managed to navigate through the world and play the game properly on their second play-through therefore, the navigation mechanics can be considered a success. Column "Points S3" in Table 2 shows the ratings of statement three.

Statement 4

It would be fun, playing the game together with my friends.

(Es würde mir Spaß machen, wenn ich das Spiel mit meinen Freunden/Freundinnen spielen könnte.)

The aim of statement four was to evaluate the need for a multi-player mode. The final score of seven points (shown in Table 2, column four) indicates that multiple players playing the game together in the same virtual world would be a good extension to the current prototype of the game. Surprisingly, one group did not like the idea of playing against other players and rated the statement with -1 points, but a multi-player mode being a good idea was the general opinion. By integrating a multi-player mode the prototype would cover the sixth DGBL principle defined by Breuer (2016), because it would add a social component to the game.

Statement 5

I would play the game at home.

(Ich würde das Spiel gerne zu Hause spielen.)

The intention of this statement was to find out, if the evaluating group members would download the game and play it at home with their own VR-headset. All groups top-rated this statement which indicates the success of the combination of game-based learning in a virtual world combined with making. The virtual reality experience motivated the children to tinker their own VR-Headset, which was done in six lessons. For the sake of completeness, the last column of Table 2 shows the final results of statement five.

Following the results of the evaluation, the next part of the evaluation section describes some observations made during the evaluation process and gives some explanations. The list follows no specific order.

Table 2. The ratings for every statement by group and the final results for every statement. The letter S in the column names stands for statement

Group No.	Points S1	Points S2	Points S3	Points S4	Points S5
1	2	2	1	-1	2
2	2	2	2	2	2
3	2	2	1	2	2
4	2	2	2	2	2
5	2	2	1	2	2
Result	10	10	7	7	10

- **The Testing Pupils Did Not See the Current Calculation to Solve:** This observation was mostly made during the first play-through. When the evaluating person was told to look up a little to see the calculation, the person lifted the head which in turn moves the game's camera upward, causing the calculation to move upward as well. The eyes have to be moved upwards to see the calculation without changing the camera angle in the game. This could take some time to get used to it.
- **Bending the Whole Body or Tilting the Head to the Left or Right was Used to Navigate the Avatar:** Due to the new way of navigation without input devices and only a brief introduction to the new navigation mechanics, some pupils needed some extra time to get used to navigation of the avatar and tried different motions to control the avatar. After short instructions during playing the game, the pupils internalized the navigation process.
- **The First Play-Through During the Evaluation Process was Worse Than the Second Play-Through:** The first play-through sometimes lasted only a few seconds because the avatar fell off the game floor due to bad navigation. If this was the case, the child was instructed again and could start the play-through one more time, which resulted in a far better first play-through. The cause of this problem is the start of the game. The avatar immediately moves forward after the game started. Until the evaluating person has adjusted the VR-Headset the avatar may have moved close to one edge of the platform.
- **Motion Sickness Did Not Occur:** This was a very positive observation because previous, adult test persons could not play the game for a long time due to motion sickness.
- **During the Evaluation-Step There was Less Discussion Between the Group Members:** The intention of forming groups to evaluate the game was, to observe the discussion during the rating of the statements. Unfortunately, there was less discussion between the group members, almost every group had some kind of leader, who took the rating-smileys and distributed them, with the other group members agreeing in most of the time.

The results of the evaluation have been very positive and they suggest that there is a high potential in game-based learning combined with making when used for educational purpose. The motivation to practice and play with the game could be increased if pupils create their own practice devices (cardboards VR-headsets), because they are proud what they made.

The next section describes issues which should be tackled, when extending the prototype and outlines potential improvements to the current prototype.

SOLUTIONS AND RECOMMENDATIONS

During the testing and evaluation of the prototype some issues were identified. This section is split into two subsections with the first discussing the issues of the prototype's gameplay and the second listing problems which occurred during the evaluation process.

Gameplay

The game design and the navigation mechanics were held simple to create a beginner friendly, Virtual Reality experience. Navigating by changing the viewing direction is an easy way to control the avatar, but was hard to explain, while introducing the game's mechanics to the pupils. After a first hands-on and a few minutes with the game, all pupils got used to it. The game objects could clearly be recognized and the children understood how to solve calculations and leverage the pickups. Changing the level after three correct solved exercises also raised the fun factor of the game and motivated the pupils in the group to solve more exercises than the previous group member to advance to the next level.

Some pupils pointed out that calculations were too easy to solve. A better Random Number Generator (RNG) has to be implemented to solve this problem to get a better distribution of the exercises. Another problem is a drop of the frame rate, which occurs when too much objects have to be rendered. This could also be prevented by a more advanced RNG, distributing the created objects equally over the gaming platform. Nevertheless, a better distribution will not completely fix this problem. If the avatar is positioned near a boundary of the game floor and the player changes the viewing direction to get back to the middle of the platform, the player gains a big field of view, containing many game objects, which, in turn, have to be rendered and a frame rate drop could occur. The engine has to be extended by more advanced rendering patterns to get a steady frame rate in all possible situations. Positively, the frame rate problem either did not occur during the evaluation nor it did not disturb the game experience for the pupils although it occurred. If the prototype is extended or revised, these problems have to be addressed.

The current calculation to solve, displayed in the upper area of the screen, was not seen by all pupils at once. A better way to display text on the screen should be found to make the players see the calculations immediately without help.

The combination of the simple game mechanics and the new experience of emerging into a virtual world via a VR-Headset was an exciting experience for the pupils. There were no complaints about the simplistic graphics or missing realism, which is rather standard in AAA-games on the new console generations or the PC nowadays.

As suggested by Kapp et al. (2014), the prototype should be extended to offer a story and proper story-telling to provide better engagement and long-term motivation. Adding new challenges as boss-fights, a multi-player mode, enabling comparison with other human players, may increase engagement as well as adding connection to social media. Showing the learning progress to friends via social media could lead to engaging friends to replay the game and progress further in the game.

Evaluation

The four-step evaluation process was an adequate method to gain user feedback of the prototype. The evaluating group size of three pupils per group and one group of two was a good decision. The groups were big enough to have the attention of every group member while introducing the game, the headset and to deal with the limited time for the whole evaluation process.

A problem of multiple groups was the introduction of the game for every group. Future groups may have gotten a more detailed introduction due to mistakes made by previous groups. Also two play-throughs have not been planned, but after the first play-through of group one, a second play-through after a second briefing, seemed to be necessary and it showed that every group needed this extra briefing and a second play-through.

Evaluating the game via rating statements using smileys was very intuitive for the children. The intention of three group members having only one vote for a statement was to observe the pupils, while discussing how to rate the statement and gain some additional information about the thoughts and feelings of the children. Unfortunately, as stated in the evaluation chapter, every group had some kind of leader, distributing the smileys at first. The other group members agreed in most cases without having a discussion about the rating. Rating the statements by pupil and a little interview on the one hand could have resulted in some additional or more precise information but on the other hand would have exceeded the time limit for this evaluation.

Despite these shortcomings, the evaluation process itself was good enough to gain some vital information and results to build atop for improvements and extensions of the prototype. It clearly showed that the pupils are willing to use and play educational games and enjoyed the game-based learning approach. Table 3 summarizes the known issues and possible solutions as listed in the above text.

FUTURE RESEARCH DIRECTIONS

The evaluation of the prototype in school showed that there is a high potential in educational games. Children are used to play games and the combination of playing and learning is a great way to break up old structures of today's teaching. A more advanced prototype could be used for testing pupils or to submit results of homework. All pupils get the same exercises and play the game at home. Their results are submitted as homework. Another step would be the integration into lessons to make the game a

Table 3. A list of the issues and possible solutions identified during the evaluation and testing of the prototype

No.	Issue	Solution
1	Too easy exercises.	Add a more advanced random number generator.
2	Possible drops of the frame rate resulting in jerking and thus could result in motion sickness.	Implement a more advanced rendering algorithm including better Level Of Detail (LOD) mechanics.
3	Current exercise is difficult to see.	Implement a better User Interface (UI) and find better ways to position text elements on screen.
4	No story was added to the game.	Add a compelling story for a higher engagement level.
5	Common game elements are missing.	Add well-known game elements as a multi-player mode, enemies, boss fights, etc.
6	Introduction of the game to every evaluating group separately, was biased.	Create a script to follow when introducing the game.
7	No discussion during the rating of the statements.	Rating the statements by pupil including a short interview.

tool for the teachers to teach new content to pupils. For that purpose, a good feedback system to better evaluate and visualize the learning progress of every pupil has to be implemented. Also, further testing of the prototype has to be done to fix unknown flaws in the gameplay.

Using the prototype during school lessons and integrating such an educational medium into learning scenarios in school demands some initial training for teachers to use this new schooling method properly. It should be evaluated how such training for teaching staff should look like and if it demands explicit training lessons for teachers or if precise guidelines on how to integrate and use such media in school will be sufficient.

The aforementioned integration of story elements should be evaluated as well. How could adding a story be done and how will the story elements fit with the math content tested by the current prototype? How can challenges of the game mask mathematical calculations and integrate them into the gameplay to blend learning and playing without being recognized by the player?

CONCLUSION

The aim of this work was to evaluate the need for educational VR computer games in combination with making. Other researchers already investigated educational VR learning modules. They confirmed the potential of VR based applications to help students learn mathematical concepts.

Wang, Yang and Lian (2009) already used VR learning modules for engineering students. They transferred real-world engineering challenges to help students to better realize and recognize mathematical concepts. They visualized a projectile in 3D space to show the parabola flight curve following the quadratic equation of a falling object. Students could freeze the flight of the projectile to show the current parameters of the equation. The feedback of the students was very good. They believe that the VR modules are helping them to better understand mathematical concepts.

Kaufmann, Schmalstieg and Wagner (2000) build a VR application for mathematics and geometry education. They used an HMD which was used to display the 3D models. Students were able to manipulate the 3D geometry of the shown objects. The students liked the experience even though the performance and accuracy caused some difficulties.

These two examples showed the potential of VR learning applications which, among other things, inspired us to build the VR math game.

The game mechanics have been described in the previous sections of this work, the making-part consisted of tinkering a VR-Headset made of cardboard. The game was evaluated by 14 pupils with one headset and a testing device. After the evaluation, the children created their own VR-Headset with the help of their teachers to play virtual reality games. In six lessons the pupils created their own personalized virtual reality headsets. This follows the maker approach. The teachers mentioned the positive attitude of the pupils during the tinkering lessons and were amazed by the high motivated children, although it was not an easy task to tinker the cardboard headsets. The lessons were a big success and the pupils were happy with their headsets. Figure 10 shows examples of the headsets created by the pupils (RQ1).

Although the game was kept very simple in its presentation due to performance bottlenecks, resulting from mobile hardware and the simple rendering system, the pupils were excited to play the game. Emerging into the virtual world via the cardboard headset was an overwhelming experience, because none of the evaluating children have ever used a VR-Headset. The results of the evaluation have been positive throughout the evaluation process and none of the pupils felt motion sickness nor criticized the

Figure 10. Images of some headsets tinkered by the pupils, which evaluated the prototype

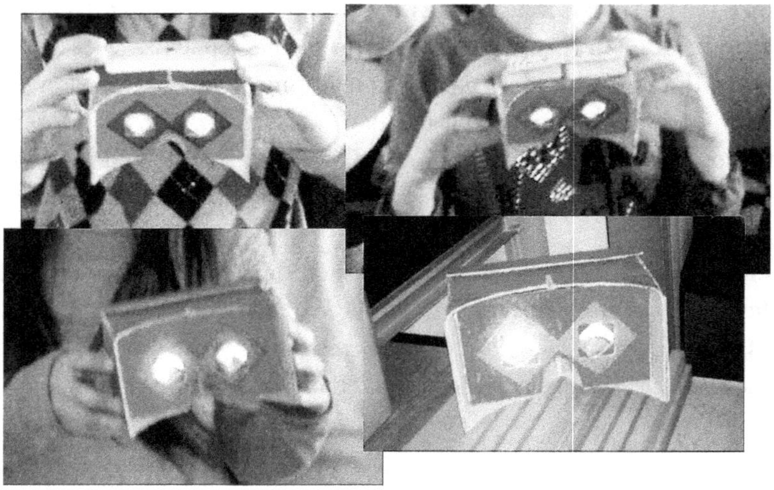

game or its mechanics. Being able to navigate within the virtual world needed some time, nevertheless, the children got used to it. The learning curve was steep and every child was able to solve exercises and therefore, was motivated to play on (RQ2).

The results of the evaluation clearly show that educational games in combination with making can be a great success. The prototype provided an extended learning experience for the pupils. Using their mathematical knowledge to advance in the game motivated the pupils to keep on playing and try again after failing to finish a level or the game. By using and playing this game, the process of repetition to tighten the mathematical skills of the pupils is enriched with a fun and motivational component, leading to long-term engagement. To improve the learning experience the mathematical content must be highly adapted to the level of education of the users. Additionally, a compelling story should be added and the user interface should be improved to keep the pupils motivated and to reduce teachers' introductions (RQ3).

The impact of a better looking, realistic 3D game using educational elements and integrating one or more new elements, mentioned in the recommendations in the last section should have an even greater impact and show great potential to be used in schools and for education of pupils. Maybe the positive and educational effects of learning by doing and game-based learning will revolutionize learning as we know it today.

REFERENCES

Abt, C. C. (1987). *Serious games*. Lanham, MD: University Press of America.

Blikstein, P., & Krannich, D. (2013). The makers' movement and FabLabs in education. In J. P. Hourcade, N. Sawhney, & E. Reardon (Eds.), *Proceedings of the 12th International Conference on Interaction Design and Children* (p. 613). Academic Press. doi:10.1145/2485760.2485884

Breuer, J. (2011). *Spielend lernen? Eine Bestandsaufnahme zum (Digital) Game-Based Learning*. Landesanstalt für Medien NRW. Retrieved May 4, 2016, from http://lfmpublikationen.lfm-nrw.de/index.php?view=product_detail&product_id=190

Burdea, G., & Coiffet, P. (2003). *Virtual reality technology*. Wiley. Retrieved April 16, 2016, from https://books.google.ca/books?id=0xWgPZbcz4AC

Cruz-Neira, C., Sandin, D. J., & DeFanti, T. A. (1993). Surround-screen projection-based virtual reality: the design and implementation of the CAVE. In M. C. Whitton (Ed.), *SIGGRAPH '93 Proceedings of the 20th annual conference on Computer graphics and interactive techniques* (pp. 135–142). New York: ACM. doi:10.1145/166117.166134

Deterding, S., Dixon, D., Khaled, R., & Nacke, L. (2011). From game design elements to gamefulness. In A. Lungmayr (Ed.), *Proceedings of the 15th International Academic MindTrek Conference Envisioning Future Media Environments* (pp. 9–15). New York, NY: ACM. doi:10.1145/2181037.2181040

Ebner, M. (2015). *Mobile Learning and Mathematics. Foundations, Design, and Case Studies*. New York: Routledge.

Ebner, M., Schön, M., Taraghi, B., & Steyrer, M. (2014). Teachers little helper: Multi-math-coach. *IADIS International Journal on WWW/Internet, 11*(3), 1-12.

Entertainment Software Association. (2005). *Essential Facts about the Computer and Video Game Industry*. Retrieved May 4, 2016, from http://www.tntg.org/documents/gamefacts.pdf

Entertainment Software Association. (2015). *Essential Facts about the Computer and Video Game Industry*. Retrieved May 4, 2016, from http://www.theesa.com/about-esa/esa-annual-report/

Gartner Press Release. (2016). *Worldwide Device Shipments to Grow 19 Percent in 2016, While End-User Spending to Decline for the First Time*. Retrieved 30 May, 2016, from http://www.gartner.com/newsroom/id/3187134

Grimus, M., & Ebner, M. (2014). Learning with Mobile Devices Perceptions of Students and Teachers at Lower Secondary Schools in Austria. In *Proceedings of World Conference on Educational Multimedia, Hypermedia and Telecommunications 2014* (pp. 1600-1609). Chesapeake, VA: AACE.

Habok, A., & Nagy, J. (2016). In-service teachers perceptions of project-based learning. *SpringerPlus, 5*(1), 83. doi:10.1186/s40064-016-1725-4 PMID:26844030

Halverson, E. R., & Sheridan, K. (2014). The Maker Movement in Education. *Harvard Educational Review, 84*(4), 495–504. doi:10.17763/haer.84.4.34j1g68140382063

Hwang, S. Y., & Kim, M. J. (2006). A comparison of problem-based learning and lecture-based learning in an adult health nursing course. *Nurse Education Today, 26*(4), 315–321. doi:10.1016/j.nedt.2005.11.002 PMID:16364510

Kapp, K. M. (2012). *The gamification of learning and instruction: Game-based methods and strategies for training and education. In Pfeiffer essential resources for training and HR professionals*. San Francisco, CA: Pfeiffer.

Kapp, K. M., Blair, L., & Mesch, R. (2014). *The gamification of learning and instruction fieldbook: Ideas into practice*. San Francisco, CA: Wiley.

Kaufmann, H., Schmalstieg, D., & Wagner, M. (2000). Construct3D: A virtual reality application for mathematics and geometry education. *Education and Information Technologies*, *5*(4), 263–276. doi:10.1023/A:1012049406877

Keshavarz, B., & Hecht, H. (2014). Pleasant music as a countermeasure against visually induced motion sickness. *Applied Ergonomics*, *45*(3), 521–527. doi:10.1016/j.apergo.2013.07.009 PMID:23957932

Kuusisto, F. (2015). Vr head-mounted displays. *XRDS: Crossroads. The ACM Magazine for Students*, *22*(1), 65. doi:10.1145/2837754

Libow Martinez, S., & Stager, G. (2013). *Invent to learn: Making, tinkering, and engineering in the classroom*. Torrance, CA: Constructing Modern Knowledge Press.

Markham, T., Larmer, J., & Ravitz, J. L. (2003). *Project based learning handbook: a guide to standards-focused project based learning for middle and high school teachers* (2nd ed.). Novato, CA: Buck Institute for Education.

Papert, S. (1986). *Constructionism: A New Opportunity for Elementary Science Education*. Academic Press.

Prensky, M. (2007). Digital game-based learning. St. Paul, MN: Paragon House.

Schön, S., Ebner, M., & Kumar, S. (2014). *The Maker Movement. Implications of new digital gadgets, fabrication tools and spaces for creative learning and teaching*. Retrieved May 6, 2016, from http://www.openeducationeuropa.eu/de/article/The-Maker-Movement.-Implications-of-new-digital-gadgets,-fabrication-tools-and-spaces-for-creative-learning-and-teaching

Solomon, C., & Papert, S. (1971). *Twenty Things to Do With a Computer*. Retrieved May 6, 2016, from http://www.stager.org/articles/twentythings.pdf

Spitzer, M., & Ebner, M. (2015). Collaborative Learning Through Drawing on iPads. In *Proceedings of World Conference on Educational Multimedia, Hypermedia and Telecommunications 2015* (pp. 633-642).Chesapeake, VA: AACE.

Stanković, S. (2015). Virtual Reality and Virtual Environments in 10 Lectures. In A. C. Bovik (Ed.), *Synthesis Lectures on Image, Video, and Multimedia Processing* (Vol. 8, pp. 1–197). Morgan & Claypool Publishers. doi:10.2200/S00671ED1V01Y201509IVM019

Sutherland, I. E. (1968). A head-mounted three dimensional display. In AFIPS (Ed.), *Proceedings of the December 9-11, 1968, fall joint computer conference, part I* (pp. 757–764). doi:10.1145/1476589.1476686

Thompson, G. (2014). The Maker Movement Conquers the Classroom. *THE Journal*. Retrieved May 6, 2016, from https://thejournal.com/Articles/2014/04/30/The-MakerMovement-Conquers-the-Classroom.aspx?Page=2

Wang, Y., Cui, S., Yang, Y., & Lian, J. A. (2009). Virtual reality mathematic learning module for engineering students. *Technology Interface Journal*, *10*(1), 1–10.

KEY TERMS AND DEFINITIONS

Avatar: A 3D model of a human, animal, thing, etc. representing the person in a virtual world enabling the person to interact with elements of the virtual world.

Digital Game-Based Learning: The approach of combining learning and teaching with playing computer games.

Gamification: Learning material is augmented with game-like elements to keep learners motivated and engaged.

Head-Mounted Display: A computer display worn like goggles to immerge into virtual worlds.

Immersion: Completely dive into a virtual world. A high immersion leads to a user's perception of being part of this virtual world.

Maker Movement: A trending movement consisting of people implementing do-it-yourself (DIY) projects using available technology to create, share and use their invention and ideas.

Project-Based Learning: An approach to use learned contents in a project to find a proper solution to an existing problem.

Virtual Reality: A computer generated virtual world the user can interact with.

Chapter 10
Mobile Augmented Reality Applications in Education

Irfan Sural
Eskisehir Osmangazi University, Turkey

ABSTRACT

The evolution and broad ownership of mobile devices has led to an increased interest to integrate the benefits of mobile learning and augmented reality applications. New possibilities for teaching and learning provided by augmented reality have been increasingly recognized by educational researchers. Mobile augmented reality can provide rich contextual learning for individuals. Currently virtual reality and augmented reality applications are used for training in fields as diverse as trades, military, entertainment, education and health. This chapter will explore different dimensions of mobile augmented reality and exemplify their potential for education. Therefore, how mobile augmented realty applies to education and training domain, and the potential impact on the future of education will be explained. Current status, opportunities and challenges of mobile augmented reality in education, mobile augmented reality applications will be included.

INTRODUCTION

Augmented reality is a growing field of technology where real life is adjusted and enhanced by computer-generated visuals and sounds. Augmented reality can be used in many different technologies, such as: computers, tablets, and smartphones. Augmented reality technology can also be utilized through wearable components, for example, glasses and helmets. Augmented reality allows us to integrate information based virtual reality and the real world physical reality. Recently as the advantages of mobile devices, mobile augmented reality concept is widely being used in many field and applications. Simply mobile augmented reality is augmented reality that users can take with them wherever they go. Most specifically, this means that the hardware required to implement an augmented reality application is something that can be taken with users wherever they go. Mobile augmented reality applications are designed to provide mobile device users with rich services, applications and functionality, which are utilized on top

DOI: 10.4018/978-1-5225-2110-5.ch010

of physical reality. Such applications use the mobile device's camera, GPS, touch screen elements, other sensors and motion detectors to integrate real images, videos or scenarios within them.

The evolution and broad ownership of mobile devices has led to an increased interest to integrate the benefits of augmented reality applications into mobile learning. New possibilities for teaching and learning provided by augmented reality have been increasingly recognized by educational researchers. Mobile augmented reality can provide rich contextual learning for individuals. Currently virtual reality and mobile augmented reality applications are used for training in fields as diverse as trades, military, education and health. The coexistence of virtual objects and real environments allows users to visualize complex spatial relationships and abstract concepts.

This chapter will explore development stage of mobile augmented reality and exemplify their potential use in education. Therefore, how mobile augmented realty applies to education and training domain, and the potential impact on the future of education will be explained. Current status, opportunities and challenges of mobile augmented reality in education, mobile augmented reality applications topics will be included.

BACKGROUND

This section will start with history of augmented reality and then continue with the emergence of mobile augmented reality. Finally, the development of mobile augmented reality for use in education domain is mentioned. Augmented reality is not a new concept. It has been used in different forms in many years. Augmented reality applications may vary from yellow first-down lines sketched over a televised football game to movies, event projector's been used to project images on the top a real setting. Those are examples of virtual graphics being placed upon a real-life situation. Since augmented reality exists about a half-century, the ancestors of this technology can be accepted as virtual reality. Concept of virtual reality is firstly used in science fiction short story named "Pygmalion's Spectacles" in 1935 (Weinbaum, 2007) (Figure 1). In this story reality system is described as a holographic recording of imaginary experiences, including smell and touch.

Later on in 1950s arcade-style theatre cabinet Sensorama which stimulate all the senses developed by cinematographer Morton Heilig. The Sensorama was able to display stereoscopic 3D images in a wide-angle view, provide body tilting, supply stereo sound also had tracks for wind and aromas to be triggered during the film (Rheingold, 1992). Morton Heiling next invited the first example of a head-mounted display (HMD) named Telesphere Mask (Figure 2). The headset provided stereoscopic 3D and wide vision with stereo sound.

In 1968 Ivan Sutherland and his student Bob Sproull created the first virtual reality / augmented reality head mounted display named Sword of Damocles that was connected to a computer (Sutherland, 1968). The device was very basic in terms of user interface and realism, and the graphics comprising the virtual environment were simple wireframe rooms. Even after all of this development in virtual reality there still wasn't a covering term for the field. In 1987 founder of virtual programming lab Jaron Lanier coined the term virtual reality. Following years the term is introduced to wider audience in movies like The Lawnmower Man in 1992, The Matrix in 1999. For example, in the movie Matrix characters that are living in a fully simulated world, were being unaware that they do not live in the real world. In the first sixteen years of the 21st century, it has seen major development of virtual reality. Development in computer technology and smartphones with high-quality displays and 3D graphics capabilities enabled a

Figure 1. Pygmalion's Spectacles, idea of a pair of goggles (Virtual-Realms, 2016)

Figure 2. Morton Heilig's Sensorama (Virtual-Realms, 2016)

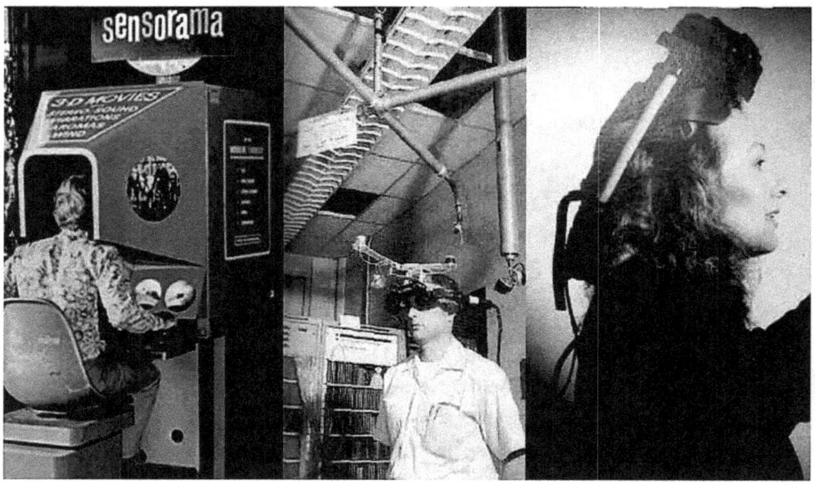

new generation of virtual reality devices. In history of augmented reality literature, most of resources use the virtual reality and augmented reality terms interchangeably. The term "augmented reality" was first coined in 1990 by Boeing researcher Tom Caudell (Rouse, 2016). Augmented reality is a combination of virtual reality and real life. In virtual reality users experience computer generated virtual environments whereas in augmented reality users are in real life and it is extended with information and imagery. Both virtual and augmented reality are similar in the goal of immersing the user, though both systems to this in different ways. With augmented reality, users continue to be in touch with the real world while interacting with virtual objects around them. On the other hand, with virtual reality, the user is isolated from the real world while immersed in a world that is completely fabricated. As it stands, virtual reality might work better for video games and social networking in a virtual environment, such as Second Life, or even PlayStation home.

Augmented reality is firstly used for entertainment purpose. Julie Martin created first 'Augmented Reality Theater production', Dancing in Cyberspace in 1994. In this theater dancers and acrobats manipulating body sized virtual object in real time, projected into the same physical space and performance plane. The acrobats appeared immersed within the virtual object and environments (Wikipedia, 2016a). In early of 2000s Hirokazu Kato and Mark Billinghurst presented an open source library ARToolKit which is a software library for building augmented reality applications. ARToolKit video tracking libraries calculate the real camera position and orientation relative to physical markers in real time (Watanuki, 2010) (Figure 3).

In 2009, ARToolKit library brings augmented reality to web browsers. This open source library allows developers to use many different languages and platforms like Android, Flash or Silverlight and being used in many augmented reality related projects. This has paved the way for many new developments in augmented reality. ATOMIC Authoring Tool a multi-platform authoring tool for creating augmented reality applications, DroidAR, GeoAR, BeyondAR augmented reality frameworks are some examples of milestones to develop augmented reality on web and mobile devices. Today augmented reality is currently being delivered in a variety of ways. One of this is mobile environments that's called mobile augmented reality.

Augmented reality has become increasingly popular in the media and entertainment and games over the last few years. Until a few years ago augmented reality applications were mostly available on workstations and personal computers. Although milestones in mobile augmented reality starts at early in 1968 it has evolved in 1993 with invention of smartphones. The first version of mobile augmented reality can be related with development of wearable technology in sense of motion. With the transformation and miniaturization of physical devices and displays, the concept of mobile augmented reality evolved towards the notion of "mobile device", aka augmented reality on a mobile device (Arth, Grasset,

Figure 3. An Application with ARToolKit (HITLab, 2016)

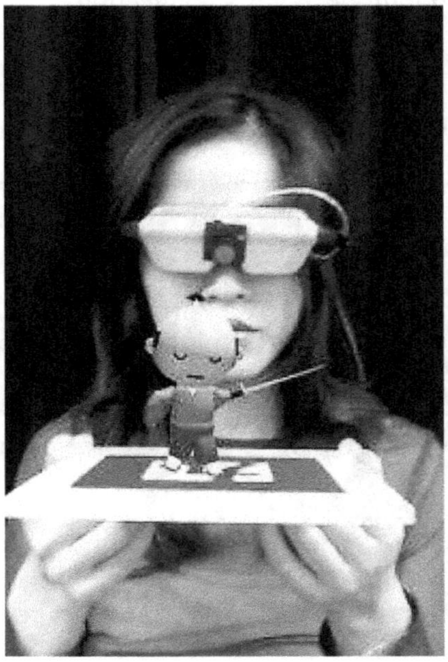

Gruber, & Wagner, 2015). Mobile augmented reality is augmented reality that you can take it with you wherever you go.

Smartphone is a mobile device that fits in a pocket and operate easily everywhere. Most tablet devices are also mobile devices that users can carry easily wherever they go. This future bring big opportunities to use mobile augmented reality and integrate in many domains. One of this domain is education. Augmented reality is a powerful and very flexible technology that can be used in educational settings.

MOBILE AUGMENTED REALITY EXAMPLES IN DIFFERENT DOMAINS

Mobile augmented reality has many application domains. In this section some example of different mobile augmented reality applications is presented. A mobile augmented reality application is a mobile application that are designed to provide mobile phone users with rich services, applications and functionality. Mobile devices bring new dimension to augmented reality systems and applications. Most of mobile augmented reality applications have turned from wearable technology to smartphone devices. The first mobile augmented reality system, which allowed outdoor and indoor users to access and manage information that is spatially registered with the real world, presented in 1999 by Columbia University (Hollerer, Feiner, Terauchi, Rashid, & Hallaway, 1999). System allowed users to walk around while they have necessary equipment mounted onto their back (Figure 4).

Vlahakis et al. (2001) presented a mobile augmented reality system Archeoguide for cultural heritage. The system provides on-site help and augmented reality reconstructions of ancient ruins, based on user'sposition and orientation in the cultural site, and real-time image rendering. Archeoguide was consist of three main components. Site information server, mobile units and WLAN connection. Site information server which store multimedia relational database supports querying according to attributes like geographic position (coordinates), item type (e.g. temple), time (e.g. chronological era), level of information (e.g. general, detail, scientific), theme (e.g. relates to sports), and versioning (e.g. stage in the restoration process). Mobile units like notebooks, tablets and PDA request material according to user preferences, position, and selected tour vie WLAN connection.

Figure 4. A user wearing prototype of Mobile Augmented Reality backpack (Feiner et al., 2016)

Daniel Wagner and Dieter Schmalstieg (2003) developed stand-alone mobile augmented reality system with self-tracking running on an unmodified personal digital assistant (PDA) with a commercial camera. The application provides the user with a three-dimensional augmented view of the environment by using marker-based tracking toolkit (ARToolKit), which runs directly on the PDA. Developers also introduced an optional client-server architecture based on wireless networking and tracking task in order to provide better performance in selected areas (Wagner & Schmalstieg, 2003).

Mathias Möhring et al. (2004) presented a running video see-through mobile augmented reality system on a consumer cell-phone. It supports the detection and differentiation of different markers, and correct integration of rendered 3D graphics into the live video stream via a weak perspective projection camera model and an OpenGL rendering pipeline. The output of this study was aimed to contribute in interactive tour guiding for museums and tourism, as well as at mobile games (Möhring, Lessig, & Bimber, 2004).

Mobile augmented reality is used in the field of building and construction too. Hakkarainen et al. (2010) developed a software prototype that provide two-way real-time augmented reality access to 4D CAD and BIM information. The system allows the user to compare scheduled building information models with the situation at the construction site in real time, as well as to attach position and time aligned visual and other feedback to the building models at the office (Hakkarainen, Woodward, & Rainio, 2010).

One of the popular mobile augmented reality app Pokémon Go is developed by Niantic in 2016. In this game, players use a mobile device's GPS capability to locate, capture, and train virtual creatures, called Pokémon (Figure 5). Players use avatar which can be created by selecting a hair, skin, and eye color, style and outfit. Player's avatar is displayed on a map using the player's current geographical location. As players move on real world, their avatar moves within game's map. There is different Pokémon species located in different areas like water-type Pokémon near to water. Players may encounter with Pokémon either in augmented reality mode or in live rendered generic background mode (Wikipedia, Pokémon Go, 2016c)

Figure 5. Pokémon Go (FigureX, 2016)

MOBILE AUGMENTED REALITY IN EDUCATION

In traditional education settings, most of activities carried out through face-to-face instructions where learning environment is arranged by the teacher. Learning materials that based on static design do not show any information in a dynamic way such as motion or continuous movement. Although sometimes these existing methods work effectively, however there is an increasing interest of educators and researchers in introducing more useful methods to improve the teaching and learning experiences (Nincarean, Alia, Halima, & Rahman, 2013). One of these recent method is augmented reality. During the last few decades, many professionals and researchers have been developing theories and applications for the adoption of mobile augmented reality into education settings.

The introduction of augmented reality applications to mobile devices such as smartphones, tablets, and mobile computers enabled augmented reality experiences for all mobile users. Because of increasingly use of mobile devices, augmented reality become a ubiquitous tool in many field and also in mobile learning. This ubiquitous availability allow many educators to design creative learning scenarios in real world environments. Most prominent mobile augmented reality examples support exploration of the physical environment with different topics of interest, e.g. history, arts, technology, biology, astronomy and others, or by enriching artifacts in the physical environment with augmented reality techniques (Specht, Ternier, & Greller, 2011). Although there are some issues like cost for development of augmented reality systems, integration of augmented reality in traditional learning environments and general resistance to new technologies, recently there is great amount research in adopting augmented reality in teaching and learning activities.

In their study Dunleavy, Dede and Mitchell (2009) designed mobile augmented reality game "Alien Contact" to teach math, language arts, and scientific literacy skills to middle and high school students. In game, there is a scenario that aliens land on earth and have the following actions; peaceful contact, invasion, plundering, or simply returning to their home planet. Students must explore augmented realty world interviewing virtual characters, collecting digital items, and solving math, language arts, and scientific literacy puzzles to determine why the aliens have landed. The purpose of this study was to understand how school teachers and students describe teaching and learning within a participatory augmented reality simulation. (Dunleavy, Dede, & Mitchell, 2009).

Chang, Wu and Hsu (2013) developed online radiation unit and focused on mobile augmented reality activity to contextualize students' learning of a socioscientific issues on nuclear energy use and radiation pollution. In this mobile augmented reality activity which situates students in the context of the nuclear accidents at the Fukushima Daiichi Nuclear Power Plant in Japan, students collects simulated radiation values using their tablet's scanning function. 22 ninth-grade students taught by a science teacher at a public middle school in South Taiwan are contributed to this study. The result of study showed that students made significant improvements from the pretest to the posttest. Moreover according to this study unit enhanced by mobile augmented reality can improve students' learning of the science concepts (Chang, Wu, & Hsu, 2013).

Adopting Mobile Augmented Reality in Teaching and Learning

Recently the rapid development of technology has offered new ways to develop applications and approaches for learning environments. Augmented Reality is such a technology that offers a new educational approach in helping learners develop critical capacity and deeper understanding of the concepts underlying scientific investigation (Lazoudis & Agogi, 2011). Also, augmented reality makes possible to study abstract concepts such as three-dimensional shapes and geometric objects, which are difficult to understand through in a text book.

Freitas and Campos (2008) developed SMART augmented reality educational system for teaching 2nd grade-level concepts, such as the means of transportation and types of animals. 54 students from different classes participated into study and three dimensional models and prototypes, such as a car, truck, and airplane was shown to the whole class. The results indicated that SMART doesn't necessarily help students to actually learn, although all students enjoy the technology but effect is much higher among weak and average students. SMART had a positive impact in their learning. The results also showed that motivation was high in every class. (Freitas & Campos, 2008).

Another project developed for mobile environment was MARQ project. Mobile Augmented Reality Quest (MARQ) aimed at developing an electronic tour guide for museums based on a self-contained, inexpensive PDA that delivers a fully interactive 3D Augmented Reality to a group of visitors. The tour was delivered in in the form of a team-oriented game. Multi-user communication system was built for interaction between groups of visitors, in particular multi-user location based games. (Wagner, Schmalstieg, Szalavári, Traxler, & Stifter, 2014).

Educational Apps

In this section, a few mobile augmented reality applications in education settings are described. The potential use and properties are introduced. Aurasma app is a powerful drag-and-drop web studio that enables anyone to easily create, manage, and track their own augmented reality experiences. Educators and students can use this open source tool to bring their learning to life. This tool can be used in classroom in several different ways (Nesloney, 2013):

- **Homework Mini-Lessons:** Students can scan a page of their homework, then the page reveals a video of their teacher helping them solve a problem.
- **Faculty Photo Wall:** Users set up a display of faculty photos near the school entrance. Visitors can scan the image of any instructor and see that figure come to life, telling more about him- or herself.
- **Book Reviews:** Students can record themselves giving a brief review of a novel that they just finished, and then attach that "aura" (assigned digital information) to a book. Afterward, anyone can scan the cover of the book and instantly access the review.
- **Parent Involvement:** Record parents giving brief words of encouragement to their child, and attach a trigger image to every child's desk. Anytime students need to hear encouraging words from their parent, they can scan the image on their desk for virtual inspiration.
- **Yearbooks:** From tributes to video profiles, from sports highlights to skits and concert footage, the ways that augmented reality can enhance a school yearbook are limitless.
- **Word Walls:** Students can record themselves providing the definitions to different vocabulary words on a word wall. Afterward, anyone can use the Aurasma app to make a peer pop up on screen, telling them the definition and using the word in a sentence.
- **Lab Safety:** Put triggers (images that activate media when scanned by an augmented reality enabled device) all around a science laboratory so that when students scan them, they can quickly learn the different safety procedures and protocols for the lab equipment.

- **Deaf and Hard of Hearing (DHH) Sign Language Flashcards:** With augmented reality, flashcards of vocabulary words can contain a video overlay that shows how to sign a word or phrase.

One of another powerful app is Anatomy 4D (Figure 6). It allows users to understand and interact with complex anatomic information with remarkable ease and speed. Anatomy 4D takes users inside the human body via a simple to use three dimensional learning environment.

ZooBurst is a digital storytelling tool that lets users to create three dimensional books easily. Also, these books can be used online via mobile app, iPad, laptop or desktop computers. Educators can share their books with students using simple hyperlink and books can be easily embedded in any blog or webpage. The interactivity in ZooBurst application gives students an opportunity for total immersion (Figure 7). The immersive in learning will increase motivation as well as allow students to improve performance (Mahadzir & Phung, 2013). This tool also provides an opportunity for students to deliver presentation, write reports, express complex ideas and tell stories.

There are plenty of mobile augmented reality application in marketspace that covers universe, stars and planet topics. Star Walk is an interactive astronomy app that shows celestial objects on the sky in the exact positions, providing detailed information about them. Whenever app users hold their mobile device up towards sky they can see all the stars and constellations (Vitotechnology, 2013). The app features are watch moon phases, meteor showers see daily sunset and sunrise times, elevation angle, and daily hours of sunlight as well as the data for all main planets. Another app similar to Star Walk is Spacecraft 3D. Users interact with many spacecraft to explore solar system. With printed augmented reality targets and camera on mobile device, user can get up close with robotic explorers and observe how they move and learn about space.

AR Dinopark app is bringing back ancient dinosaurs to life. User can move the dinosaur pictures and dinosaurs appear in various scenes taken by camera as if they are alive. Also, user can get information about dinosaur in any stage. Changing view position and taking snap shot is other features.

Another augmented reality app is a medical education. AR Liver Viewer is real-time 3D medical education and patient communication tool, featuring detailed anatomical model (Figure 8). User can place the detailed organ in any position and zoom in to any location to explore all of the anatomical structures. AR Liver is appropriate for use by secondary students, undergraduate and graduate students,

Figure 6. Anatomy 4D App (Egitim Teknoloji, 2016)

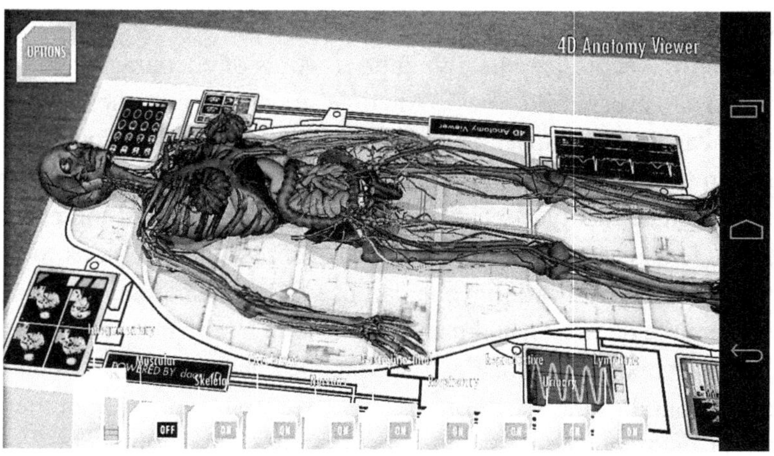

Figure 7. ZooBurst App (Apple ITunes, 2016)

and medical professionals (Edshelf, 2016). Medical augmented reality applications have great attention in marketplace. Reality of AFib is another app like AR Liver Viewer but it only allows the user to view a realistic and anatomically accurate 3D model of a heart from the exterior, and cross-sectional views.

ColarMix is a coloring app that allows users to take 2D pictures, customize their colors and textures, and then watch them come to life with animations. It is great tool for developing skills and knowledge on various topics. It allows an interaction and playing games with the animated characters by touching the screen.

Word Lens used the built-in cameras on smartphones and similar devices to quickly scan and identify foreign text, and then translate and display the words in another language on the device's display. The

Figure 8. AR Liver ViewerApp (Apple ITunes, 2016)

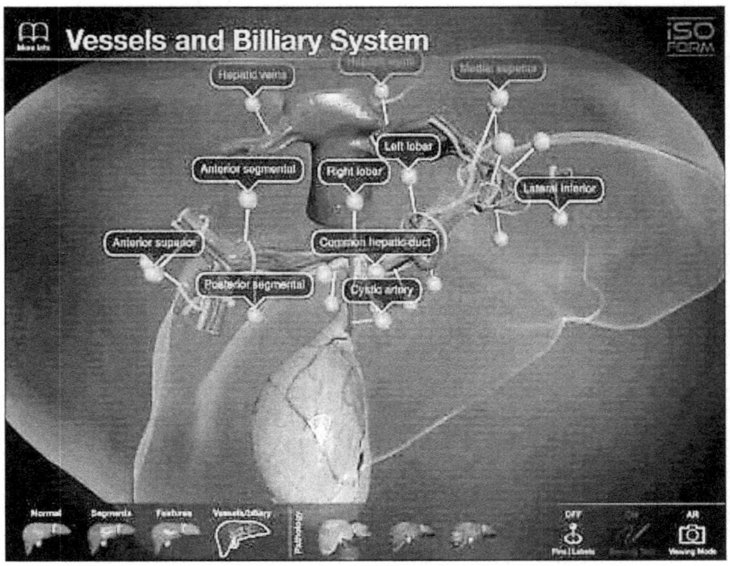

words were displayed in the original context on the original background, and the translation was performed in real-time without connection to the internet (Wikipedia, 2016b). Word Lens can be used in google glass too. Simply user should look at some printed text, hold his/her head still, and see it translated.

AR Flashcards – Animal Alphabet can be used in phonics teaching and early language development (Figure 9). The bond between animal associations to letters and phonic sounds has been a part of language learning for a long time (Learninginspired, 2016). The flashcards can be stuck in the classroom so that children can freely explore them and see the three dimensional animals pop up in front of them.

FUTURE RESEARCH DIRECTIONS

Currently augmented reality is moving to mainstream markets. Possible future uses of augmented reality are in medicine for direct training of students, military for precise location of enemy positions, fine art for mixed media and interactive artworks, industry for assistance in prototyping and teaching for visualizing learning material in interactive apps (Lass, 2015). Since mobile augmented reality merge any digital information within real-world settings, i.e. electronic data or information, in a variety of media formats not only visual and graphic media but also text, audio, video and haptic overlays, it has big research potential on educational settings. On the other hand, such a variety in media formats and lack of design may cause students to have cognitive overload in mobile augmented reality simulations. Therefore, further studies that explore the effects of augmented reality on cognitive load in brain should be done. And how much information should be displayed before it turns into a distracting technology should be examined.

Mobile augmented reality applications can be used as physical activity out of classroom for teachers and students. Interactive media, whereat augmented reality applications should support outdoor exploration and documentation of nature or that integrate physical activity and encourage children to get up and be mobile rather than sit passively in front of a screen (Naeyc & Fred Rogers Center, 2012). The

Figure 9. AR Flashcards – Animal Alphabet App (Top Best Apps for Kids, 2016)

potential of mobile augmented reality that support physical activates can be taken in consideration in teaching and learning activities.

Augmented reality applications may vary from HDM-based to handled or outdoor applications. Therefore, most suitable platform and application environment should be carefully reviewed and research should be done to use best appropriate solution in learning environment. Additional studies on effects of augmented reality in terms of student communication can be done. If students who use augmented reality communicate more effectively compared to students who is not, this would be a chance to improve student's communication capabilities.

CONCLUSION

Augmented reality is one of the growing technologies that have been increasing recognized by educational researchers. The effectiveness of augmented reality can be extended with other type of technologies such as mobile device. The recent rise of smartphones has sparked the development of mobile augmented reality applications in almost any field of study. Augmented reality has big potential use in schools because it brings real life experiences to the classrooms. Moreover, augmented reality applications are great at providing metacognition information to students. This kind of apps may help students to focus on key information normally they miss and provide just in time feedback.

Using augmented reality in education brings many benefits. One of this is increasing the accessibility of virtual educational content. Students can access to virtual content with mobile devices when traditional access is not readily available such as when walking on the street. Augmented reality based educational applications and the topics covered also increase day by day. In mobile marketplace augmented reality applications mostly focus on games and simulations. When the potential use of augmented reality in education is fully discovered, the useful functions of augmented reality can begin to be used widely in all fields of education. Mobile augmented reality applications will become an easy and widely used information retrieval and interaction technologies. It is an important part of education, and will continue to grow.

REFERENCES

Apple ITunes. (2016). *AR Liver ViewerApp*. Retrieved from https://itunes.apple.com/us/app/ar-liver-viewer, 2016

Apple ITunes. (2016). *ZooBurswt App*. Retrieved from https://itunes.apple.com/us/app/zooburst

Arth, C., Grasset, R., Gruber, L., & Wagner, D. (2015). *The History of Mobile Augmented Reality*. arXiv preprint arXiv:1505.01319

Chang, H.-Y., Wu, H.-K., & Hsu, Y.-S. (2013). Integrating a mobile augmented reality activity to contextualize student learning of a socioscientific issue. *British Journal of Educational Technology*, *44*(3), E95–E99. doi:10.1111/j.1467-8535.2012.01379.x

Dunleavy, M., Dede, C., & Mitchell, R. (2009). Affordances and Limitations of Immersive ParticipatoryAugmented Reality Simulations for Teaching and Learning. *Journal of Science Education and Technology*, *18*(1), 7–22. doi:10.1007/s10956-008-9119-1

Edshelf. (2016). *AR Liver Viewer*. Retrieved August 2016, from edshelf: https://edshelf.com/tool/ar-liver-viewer/

Egitim Teknoloji. (2016). *Anatomy 4D App*. Retrieved from http://egitimteknoloji.net/

Feiner, S., Höllerer, T., Gagas, E., Hallaway, D., Terauchi, T., Güven, S., & MacIntyre, B. (2016). *MARS - Mobile Augmented Reality Systems*. Columbia University. Retrieved from http://graphics.cs.columbia.edu/projects/mars

Figure, X. (2016). *Pokémon Go*. Retrieved from http://figurex.net/

Freitas, R., & Campos, P. (2008). SMART: a System of augmented reality for teaching 2nd grade students.*Proceedings of the 22nd British Computer Society Conference on Human-Computer Interaction (HCI 2008)* (pp. 27-30). Liverpool, UK: John Moores University.

Hakkarainen, M., Woodward, C., & Rainio, K. (2010). Mobile Augmented Reality for Building and Construction.*Mobile World Conference Barcelona*, Barcelona, Spain.

HITLab. (2016). *ARToolKit*. Retrieved from https://www.hitl.washington.edu/artoolkit/

Hollerer, T., Feiner, S., Terauchi, T., Rashid, G., & Hallaway, D. (1999). Exploring MARS: Developing Indoor and Outdoor User Interfaces to a Mobile Augmented Reality System. *Computers & Graphics*, *23*(6), 779–785. doi:10.1016/S0097-8493(99)00103-X

Lass, W. (2015, July 13). *The Future of Augmented Reality: Limitations, Possibilities and Hopes*. Retrieved September 02, 2016, from Emergingedtech: http://www.emergingedtech.com/2015/07/future-of-augmented-reality-limitations-possibilities-hopes/

Lazoudis, A., & Agogi, E. (2011). The "Science Center To Go" project. EDEN - 2011 Open Classroom Conference, 7-13.

Learninginspired. (2016). *My Top 5 Educational Augmented Reality Apps*. Retrieved September 1, 2016, from Learninginspired: http://www.learninginspired.com/my-top-five-educational-augmented-reality-apps

Mahadzir, N. N., & Phung, L. F. (2013). The Use of Augmented Reality Pop-Up Book to Increase Motivation in English Language Learning For National Primary School. *IOSR Journal of Research & Method in Education, 1*(1), 26-38.

Möhring, M., Lessig, C., & Bimber, O. (2004). Video See-Through AR on Consumer Cell-Phones.*IEEE International Symposium on Mixed and Augmented Reality (ISMAR)*. doi:10.1109/ISMAR.2004.63

Naeyc & Fred Rogers Center. (2012, January). *Technology and Interactive Media as Tools in Early Childhood Programs Serving Children from Birth through Age 8*. Retrieved September 20, 2016, from Naeyc.org: http://www.naeyc.org/files/naeyc/file/positions/PS_technology_WEB2.pdf

Nesloney, T. (2013, April 11). *Augmented Reality Brings New Dimensions to Learning*. Retrieved August 2016, from Edutopia: http://www.edutopia.org/blog/augmented-reality-new-dimensions-learning-drew-minock

Nincarean, D., Alia, M. B., Halima, N. D., & Rahman, M. H. (2013). Mobile Augmented Reality: the potential for education. *13th International Educational Technology Conference*, 103, 657 – 664.

Rheingold, H. (1992). *Virtual Reality*. Simon & Schuster. Retrieved August 20, 2016, from Morton Heiling: http://www.mortonheilig.com/SensoramaPatent.pdf

Rouse, M. (2016, February). *Augmented Reality (AR)*. Retrieved August 2016, from Whatis.com: http://whatis.techtarget.com/definition/augmented-reality-AR

Specht, M., Ternier, S., & Greller, W. (2011). Dimensions of Mobile Augmented Reality for Learning: A First Inventory. *Journal of the Research Center for Educational Technology*, *7*(1), 117–127.

Sutherland, I. I. (1968). A head-mounted three dimensional display. *Proceedings of AFIPS*, *68*, 757–764. Retrieved from http://design.osu.edu/carlson/history/PDFs/p757-sutherland.pdf

Top Best Apps for Kids. (2016). *Best Kids Apps – AR Flashcards*. Retrieved from http://topbestapps-forkids.com/best-kids-apps-ar-flashcards/, 2016

Virtual-Realms. (2016). *New Gallery 2016*. Retrieved from http://www.virtual-realms.com/new-gallery/2016

Vitotechnology. (2013, December). *User Manual Star Walk*. Retrieved August 2016, from Vitotechnology: http://vitotechnology.com/manual_/pdf/StarWalk_manual_en.pdf

Vlahakis, V., Karigiannis, J., Tsotros, M., Gounaris, M., Almeida, L., Stricker, D., & Ioannidis, N. et al. (2001). Archeoguide: First results of an augmented reality, mobile computing system in cultural heritage sites.*Proceedings of the 2001 Conference on Virtual Reality, Archeology, and Cultural Heritage*. New York: ACM. doi:10.1145/584993.585015

Wagner, D., & Schmalstieg, D. (2003). *First Steps Towards Handheld Augmented Reality. IEEE International Symposium on Wearable Computers*, 127-135.

Wagner, D., Schmalstieg, D., Szalavári, Z., Traxler, C., & Stifter, M. (2014). *Mobile augmented reality quest (MARQ) 2005-2007*. Retrieved August 2016, from Handheld Augmented Reality: http://handheldar.icg.tugraz.at/marq.php

Watanuki, K. (2010). A Mixed Realitybased Emotional Interactions and Communications for Manufacturing Skills Training. In S. Fukuda (Ed.), Emotional Engineering: Service Development (p. 58). Springer Science & Business Media.

Weinbaum, G. S. (2007, October 5). *Project Gutenberg's Pygmalion's Spectacles*. Retrieved Agust 2016, from Gutenberg: http://www.gutenberg.org/files/22893/22893-h/22893-h.htm

Wikipedia. (2016a, August). *Augmented Reality*. Retrieved from Wikipedia: https://en.wikipedia.org/wiki/Augmented_reality

Wikipedia. (2016b). *Word Lens*. Retrieved August 20, 2016, from Wikipedia: https://en.wikipedia.org/wiki/Word_Lens

Wikipedia. (2016c, September). *Pokémon Go*. Retrieved September 23, 2016, from Wikipedia: https://en.wikipedia.org/wiki/Pok%C3%A9mon_Go

KEY TERMS AND DEFINITIONS

App: An application, typically a small, specialized program downloaded onto mobile devices.

Augmented Reality: The integration of digital information with the user's environment in real time. It uses the existing environment and overlays new information on top of it.

Cognitive Overload: A situation where the teacher gives too much information or too many tasks to learners simultaneously, resulting in the learner being unable to process this information.

Head-Mounted Display: A computer displays worn like goggles to immerge into virtual worlds.

Mobile Augmented Reality: It applies the concept of Augmented Reality in truly mobile settings; that is, away from the carefully conditioned environments of research laboratories and special-purpose work areas.

Mobile Learning: Learning via mobile devices such as smartphones, MP3 players, laptops and tablets.

Virtual Reality: A computer generated virtual world the user can interact with.

Chapter 11
Design Principles for an Intelligent-Augmented-Reality-Based M-Learning Application to Improve Engineering Students' English Language Skills

Derya Bozdogan
TED University, Turkey

Buket Kasap
Ankara University, Turkey

Utku Kose
Usak University, Turkey

ABSTRACT

Introducing an intelligent augmented reality based M-learning application designed and developed for improving engineering students' English language skills, this chapter reports a work-in-progress that focuses on system design procedure. The application consists of Artificial Intelligence (AI) based functions to ensure an effective learning flow while taking advantages of game-based learning by providing a story board structure with a content tree. Four design principles "fair use, flexible use, fault tolerance, educational climate" by Elias in addition to Stockwell and Hubbard's principles for mobile supported language learning have been taken into account. Furthermore, the proposed system here employs an effective approach combining both real and virtual environments to achieve an Augmented Reality based learning experiences for students. After the introduction of the application, the chapter outlines how it will be processed in the future.

DOI: 10.4018/978-1-5225-2110-5.ch011

INTRODUCTION

Rapid developments and changes in technology have affected mostly the way of providing education. Since the introduction of computer technology and also Internet, especially educational research has gained a remarkable momentum and many different approaches, methods, and techniques on delivering educational processes thanks to computer and Internet environments were introduced to the associated literature. Today, e-learning is the most popular technology-oriented educational technique as an effective and efficient way of distance education approach. In time, e-learning has also gone through many developments and it is currently used in different forms according to environments, platforms, or devices over which teaching-learning is experienced.

It can be said that because there are many dynamic environmental factors affecting the whole M-Learning process, there will be always an effort on searching for better language education supported by mobile technologies. Furthermore, it is also a good way to adapt the literature of M-Learning with the latest technological improvements by taking the real-world based problems into consideration. Here, it is clear that supporting M-Learning based language teaching – learning processes with the latest technological components will take the effectiveness and efficiency of M-Learning one step away, by shaping also the future of this research area. In this context, two popular approaches: Augmented Reality and Artificial Intelligence are strong candidates to support M-Learning and so Mobile Assisted Language Learning studies.

Augmented Reality (AR) is an effective interaction oriented approach, which employs both virtual environment and real world in a common platform. Since its first introduction AR and its applications have become popular among different fields and employed in the context of especially multimedia based platforms to ensure highly interactive, effective using solutions for different fields like education, health, natural sciences, economy, military...etc (Chen, & Chung, 2007; Kose, 2015; Kose, Koc, & Yucesoy, 2013). Because AR provides interactive aspects and an adaptive working function interacting with the real time environments, there is a remarkable interest on performing scientific research studies in which practical, immediate solutions for real-world based problems are tried to be achieved (Kose, 2015; Kose et al., 2013). Because of its effectiveness on enabling people to interact with the real-world while still in contact with digital elements, AR has been a popular approach for educational studies.

Regarding to educational studies, Artificial Intelligence (AI) is another research approach, which is widely used. Because AI comes with a multidisciplinary approach by achieving a logical and mathematical way of enabling computer based systems to solve real-world problems, it employs a great potential for even future of research studies. So, adding some ‘intelligent’ working mechanisms to educational software systems has become a popular trend over the intersection of education and AI.

Associated with the explanations above, this chapter is based on an application combining some popular approaches to provide a better way of teaching – learning English. Introducing an intelligent Augmented Reality based M-Learning application designed and developed for improving engineering students’ English language skills, this chapter reports a work-in-progress that focuses on system design procedure. The application consists of Artificial Intelligence based functions to ensure an effective learning flow while taking advantages of game-based learning by providing a story board structure with a content tree. Four design principles “fair use, flexible use, fault tolerance, educational climate” by Elias (2011) in addition to Stockwell and Hubbard’s (2013) principles for Mobile Assisted Language Learning have been taken into account. Furthermore, the proposed system here employs an effective approach combining both real and virtual environments to achieve an Augmented Reality based learning experiences for students. In this way, it is aimed to improve students' learning experiences with two approaches based on intelligent working mechanism and interaction with both real and virtual world. Some starting points of the application is introduced in this chapter. After the introduction of the application, the chapter outlines how it will be processed in the future.

In the context of the chapter subject, remaining sections of the paper are organized as follows: The next section is devoted to essential educational motivations considered along the application introduced in this study. After this section, the third section explains general setting of the application and its importance. Following to this section, the fourth section explains the further methodology that was planned to be followed along next steps of the study. After this section, design principles regarding to the application have been explained under the fifth section. Under next two sections, potential risks – challenges and also future research regarding to the study are discussed. Finally, the paper ends with the final section discussing about conclusions.

EDUCATIONAL MOTIVATIONS

Broadly in education, defined as non-spatial, personalized learning technique engaging learners via mobile devices such as smart phones or tablets, mobile learning (M-Learning) gets the attention of researchers (Ally, 2009; Kukulska-Hulme, 2005; Motiwalla, 2007). Specifically, in language learning, Mobile Assisted Language Learning (MALL) integrates Mobile Learning Environments (MLEs) into language teaching – learning (Beatty, 2003; Bozdogan, 2015; Colpaert, 2004) with features of mobility, content flexibility, immediacy, individuality, and interaction opportunities (Attewell, & Webster, 2005; Chinnery, 2006). So, it is an important way to employ M-Learning for language teaching – learning. But on the other hand, there is still a remarkable interest in improving the literature associated with this subject.

Introducing learners to learning opportunities via mobile devices has already proven its educational benefits. The theoretical framework of this study links MALL to game-based learning in an ESP (English for Specific Purposes) context with attempts to convert it to a CLIL (Content and Language Integrated Learning) setting. While MALL allows learners to study in a contextualized learning environment in their own pace (Silberman, & Biech, 2015), Game-Based Learning (GBL) enables interaction with the subject in a fun and intriguing way (Demirel, Seferoglu, & Yagci, 2003). Language-wise, CLIL "a dual-focused teaching approach where content is taught through a language other than the mother-tongue." (Marsh, 2002) promises necessary language support in content classes.

Games in M-Learning function as a new world where learner ubiquitously play and learn within the scope of defined learning objectives. What attracts learners the most as listed by Bayirtepe and Tuzun (2007) are the competitiveness and mystery to unlock. Games provide learners with the opportunity to practice knowledge transferring their prior knowledge into a new reality (Bottino, Ferlino, Ott, & Travella, 2007; Ebner, & Holzinger, 2007). Additionally, GBL poses a challenge to mobile application developers as to include learners into learning process in an enjoyable, challenging and encouraging (Gee, 2005). One of its basic principles portrays the nature of gaming: The more challenging games are, the better learning outcomes will be. So far, research on game-based M-Learning presents positive outcomes (Dabbagh et al., 2016; Hamari et al., 2016; Chee, 2016; Van der Wal, de Kraker, Kroeze, Kirschner, & Valkering 2016).

Adopted by the European Union as part of the multilinguality principle, any content can be taught using CLIL; however, the focus may shift slightly between content and language according to the objectives in the curriculum. Studies indicate CLIL success in Europe (O'Driscoll, 2004; Schreppegrell, Achugar, & Oteiza 2004; Shamsudin, & Nesi, 2006) supporting its claim to facilitate meaningful contextualized language learning (Kincheloe, Slattery, & Steinberg, 2000). CLIL in higher education needs attention as learners seek language guidance and support in content classes (Hanesova, 2015). In the Turkish higher education context, British Council (2015) reports the issues and recommends CLIL incorporation as an aid to boost both content and language instruction. However, studies from Turkey focus mainly on project based, CBI (Content-Based Instruction) and specifically theme-based approach at the primary and secondary levels (Arslan, & Saka, 2010; Kiziltan, & Ersanli, 2007).

Task-based Language Teaching (TBLT) promotes a student-centered learning environment where meaningful language learning takes place (Willis, 1996). Language becomes a concrete means to accomplish a real task from students' lives. To do so, students improve their thinking skills such as inference, analytical thinking, creative thinking, planning (Gunes, 2012).

The framework of TBLT is shown in Figure 1.

In addition to the mentioned educational approaches, methods, and techniques explained under previous paragraphs, Augmented Reality (AR) has the main mechanism of the application by enabling students to interact with the 'mixed reality' (Milgrami, & Kishino, 1994) along their learning processes. In the literature, Chang, Morreale, & Medicherla (2010) indicates that learners can improve their motivation for learning and it is also possible for them to enhance their educational practices via AR. Also, Johnson, Levine, Smith and Stone (2010) think that the AR can provide strong, contextual, onsite learning experiences with also serendipitous exploration and discovery of the connected nature of the information over the real world. On the other hand, Cascales, Laguna, Pérez-López, Perona, & Contero (2013) explain that the AR is an emerging technology for educational activities because it affects students' many senses. Briefly, success of AR in educational studies has encouraged the authors to include in the application introduced here.

Nowadays, it is also a remarkable interest to employ Artificial Intelligence in many fields of modern life including education. Running intelligent mechanisms in educational applications have improved effectiveness and efficiency of educational works and made it easier for both teachers and students to take active part in teaching-learning processes. At this point, using intelligent educational software systems is an important approach that is followed by educational institutions. It can be observed from the as-

Figure 1. Task-based Language Teaching framework (Willis, 1996)

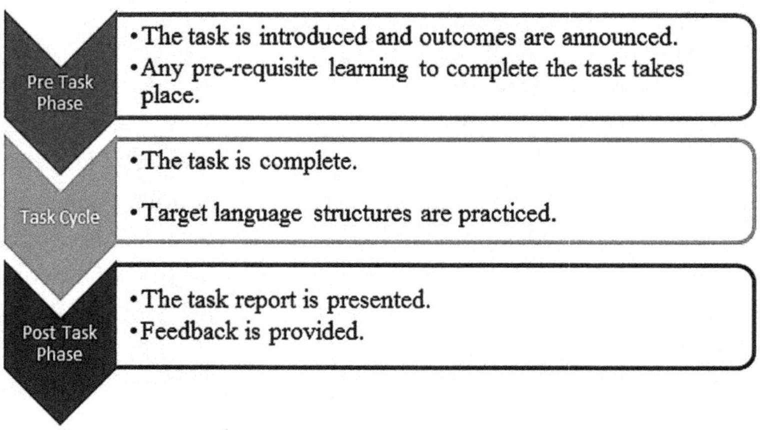

sociated literature that Artificial Intelligence is widely employed to develop such educational software systems (Aleven, 2013; Carpio Cañada, Mateo Sanguino, Merelo Guervós, & Rivas Santos, 2015; Nye, 2015; Peña-Ayala, 2013; Vera, Breis, Sánchez, & Espinosa, 2013; Zatarain-Cabada, Barron-Estrada, Alor-Hernandez, & Reyes-Garcia, 2014). In addition to that, improving such intelligent educational software systems more with additional interactive components (i.e., Augmented Reality) is a remarkable key factor in forming more advanced educational processes.

SETTING OF THE APPLICATION AND IMPORTANCE

In order to understand more about design principles of the application, it is a good way to explain general setting though for it and shaping the importance of the research done with the application.

General Setting

This study aims to analyze the content instruction context following the EMI policy and to align higher education language instruction with the 21th century learning skills and requirements. Though numerous reforms have been initiated in Turkish higher education, undergraduate students in Turkey were ranked among the last ten countries in Europe for language competence (British Council, 2015). Such a devastating failure in higher education can be associated with the foreign language teaching policy and low student motivation.

Undergraduate programmes in Turkey display classification in three groups according to their medium of instruction: English as a medium of instruction (EMI), Turkish as a medium of instruction (TMI), and Turkish and English as a medium of instruction (T-EMI). In practice, students' proficiency level and lack of interest, and faculty's inefficiency and reluctance to deliver EMI classes pose challenges. Moreover, monotonous and difficult to access instructional materials and the scarcity of international students/relations add to these issues that can be handled with a needs analysis for identification purposes, and a curriculum and material revision (Cheon & Reeve, 2014). Hence, a transition from EMI to CLIL with dual focus on language and content improvement seem to meet such a demand and challenge (Arkin, 2013; British Council, 2015).

Ankara University, where the study takes place, promotes EMI policy in some of its programmes including Electrical and Electronical Engineering department. The 4-year program includes basic sciences, engineering, English language (general and genre-specific) and a range of elective courses. The number of students per class is approximately 30-40 and based on their preparatory class exit level, B1-B2 identify their language proficiency. This study concentrates on a second-year, fall semester content course and develop a game-based mobile app in support of their course and foreign language proficiency.

Importance of the Application

A needs analysis based curriculum improvement and game-based MALL integrated CLIL approach form the strengths and pillars of the study. This attempt to integrate game-based MALL and CLIL in higher education level is believed to contribute to the content classes in terms of affective factors like motivation and attitude and achievement both in content and language (Dornyei, MacIntyre, & Henry, 2015; Finn, 2015). A thorough needs analysis triangulated with qualitative and quantitative data collection identified student and teacher perspectives on EMI, learning styles, course material evaluation and the content and language balance in the classrooms. In a complementary manner, a game-based mobile application is being developed believing its contribution in the curricular and pedagogical issues. Additionally, in the Turkish context where language and content learning are treated separately, "a dual focus" (Marsh, 2002) would promote both.

METHODOLOGY

Grounded on the needs analysis executed in the previous research project in the same setting, this system design of the mobile application, utilized both qualitative and quantitative methods. The second year undergraduate course ELE218 Microprocessor materials and syllabus will be revised with the integration mobile app. Initially, both the EEE academic staff including the course instructor and students will be informed about the study, its theoretical background and potential outcomes while getting their consent to participate. First, corpus concordance will be performed to form the professional English bilingual dictionary app providing word meanings, examples, connotations and visuals. In a similar context, to underline the power of visuals, the study of Saran, Cagiltay, and Seferoglu (2008) suggests including pronunciation and usage in sentence as well as visuals to clarify meaning.

The textbook 'Digital Design with an Introduction to Verilog HDL (5th edition)' (Mano, & Ciletti, 2013) is chosen to be analyzed to identify content-specific vocabulary. For reference purposes, monolingual and bilingual hardcover and mobile dictionaries are taken into consideration. Along the further steps of the research – study, it is planned to gather some data about M-Learning with a questionnaire developed by the authors focusing on general and educational uses of mobile phone and related perceptions. After consulting the course instructor and students, the game based themes will be conceptualized for the tasks following the framework for MALL by Stockwell and Hubbard (2013) that offer key principles like creating short and clear activities considering characteristics of students' smart phone use, user environments, and learning styles. All in all, ten tasks will be developed for students to interact with artificial intelligent-expert systems in the M-Learning application. In order to better infuse tasks into MALL, four characteristics of tasks by Ellis (2003) for Task-Based Language Teaching (TBLT) are taken into consideration:

- Each task has a meaningful primary focus.
- Each task aims to address a need.
- Students are required to use certain language structures to accomplish each task.
- Each task has a real life-like outcome other than language learning.

Kalantzis and Cope (2005) suggest Performance- Play- Dialog model (PPD) for GBL. With PPD, students become active players rather than passive doers in the MLEs. The basic principles of PPD are these (Figure 2):

In addition to the dictionary and tasks, the third component of the application includes an avatar section, where students edit their avatars, store their achievement tokens and interact with other online users simultaneously. Students identify with their avatars and struggle to complete the tasks while performing

Figure 2. Basic principles of PPD (Kalantzis & Cope, 2005)

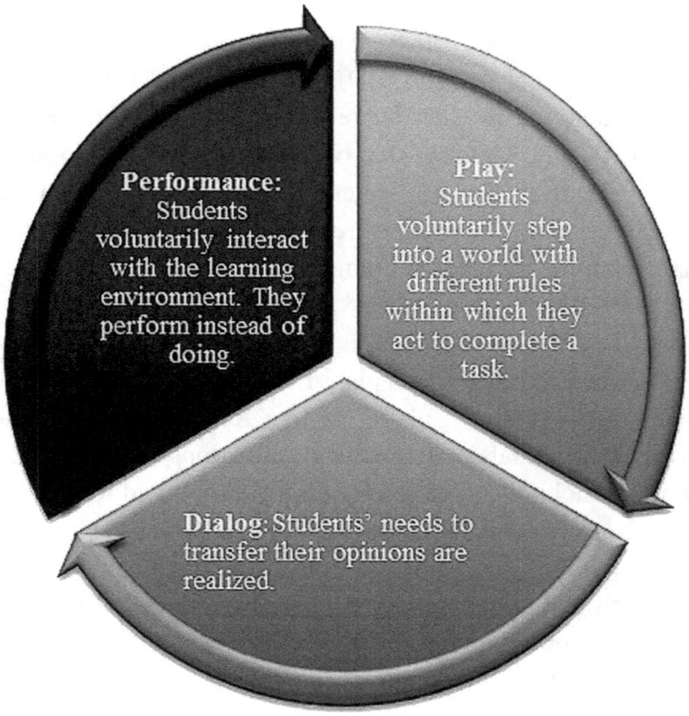

meaningful language use in context. Ab Hamid and Yu Fung (2007) in a similar game-based M-Learning platform identify user attentiveness in the game-based tasks reporting an 80% MLE interaction every day. Student achievement correlates highly the amount of MLE interaction and the extent of motivation out of game-based challenging tasks (Schawabe, & Göth, 2005). In a similar experimental design, Chen and Chung (2007) reports students' motivation and achievement are improved when they are engaged in an M-Learning environment on their own pace.

To facilitate out-of-class access to the M-Learning environment, task completion requirements are manageable but incessant. Based on the rubric prepared with the course instructor and provided to the students at the end of the implementation, summatively students are evaluated 10% out of their course grade. Each task is starred by the students upon completion; students also post comments and reviews, such a need emphasized by Moreno-Ger, Burgos and Torrente (2009). Furthermore, frequent teacher feedback increases value attached and improves the active learning process. The system will be left open for any updates and improvements for further uses.

Augmented Reality and Artificial Intelligence oriented mechanisms employed in the application are for improving effectiveness and efficiency of the learning process. By using intelligent mechanism in the background, it will be possible to run a more accurate learning flow in the context of changing nature of learning affected by many factors. On the other hand, use of Augmented Reality will enable students to interact with the real world in order to complete activities run over a mixed reality.

DESIGN PRINCIPLES OF THE APPLICATION

Designing mobile learning environments (MLEs) for smart phones is a meticulous process. Elias (2011) suggests four universal instructional design (UID) principles for MLEs: equitable use, flexible use, tolerance for error, and instructional climate. These UID principles are followed to develop the mobile learning environment in this study. The first step of the design and development process of the MLE is to formulate the content tree in the light of the curriculum and learning objectives. The next step is to develop the story board based on the content tree. Whole process is supervised by a team of field experts and educational technologists whose comments are taken into consideration to edit and update the story board.

Upon forming the story board, coding process is initiated for two different operational systems: Android and iOS. The application is developed via platforms like Android Studio and Eclipse for Android, and XCode and Swift for iOS. Similarly, platforms like Adobe Fireworks, Illustrator and Flash are utilized to design interfaces and visuals (e.g. graphics, animations). It is of vital importance to develop two identical versions of the mobile application for both Android and iOS in terms of design, features and functions. SQLite, which runs with both operational systems, is used for database support. Data for schematic and functional structures, including the story board are stored in the data base. On the other hand, codes and data for artificial intelligent- expert systems approaches are included for pedagogical operation mechanisms. At this point, necessary framework for the Augmented Reality is included by keeping touch with other elements of the mobile application.

The feedback for the prototype is obtained from field experts and educational technologists. Any required changes are included according to the constructive feedback. The resulting prototype is submitted to random student feedback and is edited accordingly. In order to achieve the optimum format, each version of the prototype is tested, updated, and edited accordingly before it is put to use.

The proposed system design processes are shown briefly in Figure 3:

Technical structure of the application focuses on the following aspects:

Figure 3. System design processes

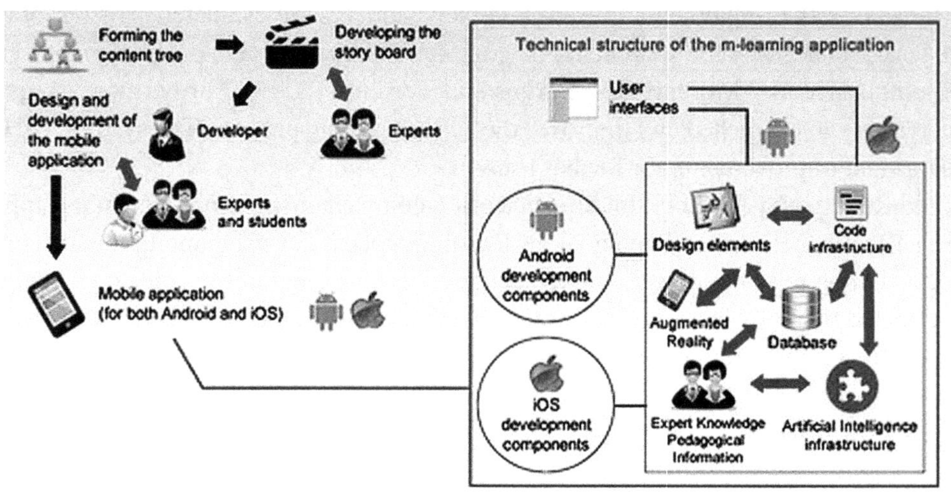

- Key using features and functions of the application are adapted to the different environments of Android and iOS operating systems. Android and iOS development components are employed within two separate development processes to obtain accurate versions of the mobile application.
- Database of both different versions is designed in the same structure. It is aimed to make development processes of user interfaces and code infrastructures easier and faster. In addition, design elements (i.e., icons, images, style files) are used directly in both Android and iOS versions.
- Interrelations among Design Elements, Code Infrastructure, Database, Expert Knowledge, Pedagogical Information, and Artificial Intelligence Infrastructure of the mobile application can be seen in Figure 3. This is because of running Expert Knowledge, Pedagogical Information and also Artificial Intelligence approaches under user-oriented sides of the application. AI has an important role on making the application adaptive to the learning process without the need for face-to-face teacher interaction. In detail, some well-known AI techniques like Artificial Neural Networks and also some optimization oriented techniques under Swarm Intelligence are used in order to form an intelligent working mechanism affecting the whole application in terms of experiencing an adaptive learning process (Cochocki, & Unbehauen, 1993; Hassoun, 1995; Kose, 2015; Miller, Sutton, & Werbos 1995).
- Augmented Reality infrastructure of the application has been formed to ensure a flexible approach, which is connected with the whole elements included like design elements, code infrastructure and of course, Artificial Intelligence infrastructure.

Potential Risks and Challenges

Before, during and after the implementation, some problems are likely to emerge; first predicable problem is related to interface: interfaces are partially monitored or run because different smart phones do not support the platform. In such cases, source codes are updated to support all the smart phones. Another relates to user interface freezes: after checking the compatibility, source codes are edited according to user smart phones. Moreover, visuals may appear partially or differently on different smart phones; in such a case, source codes of those schematic images are resized and relocated. To avoid such interface grounded pitfalls, design and coding process need to consider possible difficulties with instant and appropriate actions to be taken.

As an important component of the application designed, Augmented Reality may be also another challenging factor for students. In this sense, it may be a little difficult to interact with the real-world under the control of Augmented Reality. But this will transform itself in a short time to a great advantage of interaction and an enjoyable learning experience, after students learn how to benefit from Augmented Reality.

Another possible problem is regarding the mobile learning skills of students; they might have weak mobile learning skills. With exposure to supplementary training and support, they will be merged into the world of M-Learning in a comprehensive and contextualized way. Finally, students' perception of M-Learning might cause problems; they may perceive educational use of smart phones pointless. To overcome such a barrier, they will be encouraged to use small-scale educational applications and to reflect on their experiences. Positive experiences are believed to motivate them to change their perspectives.

FUTURE RESEARCH DIRECTIONS

As it can be understood, future research directions regarding to the application considered in this chapter includes running of some project phases. In this context, the project will include four phases at total, called as Preparation, System Design, Application, and Observing and Evaluation respectively. Principles considered in this chapter will take place as the second phase of the whole project flow. Before it, there is a preparation phase in which all pedagogical infrastructure needed for the project will be formed. After the system design phase, the application will be run by employing lots of statistical observations in order to have idea about the process. Finally, the application phase will be followed by an observing and evaluation phase in which students will have the responsibility to evaluate the application.

By including the proposed system design processes mentioned under Figure 3 as the second phase, a bigger picture of the project (project flow) is shown in Figure 4:

In addition to the mentioned project flow, the application will also include further steps including development of newer versions of the mobile application with more advanced Augmented Reality functions and also use of alternative Artificial Intelligence based approaches to achieve more educational outputs to be evaluated. Of course, future research will always include the related phases of statistical observations, group meetings, and evaluation works done by students.

Figure 4. Project flow

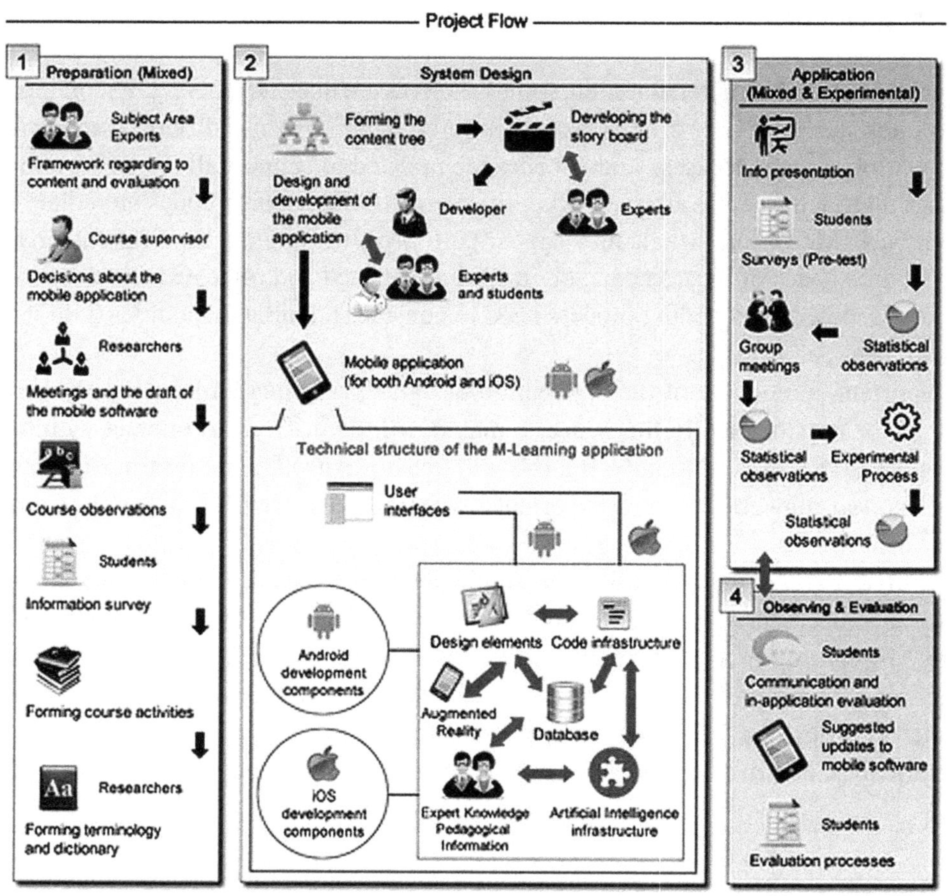

CONCLUSION

This chapter has provided a brief view on preparation of an improved educational approach for improving engineering students' English language skills. Because of the detailed background regarding to the infrastructure, more focus was given to the design of the application. Briefly, the study aims to analyze the content instruction context following the English as a Medium of Instruction (EMI) policy and to align higher education language instruction with the 21th century learning skills and requirements. Though numerous reforms have been initiated in Turkish higher education, undergraduate students in Turkey were ranked among the last ten countries in Europe for language competence (British Council, 2015). Students' proficiency level and lack of interest, and faculty's inefficiency and reluctance to deliver EMI classes pose challenges. Moreover, monotonous and difficult to access instructional materials and the scarcity of international students – relations add to these issues that can be handled with a needs analysis for identification purposes, and a curriculum and material revision (Cheon, & Reeve, 2015). Hence, a transition from EMI to CLIL with dual focus on language and content improvement seem to meet such a demand and challenge (Arkin, 2013; British Council, 2015). As the technological infrastructure, the application is supported with the most remarkable and effective approaches like Augmented Reality, M-Learning, and Artificial Intelligence.

The proposed study merging ICT skills through Augmented Reality, Artificial Intelligence supported M-Learning and language learning competences of the engineering students is believed to contribute to the pedagogical processes, student and teacher attitudes to M-Learning and practicing English through mobile phones. The program curriculum might benefit in diversifying and individualizing the content education with an application piloted and developed considering the students' needs in terms of language and content.

REFERENCES

Ab Hamid, S. H., & Yu Fung, L. (2007). Learn Programming by Using Mobile Edutainment Game Approach. In *Proceedings of DIGITEL, 1st IEEE International Workshop*. IEEE. doi:10.1109/DIGITEL.2007.31

Aleven, V. (2013). Help seeking and intelligent tutoring systems: Theoretical perspectives and a step towards theoretical integration. In *International Handbook of Metacognition and Learning Technologies* (pp. 311–335). Springer New York. doi:10.1007/978-1-4419-5546-3_21

Ally, M. (2009). Introduction. In Mobile learning: Transforming the delivery of education and training (pp. 1-8). Athabasca: Athabasca University Press.

Ally, M., Schafer, S., Cheung, B., McGreal, R., & Tin, T. (2007). Use of mobile learning technology to train ESL adults.*Proceedings of the 6th Annual International Conference on Mobile Learning.*

Arkin, I. E. (2013). *English-Medium Instruction in Higher Education: A Case Study in a Turkish University Context* (Unpublished doctoral dissertation). Eastern Mediterranean University.

Arslan, R. S., & Saka, C. K. (2010). Teaching English to science students via theme-based model of content-based instruction. *Journal of Turkish Science Education*, *7*(4), 26.

Attewell, J., & Webster, T. (2005). Engaging and supporting mobile learners. In *Proceedings of M-Learning 2004: Mobile learning anytime everywhere (A book of papers from M-LEARN 2004, Italy)*. Learning and Skills Development Agency.

Bayirtepe, E., & Tuzun, H. (2007). The effects of game-based learning environments on students' achievement and self-efficacy in a computer course. *Hacettepe University Journal of Education*, *33*, 41–54.

Beatty, K. (2003). *Teaching and Researching Computer-Assisted Language Learning*. Essex, UK: Pearson Education Limited.

Beauchemin, R. W. (2016). Augmenting education: Using augmented reality technologies. In D. Mentor (Ed.), *Handbook of Research on Mobile Learning in Contemporary Classrooms* (pp. 160–180). Hershey, PA: IGI Global. doi:10.4018/978-1-5225-0251-7.ch008

Bottino, R. M., Ferlino, L., Ott, M., & Travella, M. (2007). Developing strategic and reasoning abilities with computer games at primary school level. *Computers & Education*, *49*(4), 1272–1286. doi:10.1016/j.compedu.2006.02.003

Bozdogan, D. (2015). MALL Revisited: Current Trends and Pedagogical Implications. *Procedia: Social and Behavioral Sciences*, *195*, 932–939. doi:10.1016/j.sbspro.2015.06.373

British Council. (2015). *The State of English in Higher Education in Turkey*. British Council Turkey. Retrieved from http://www.britishcouncil.org.tr/sites/default/files/turkiyede_yuksekogretim_kurumlarindaki_ingilizce_egitimi.pdf

Carpio Cañada, J., Mateo Sanguino, T. J., Merelo Guervós, J. J., & Rivas Santos, V. M. (2015). Open classroom: Enhancing student achievement on artificial intelligence through an international online competition. *Journal of Computer Assisted Learning*, *31*(1), 14–31. doi:10.1111/jcal.12075

Cascales, A., Laguna, I., Pérez-López, D., Perona, P., & Contero, M. (2013). An experience on natural sciences augmented reality contents for preschoolers. In R. Shumaker (Ed.), *Virtual, Augmented and Mixed Reality. Systems and Applications* (pp. 103–112). Berlin: Springer-Verlag. doi:10.1007/978-3-642-39420-1_12

Chang, G., Morreale, P., & Medicherla, P. (2010). Applications of augmented reality systems in education. *Proceedings of Society for Information Technology & Teacher Education International Conference*.

Chee, Y. S. (2016). *Games-to-teach or games-to-learn: Unlocking the power of digital game-based learning through performance*. Dordrecht: Springer. doi:10.1007/978-981-287-518-1

Chen, C. M., & Chung, C. J. (2007). Personalized Mobile English Vocabulary Learning System Based on Item Response Theory and Learning Memory Cycle. *Computers & Education*, *51*(2), 624–645. doi:10.1016/j.compedu.2007.06.011

Chen, P., Liu, X., Cheng, W., & Huang, R. (2017). A review of using Augmented Reality in Education from 2011 to 2016. In *Innovations in Smart Learning*. Singapore: Springer. doi:10.1007/978-981-10-2419-1_2

Cheon, S. H., & Reeve, J. (2015). A Classroom-Based Intervention to Help Teachers Decrease Students Amotivation. *Contemporary Educational Psychology*, *40*, 99–111. doi:10.1016/j.cedpsych.2014.06.004

Chinnery, G. (2006). Emerging Technologies - Going to the MALL: Mobile Assisted Language Learning. *Language Learning & Technology*, *10*(1), 9–16.

Cochocki, A., & Unbehauen, R. (1993). *Neural networks for optimization and signal processing*. John Wiley & Sons, Inc.

Colpaert, J. (2004). From courseware to coursewear? *Computer Assisted Language Learning*, *17*(3-4), 261–266. doi:10.1080/0958822042000319575

Dabbagh, N., Benson, A. D., Denham, A., Joseph, R., Al-Freih, M., Zgheib, G., & Guo, Z. et al. (2016). *Learning Technologies and Globalization: Pedagogical Frameworks and Applications*. Springer International Publishing. doi:10.1007/978-3-319-22963-8

Demirel, O., Seferoglu, S., & Yagci, E. (2003). Developing Instruction Technologies and Material. Öğretim Teknolojileri ve Materyal Geliştirme. Ankara: PegemA. (in Turkish)

Dörnyei, Z., MacIntyre, P., & Henry, A. (Eds.). (2015). Motivational dynamics in language learning. Bristol: Multilingual Matters.

Ebner, M., & Holzinger, A. (2007). Successful implementation of user-centered game based learning in higher education: An example from civil engineering. *Computers & Education*, *49*(3), 873–890. doi:10.1016/j.compedu.2005.11.026

Elias, T. (2011). Universal Instructional Design Principles for Mobile Learning. *International Review of Research in Open and Distance Learning*, *12*(2), 143–156. doi:10.19173/irrodl.v12i2.965

Ellis, R. (2003). *Task-Based Language Learning and Teaching*. Oxford.

Finn, B. (2015). Measuring motivation in low-stakes assessment. *ETS Research Report Series*, *2*(2), 1–17. doi:10.1002/ets2.12067

Gee, J. P. (2005). What would a state of the art instructional video game look like? *Innovate: Journal of Online Education*, *1*(6), 1.

Gunes, F. (2012). Turkish instruction from tests to activities. (In Turkish) Testlerden etkinliklere Türkçe öğretimi. *Journal of Language and Literature Education*, *1*(1), 31–42.

Hamari, J., Shernoff, D. J., Rowe, E., Coller, B., Asbell-Clarke, J., & Edwards, T. (2016). Challenging games help students learn: An empirical study on engagement, flow and immersion in game-based learning. *Computers in Human Behavior*, *54*, 170–179. doi:10.1016/j.chb.2015.07.045

Hanesova, D. (2015). Development of CLIL teacher education in UMB (Introduction of European project). In Learning Together: To be a better CLIL Teacher. Banskå Bystrica.

Hassoun, M. H. (1995). *Fundamentals of Artificial Neural Networks*. Cambridge, MA: MIT Press.

Johnson, L., Levine, A., Smith, R., & Stone, S. (2010). *Simple augmented reality. The 2010 Horizon Report*. Austin, TX: The New Media Consortium.

Kalantzis, M., & Cope, B. (2005). *Learning by Design, Common Ground*. Melbourne, Australia: Common Grounds.

Kincheloe, J. L., Slattery, P., & Steinberg, S. R. (2000). *Contextualizing Teaching Introduction to Education and Educational Foundations*. Longman.

Kiziltan, N., & Ersanli, C. Y. (2007). The contributions of theme-based CBI to Turkish young learners' language development in English. *Journal of Language and Linguistic Studies*, *3*(1).

Kose, U. (2015). Present state of swarm intelligence and future directions. In *Encyclopedia of Information Science and Technology* (pp. 239–252). IRMA International. doi:10.4018/978-1-4666-5888-2.ch023

Kose, U., Koc, D., & Yucesoy, S. A. (2013). An augmented reality based mobile software to support learning experiences in computer science courses. *Procedia Computer Science*, *25*, 370–374. doi:10.1016/j.procs.2013.11.045

Kukulska-Hulme, A. (2005). *Mobile learning: A Handbook for Educators and Trainers*. Oxon, UK: Routledge.

Mano, M. R., & Ciletti, M. D. (2013). *Digital Design with an Introduction to Verilog HDL*. Pearson.

Marsh, D. (2002). *CLIL/EMILE - The European Dimension: Actions, Trends and Foresight Potential. DG Education & Culture*. European Commission.

Milgram, P., & Kishino, F. (1994). A taxonomy of mixed reality visual displays. *IEICE Transactions on Information and Systems*, *77*(12), 1321–1329.

Miller, W. T., Sutton, R. S., & Werbos, P. J. (1995). *Neural Networks for Control*. Cambridge, MA: MIT Press. doi:10.1007/978-3-642-57760-4

Motiwalla, L. F. (2007). Mobile learning: A framework and evaluation. *Computers & Education*, *49*(3), 581–596. doi:10.1016/j.compedu.2005.10.011

Nye, B. D. (2015). Intelligent tutoring systems by and for the developing world: A review of trends and approaches for educational technology in a global context. *International Journal of Artificial Intelligence in Education*, *25*(2), 177–203. doi:10.1007/s40593-014-0028-6

O'Driscoll, J. (2004). The discourse of English as T-Rex. In *Language and the Future of Europe: Ideologies Policies and Practices*. University of Southampton.

Peña-Ayala, A. (2013). Intelligent and Adaptive Educational-Learning Systems. *Smart Innovation, Systems and Technologies*, 17.

Saran, M., Cagiltay, K., & Seferoglu, G. (2008). Use of Mobile Phones in Language Learning: Developing Effective Instructional Materials. In *Proceedings of the Fifth IEEE International Conference on Wireless, Mobile, and Ubiquitous Technology in Education*. IEEE. doi:10.1109/WMUTE.2008.49

Schawabe, G., & Goth, C. (2005). Mobile Learning with a Mobile Game: Design and Motivational Effects. *Journal of Computer Assisted Learning*, *21*(3), 204–216. doi:10.1111/j.1365-2729.2005.00128.x

Schreppegrell, M., Achugar, M., & Oteiza, T. (2004). The grammar of history: Enhancing content based instruction through a functional focus on language. *TESOL Quarterly*, *38*(1), 67–93. doi:10.2307/3588259

Shamsudin, S., & Nesi, H. (2006). Computer-mediated communication in English for specific purposes: A case study with computer science students at Universiti Teknologi Malaysia. *Computer Assisted Language Learning*, *19*(4-5), 317–339. doi:10.1080/09588220601043164

Silberman, M. L., & Biech, E. (2015). *Active Training: A Handbook of Techniques, Designs, Case Examples, and Tips*. John Wiley & Sons. doi:10.1002/9781119154778

Stockwell, G., & Hubbard, P. (2013). *Some emerging principles for mobile-assisted language learning*. The International Research Foundation for English Language Education. Retrieved from http://www.tirfonline.org/wp-content/uploads/2013/11/TIRF_MALL_Papers_StockwellHubbard.pdf

Van der Wal, M. M., de Kraker, J., Kroeze, C., Kirschner, P. A., & Valkering, P. (2016). Can computer models be used for social learning? A serious game in water management. *Environmental Modelling & Software*, *75*, 119–132. doi:10.1016/j.envsoft.2015.10.008

Vera, M. D. M. S., Breis, J. T. F., Sánchez, J. L. S., & Espinosa, M. P. P. (2013). Practical experiences for the development of educational systems in the semantic web. *Journal of New Approaches in Educational Research*, *2*(1), 23–31. doi:10.7821/naer.2.1.23-31

Willis, J. (1996). *A Framework for Task-Based Language Teaching*. New York: Longman.

Zatarain-Cabada, R., Barrón-Estrada, M. L., Alor-Hernández, G., & Reyes-García, C. A. (2014). Emotion Recognition in Intelligent Tutoring Systems for Android-Based Mobile Devices. In *Mexican International Conference on Artificial Intelligence* (pp. 494-504). Berlin: Springer International Publishing. doi:10.1007/978-3-319-13647-9_44

ADDITIONAL READING

Alpaydin, E. (2014). *Introduction to Machine Learning*. MIT press.

Anaraki, F. B. (2009). A flash-based mobile learning system for learning English as second language. In *Computer Engineering and Technology, 2009. ICCET'09. International Conference on* (pp. 400-404). USA: IEEE. doi:10.1109/ICCET.2009.183

Andersson, R., Anker, M., Dunford, A., Lundqvist, J., & Weiss, A. (2016). HAMLIN: An augmented reality solution to visualize abstract concepts for science education. In *Proceedings of SIDeR'16–Student Interaction Design Research Conference*, Sweden: Chalmers Publication.

Brusilovsky, P. (1999). Adaptive and intelligent technologies for web-based education. *KI, 13*(4), 19-25.

Brusilovsky, P., & Peylo, C. (2003). Adaptive and intelligent web-based educational systems. *International Journal of Artificial Intelligence in Education*, *13*(2), 159–172.

Cavus, N. (2016). Development of an Intellegent Mobile Application for Teaching English Pronunciation. *Procedia Computer Science*, *102*, 365–369. doi:10.1016/j.procs.2016.09.413

Cavus, N., & Ibrahim, D. (2016). Learning English using childrens stories in mobile devices. *British Journal of Educational Technology*, n/a. doi:10.1111/bjet.12427

Chen, D. M., Tsai, S. S., Vedantham, R., Grzeszczuk, R., & Girod, B. (2009). Streaming mobile augmented reality on mobile phones. In *Mixed and Augmented Reality, 2009. ISMAR 2009. 8th IEEE International Symposium on* (pp. 181-182). USA: IEEE. doi:10.1109/ISMAR.2009.5336472

Chinnery, G. M. (2006). Emerging technologies: Going to the MALL (Mobile Assisted Language Learning). *Language Learning & Technology*, *10*(1), 9–16.

Dashtestani, R. (2016). Moving bravely towards mobile learning: Iranian students use of mobile devices for learning English as a foreign language. *Computer Assisted Language Learning*, *29*(4), 815–832. doi:10.1080/09588221.2015.1069360

Ekengren, B. (2009). *Mobile augmented reality*. MSc. Thesis, Royal Institute of Technology - School of Computer Science and Communication, Sweden: Royal Institute of Technology.

Fedosov, A., Elhart, I., Niforatos, E., North, A., & Langheinrich, M. (2016). SkiAR: Wearable Augmented Reality System for Sharing Personalized Content on Ski Resort Maps. In *Proceedings of the 7th Augmented Human International Conference* (pp. 46). USA: ACM. doi:10.1145/2875194.2875234

Feiner, S., Macintyre, B., & Seligmann, D. (1993). Knowledge-based augmented reality. *Communications of the ACM*, *36*(7), 53–62. doi:10.1145/159544.159587

Henrysson, A., & Ollila, M. (2004). UMAR: Ubiquitous mobile augmented reality. In *Proceedings of the 3rd International Conference on Mobile and Ubiquitous Multimedia* (pp. 41-45). USA: ACM. doi:10.1145/1052380.1052387

Hollerer, T., & Feiner, S. (2004). *Mobile Augmented Reality. Telegeoinformatics: Location-Based Computing and Services*. London: Taylor and Francis Books Ltd.

Hollerer, T., Feiner, S., Hallaway, D., Bell, B., Lanzagorta, M., Brown, D., & Rosenblum, L. (2001). User interface management techniques for collaborative mobile augmented reality. *Computers & Graphics*, *25*(5), 799–810. doi:10.1016/S0097-8493(01)00122-4

Hollerer, T., Feiner, S., Terauchi, T., Rashid, G., & Hallaway, D. (1999). Exploring MARS: Developing indoor and outdoor user interfaces to a mobile augmented reality system. *Computers & Graphics*, *23*(6), 779–785. doi:10.1016/S0097-8493(99)00103-X

Huang, T. C., Chen, C. C., & Chou, Y. W. (2016). Animating eco-education: To see, feel, and discover in an augmented reality-based experiential learning environment. *Computers & Education*, *96*, 72–82. doi:10.1016/j.compedu.2016.02.008

Ibanez, M. B., Di-Serio, Á., Villaran-Molina, D., & Delgado-Kloos, C. (2016). Support for Augmented Reality Simulation Systems: The Effects of Scaffolding on Learning Outcomes and Behavior Patterns. *IEEE Transactions on Learning Technologies*, *9*(1), 46–56. doi:10.1109/TLT.2015.2445761

Jennings, N. R. (2000). On agent-based software engineering. *Artificial Intelligence*, *117*(2), 277–296. doi:10.1016/S0004-3702(99)00107-1

Kaya, Z. (2002). *Distance Education*. Ankara: Pegem.

Keegan, D. (2004). *Foundations of Distance Education*. New York: Routledge.

Kose, U. (2014). On the State of Free and Open Source E-Learning 2.0 Software. International[IJOSSP]. *Journal of Open Source Software and Processes*, *5*(2), 55–75. doi:10.4018/ijossp.2014040103

Kose, U., & Koc, D. (2014). *Artificial Intelligence Applications in Distance Education*. Hershey, PA, USA: IGI Global.

Kose, U., & Tufekci, A. (2015). On the Future Possibilities of Artificial Intelligence Based M-Learning Content Development. *The International Journal of Human, Community & Technology, 1*(2).

Kounavis, C. D., Kasimati, A. E., & Zamani, E. D. (2012). Enhancing the tourism experience through mobile augmented reality: Challenges and prospects. *International Journal of Engineering Business Management, 4*.

Lin, Y. T., Kao, C. L., & Lan, Y. J. (2016). The effects of mobile learning on students oral performance in Mandarin Chinese and their attitudes. *International Journal of Mobile Learning and Organisation*, *10*(1-2), 78–101. doi:10.1504/IJMLO.2016.076191

Liu, P. L. (2016). Mobile English Vocabulary Learning Based on Concept-Mapping Strategy. *Language Learning & Technology*, *20*(3), 128–141.

McMahon, D. D., Cihak, D. F., Wright, R. E., & Bell, S. M. (2016). Augmented Reality for Teaching Science Vocabulary to Postsecondary Education Students With Intellectual Disabilities and Autism. *Journal of Research on Technology in Education*, *48*(1), 38–56. doi:10.1080/15391523.2015.1103149

O'Shea, P. M., & Elliott, J. B. (2016, June). Augmented Reality in Education: An Exploration and Analysis of Currently Available Educational Apps. In *International Conference on Immersive Learning* (pp. 147-159). Berlin-Heidelberg: Springer International Publishing. doi:10.1007/978-3-319-41769-1_12

Ponce, J., Ornelas, F., Álvarez, F., & Toscano, B. (2016). Use of Augmented Reality a New Vision on the Massive Open Online Courses. *User-Centered Design Strategies for Massive Open Online Courses (MOOCs),* 254.

Rekimoto, J., & Ayatsuka, Y. (2000). CyberCode: designing augmented reality environments with visual tags. In *Proceedings of DARE 2000 on Designing Augmented Reality Environments* (pp. 1-10). ACM. doi:10.1145/354666.354667

Russell, S. J., Norvig, P., Canny, J. F., Malik, J. M., & Edwards, D. D. (2003). *Artificial Intelligence: A Modern Approach*. Upper Saddle River: Prentice Hall.

Specht, M. (2014). Design of Contextualised Mobile Learning Applications. *Increasing Access,* 61.

Stead, G. (2014). Open Formats for Mobile Learning. *Increasing Access*, 99.

Stockwell, G. (2016). Mobile language learning. In *The Routledge Handbook of Language Learning and Technology* (pp. 296–307). UK: Taylor and Francis Inc.

Tan, T. H., & Liu, T. Y. (2004). The mobile-based interactive learning environment (MOBILE) and a case study for assisting elementary school English learning. In *Proceedings of Advanced Learning Technologies. IEEE International Conference* (pp. 530-534). USA: IEEE.

Thornton, P., & Houser, C. (2005). Using mobile phones in English education in Japan. *Journal of Computer Assisted Learning*, *21*(3), 217–228. doi:10.1111/j.1365-2729.2005.00129.x

Tufekci, A., Ekinci, H., & Kose, U. (2013). Development of an internet-based exam system for mobile environments and evaluation of its usability. *Mevlana Int. J. of Education*, *3*(4), 57–74. doi:10.13054/mije.13.59.3.4

Van Eck, R. (2008). Building Artificially Intelligent Learning Games. In V. Sugumaran (Ed.), *Intelligent Information Technologies: Concepts, Methodologies, Tools, and Applications* (pp. 793–825). Hershey, PA, USA: IGI Global. doi:10.4018/978-1-59904-941-0.ch048

Wagner, E. D. (2005). Enabling mobile learning. *EDUCAUSE Review*, *40*(3), 41–42.

Woolf, B. P. (2010). *Building Intelligent Interactive Tutors: Student-Centered Strategies for Revolutionizing E-Learning*. USA: Morgan Kaufmann.

Yilmaz, R. M. (2016). Educational magic toys developed with augmented reality technology for early childhood education. *Computers in Human Behavior*, *54*, 240–248. doi:10.1016/j.chb.2015.07.040

KEY TERMS AND DEFINITIONS

Artificial Intelligence: (1) A term that is used to describe the feature, function or characteristic of computer systems or machines that try to simulate human-thinking behavior or human intelligence. (2) A field of Computer Science, which is based on research studies or developments on providing intelligent systems simulating the human-thinking behavior or human intelligence.

Augmented Reality: Some kind of reality in which virtual world based elements – objects are added to the real world based environment.

E-Learning: Learning activities, which are supported and performed via electronic media – sources, by enabling individuals to experience educational processes on anytime and anywhere.

Intelligent M-Learning: M-Learning supported with Artificial Intelligence based approaches, methods, or techniques.

Mixed Reality: The reality including elements – objects from both real world and virtual world based approaches.

Mobile Assisted Language Learning: An approach of language learning supported with mobile technologies.

M-Learning: A type of E-Learning done via mobile technologies including both software and hardware oriented factors.

Mobile Learning Environments: Software oriented platforms (especially over Web) that is used for M-Learning purposes.

Chapter 12
Integration of Augmented Reality and Virtual Reality in Building Information Modeling:
The Next Frontier in Civil Engineering Education

Sai Rohit Chenchu Boga
VIT University, Chennai, India

Bhargav Kansagara
VIT University, Chennai, India

Ramesh Kannan
VIT University, Chennai, India

ABSTRACT

In an educational perspective, unlike other disciplines, hands-on practice is difficult to come by in Civil Engineering. By providing a student with a realistic 3D simulation, we propose a concept that improves the understanding of the individual and eliminate guess-work entirely. Our platform that makes education fun and interactive by eliminating the constraints of a conventional teaching environment by incorporating Virtual Reality (VR) or Augmented Reality (AR) as a tool. AR can help the target audience visualize a model (to scale) in all three dimensions in the palm of their hand. This chapter explores the use of interactive 3D game environments in design visualization in Building Information Modeling (BIM) by adopting various available software packages and APIs. VR will allow the prospective customer to enter and explore a structure before it is constructed. This can be achieved by making use of a powerful game engine, in this case, Unity3D. In this chapter, we will describe ways to pivot Unity's functions towards the benefit of civil engineering.

DOI: 10.4018/978-1-5225-2110-5.ch012

INTRODUCTION

As defined by Wikipedia, "Augmented reality (AR) is a live direct or indirect view of a physical, real-world environment whose elements are augmented (or supplemented) by computer-generated sensory input such as sound, video, graphics or GPS data" (Wikipedia, 2016a). Augmented Reality helps improve the person's optical sense by superimposing a computer generated image over the user's view of the real world. By contrast, virtual reality replaces the real world with a simulated one. With newer AR technology like adding computer vision and object recognition, the surrounding world of the user becomes interactive and docile. Information about the environment and its objects is overlaid on the real world. See Figure 1.

In technical terms, virtual reality or virtual realities (VR), also known as immersive multimedia or computer-simulated reality, is a computer technology that replicates an environment, real or imagined, and simulates a user's physical presence and environment in a way that allows the user to interact with it (Wikipedia, 2016b). Simply put, Virtual Reality can be defined as the digital emulation of an environment that can be experienced by a person's multiple senses. An individual has more senses than the standard five of taste, touch, smell, sight and hearing. For example, the sense of balance is essential to human locomotion.

Our entire perception of reality is a combination of sensory information processed by our brain. With virtual reality, we present our brain with made-up information, and trick it into believing in a world that doesn't actually exist.

The pace of technological evolution and adoption is growing exponentially in the modern age. Every major stream of science and art rushes to incorporate the next "big thing" developed by technology. We have seen this in the past. We will continue to see the same trends in the future. Beginning before 1900, it took a couple of decades for the telephone to touch 50% of households. It took five years or less for cellphones to accomplish the same level of penetration in 1990. These statistics are reported from the Harvard Business Review (McGrath, 2016). HBR goes on to state that, by analogy, firms with competitive advantages in those areas will need to move faster to capture those opportunities that present

Figure 1. Consumption spreads faster today -Harvard Business Review

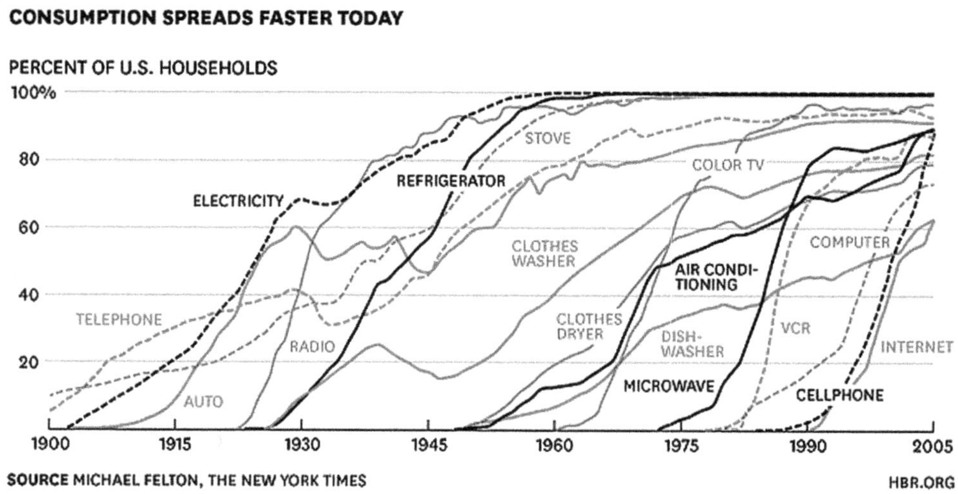

themselves. New innovations are being adopted faster than ever. What is certain is that the smartphone population is rising, and with this, the level of processing power is too. More and more consumers are carrying phones capable of displaying AR/VR content.

We would like to go on record and make a wild assertion that augmented and virtual reality are the next big things to hit all fields. On closer inspection, the statement will prove to be quite sound. VR/AR aren't new kids on the block but they are the most 'popular'. Panoramic paintings in the 1800s were the first attempt at virtual reality aimed at surrounding the viewer making them feel present in an historic event. In 1838, the first stereoscope viewers were created. In the 1930s, science fiction writer, Stanley G. Weinbaum, described devices very similar to the modern day VR headsets. The mid-1950s creation, "Sensorama" was a major step for consumer virtual reality. It presented an immersive experience in an arcade-style setup featuring stereo speakers, a stereoscopic 3D display, fans, smell generators and a vibrating chair. VR and AR has only picked up mainstream traction in recent years.

Fast forward to modern times, CES2016 finally ushers in an explosion in consumer AR/VR devices and applications. Shawn DuBravac, chief economist for the Consumer Technology Association (CTA), which puts on CES, predicts sales of VR and AR headsets will grow from approximately 200,000 in 2015 to over 1.2 million in 2016, which marks a 500% increase (Harrison, 2003). Major companies like Samsung, Microsoft, Oculus VR, Intel, Qualcomm and many more, have made significant investments and contributions to augmented and virtual reality platforms. VR/AR has many future applications. It is generally accepted that early adoption will be majorly done by the gaming industry because of the massive backing from Playstation VR and Microsoft. In our opinion, it is important for the construction and real estate industries to be quick in the adaptation of this technology.

In the construction industry, AR/VR can help visualize the entire construction process at every step. "Hospitality will never be the same," DuBravac says. "Before booking a cruise you'll be able to walk into the cabin and tour the entire ship. You'll no longer be surprised when you walk into a hotel room because you'll have already experienced it in a VR setting" (Gaudiosi, 2016). A customer can visualize his structure in its entirety before construction. A structure in the palm of their hand is very lucrative for many people. Augmented and Virtual Reality are the next logical steps to building information modelling.

A few classrooms were chosen to experiment the compatibility of modern educational devices into the existing syllabi. The assimilation of the created apps into the target classrooms had a few setbacks. One noteworthy case was the Virtual Tour application that was showcased before. Many students claimed to have experienced nausea and headaches after their time using the Virtual Reality application. This can be attributed to the well documented, Virtual Reality Sickness. It is a caveat to the technology itself. More advanced Virtual Reality systems like the Oculus Rift are better equipped and have improved in their ability to handle this limitation. However, the more primitive Google Cardboard system that we utilized is more susceptible to this motion sickness. It has been observed that the students were more comfortable with higher resolution screens and AMOLED displays as the image was crisper and clearer. Turning down brightness was one immediate partial solution to the headaches. It was also suggested that the VR experience be more realistic. As technology evolves, we will begin to witness the emergence of photorealistic virtual worlds.

User Interface is a crucial element in AR and VR app development. The interface must be non-intrusive and easy to adapt to. Our AR applications did not receive any complaints in this regard. In Virtual Reality, the navigation interface is crucial. In our Virtual Tour app, we used a gaze based movement system which made a few users uncomfortable. In future applications, we have decided to incorporate an optional controller-based movement input to make movement more responsive, like in a video game. A controller would allow for more user interaction with the environment. Google Cardboard's single physical input was a major limitation to our application. It has been observed that a few outdated smartphones were incompatible with these apps. This is a known problem. Most new smartphones running Android can easily run these apps. Apps can also be developed for iOS with minimal additional effort.

The integration of Augmented Reality into an educational environment was more seamless. The "Architecture and Town Planning" class was chosen and an app was developed with the professor to demonstrate the different architectural styles. Handouts of the reading material with additional image trackers for the application were circulated to the students. The app was later distributed in the class. Students simply had to point at the image trackers and 3D models popped out of the page. For example, below the information corresponding to the Corinthian column was a diagram of the said-column which was a disguised tracker. Students point their cameras at the diagram through their phones and a labelled three-dimensional model of the column appeared before them. Students could get a look at the column from all directions and truly understand the style of architecture that the Corinthian order dictated. Unlike the 2D pictures that were utilized earlier, the 3D models that they were now experiencing added significant value to their understanding of the subject. It can be concluded that the experiments returned a positive and students were eager for the adaptation of similar tools to aid the curriculum.

The greatest challenge is in convincing professors to utilize this new medium of knowledge dispersion. A few professors we spoke to were under the opinion that this new platform was an additional burden. Why fix something that is not broken? PowerPoint and other traditional teaching tools are convenient and offer less hassle. It can be reasoned that this resistance is observed for any transition into new technology. Time and an understanding of the massive benefits will be the remedy to this problem. Another obstacle is that most professors do not have any prior coding experience. There will be a dependency on a third party for content creation. Another roadblock is the development time of a functional application. Professors would need to commission an app two weeks in advance for timely delivery. It can be argued that this is not a problem as professors often have such visibility of the curriculum and these apps can be re-used every semester. Despite a few shortcomings, the usage of AR/ VR in education seems to be absolute. The dream of a quintessential teaching tool will drive the technology and its adaptation forward.

This chapter has been divided into 2 different equally important segments. The first segment is augmented reality of which a majority will be completed by the first review. The second segment will mostly deal with VR applications. Working prototypes will be showcased in each review.

The major objectives of this chapter are as follows:

- To utilize a gaming engine like Unity3D to develop interactive and immersive game environments
- To develop applications that will aid education in Civil Engineering
- To expand the domain and reach of civil engineering by adopting the latest trends of technology
- To bring massive structures to the "palm" of the end user
- To develop easy-to-use smartphone applications that help visualize a structure and/or objects
- To develop a platform that makes education fun and interactive by eliminating the constraints of a 2D teaching environment by incorporating VR/AR
- To develop a tool for the real estate/construction industry that helps visualize the end product before a massive investment
- To develop a strong marketing tool for the end user that allows them to see, feel, and customize a product before purchase

REVIEW OF LITERATURE

The literature review process was initiated in the December of 2015. The best understanding of workflow came from the conference paper "Using 3D Design software, BIM and game engines for architectural historical reconstruction" by Stefan Boeykens (2011). A quick skim through the paper provided a valuable infographic that showcased various approaches to our problem statement. See Figure 2. It became apparent that utilizing a game engine would be required.

In "Integrating BIM and gaming for real-time interactive architectural visualization", author Wei Yan et al. describes the difficulties in working with a gaming engine (Yan, Culp, & Graf, 2011). We refer to multiple citations from this paper. In 2002, S. Lehtinen stated the hitches in importing and exporting geometry to and from common CAD software (Lehtinen, 2002). An interesting finding is the use of CryENGINE® by Reynes and Yerli to present architectural visualizations (Reynes & Yerli, 2007). Evidence was found that a certain A.S. Indraprastha utilized the Unity® 3D game engine to develop a virtual urban environment (Wang et al., 2013). At this point, along with research on the side, we singled in on Unity® 3D as the game engine to work on among engines like CryENGINE®, Unreal Engine 4®, Source Engine etc.

It is possible to import models from CAD or BIM tools into games through complicated methodologies as depicted by Uddin and Yoon (2002), but thorough internet research shows that this is no longer the case. Technology has evolved. Most modern CAD tools now export into formats that game engines like Unity can directly import and process. Another challenge we realized through "The Investigation on Using Unity3D Game Engine in Urban Design Study" by Indraprastha and Shinozaki (2008) is the requirement of C# and Javascript for manipulating the game environment and engine (Indraprastha & Shinozaki, 2008). We recognized the necessity and began the tedious quest of becoming familiar with the two programming languages. Our personal goal of creating Android applications required us to be acquainted with the Android SDK and development techniques. Various off-the-counter and online resources helped us achieve this.

For illustrating building construction processes, BIM and 4D (3D space and time) visualization have been studied. For example, an approach is presented by J. Tulke in "4D construction sequence planning—new process and data model" for the preparation of 4D construction process simulations by using both 3D models and BIM (Tulke, Nour, & Beucke, 2008). Revit models are used in conjunction with Naviswork to add the extra dimension of time. The possibility of exporting of such 4D models into the augmented reality format is being pursued. An interesting real life application of augmented reality that was found through the literature review is a case study, "Collaborative BIM-based Markerless Mixed Reality Framework for Facilities Maintenance" by Khaled El Ammari and Amin Hammad which proved to be inspiring (Ammari & Hammad, 2014).

RESEARCH METHODOLOGY

After research, and many trial-and- errors, we have perfected a methodology to develop augmented reality applications using a combination of multiple software. In this report, we will be describing the best streamlined way to develop a basic augmented reality app that showcases the design of an apartment.

Figure 2. Infographic showing the various workflows to achieve our problem statement from Stefan Boeykens' paper

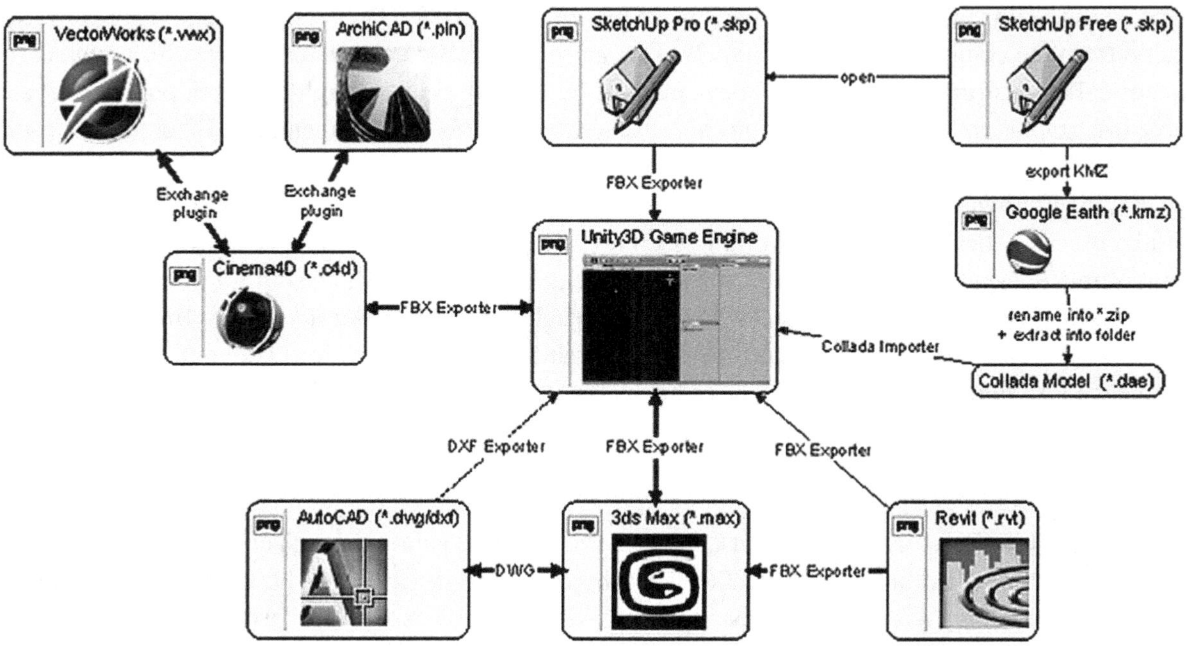

Step 1: Using any preferable technique, model the apartment in any CAD software. In this example, we use AutoCAD for simplicity. It is recommended that Revit is used. See Figure 3.

Step 2: Export the 3D file into a universal 3D format like ".fbx" (**F**ilm**box**). This format is acceptable in most game engines and software like 3ds Max. It is infact recommended to use 3ds Max for texturing as we did, as textures often aren't recognized from a direct AutoCAD import. See Figure 4.

Figure 3. Different views of an apartment in AutoCAD

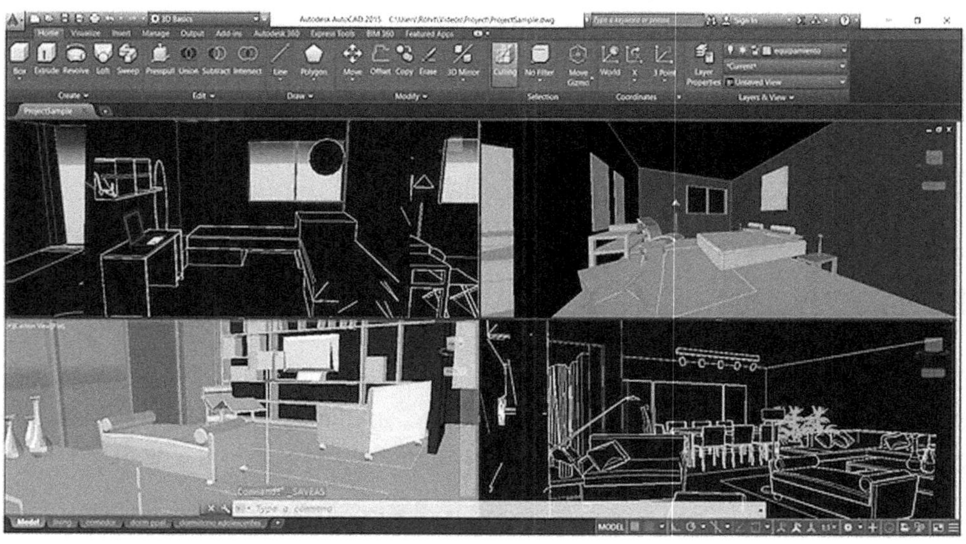

Figure 4. AutoCAD allows files to be exported into multiple file formats. A view of the export manager is shown

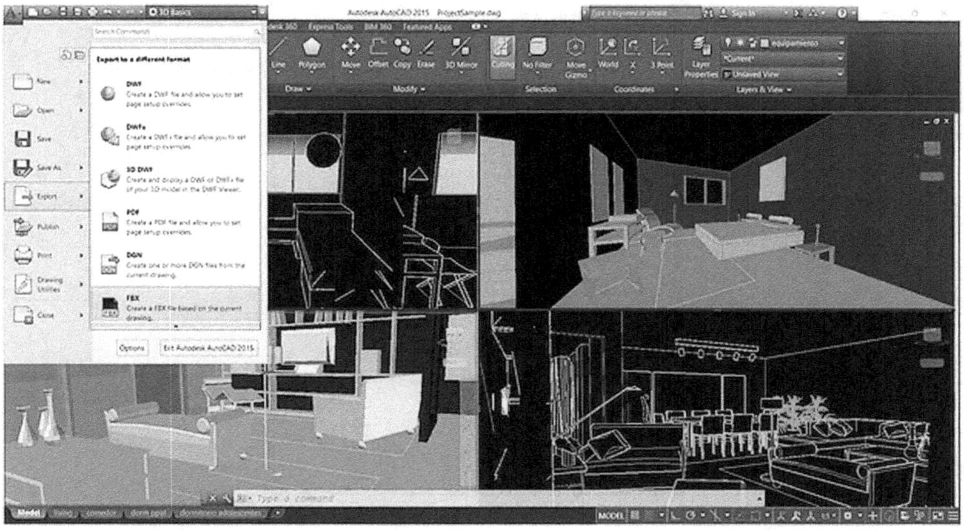

Step 3: Log on to the Vuforia™ Developer Portal, and download the required unity packages. This task is basic and hence we skip forward to License Key. See Figure 5.

Step 4: Next, select the image that will serve as the tracker for your model. Ideally, the image should be of high quality and trackable details. Once uploading the image, Vuforia™ will show you a rating of the tracker. Download the dataset of your trackers. See Figure 6.

Figure 5. A view of the Vuforia™ Developer Portal's License Manager

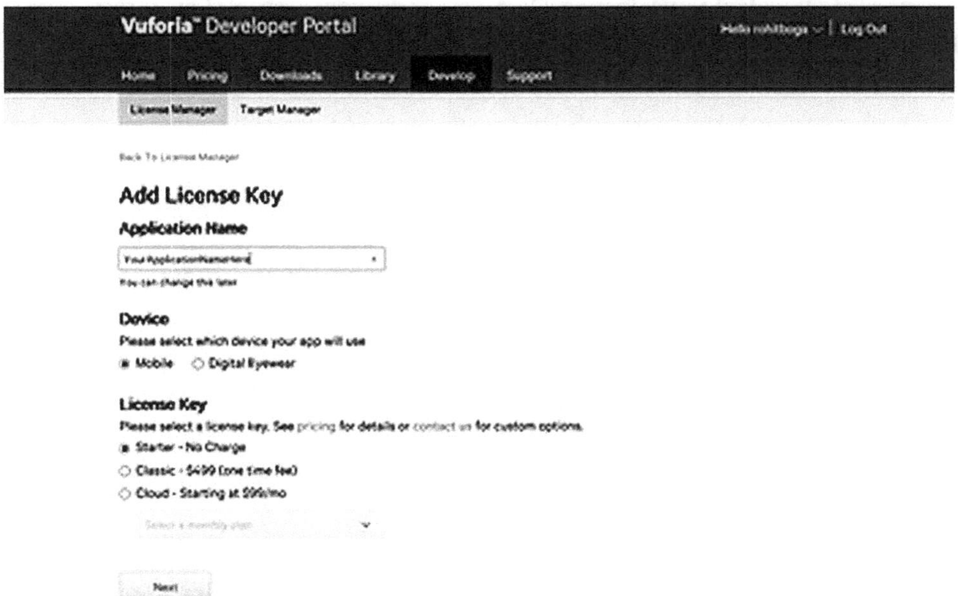

Figure 6. A view of Vuforia™ Developer Portal's Target Manager

Step 5: Proceed to open Unity3D, create a new project with any assets you may require. It is advised you open a 3D project. See Figure 7.

Step 6: Import the Vuforia™ Package from the Developer Portal as well as the Tracker Unity package from earlier. Remove the main camera. You should end up with something like this, a blank slate. See Figure 8.

Step 7: Apply the ARCamera and ImageTarget prefab into the scene. The result is as follows. The white surface represents the surface below your target. Once, you select your target the surface will disappear. See Figure 9.

Step 8: Add your license key created previously to the ARCamera, and allow the data to load as shown in the inspector. Refer to the picture below. See Figure 10.

Step 9: In ImageTarget, select your desired Data Set, and Image Target. Our target, here, is currency. See Figure 11.

Figure 7. Unity3D's New Project Window

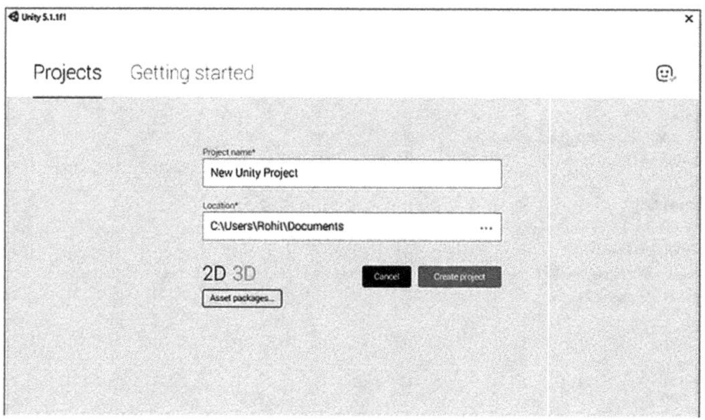

Figure 8. A visualization of a fresh Unity Project ready for development

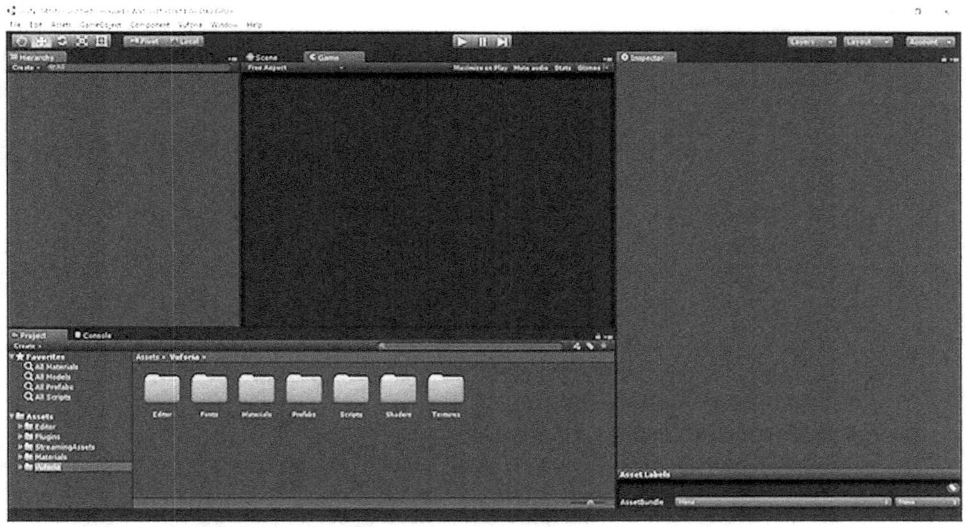

Figure 9. A visualization of the Unity scene with ARCamera and ImageTarget prefabs applied

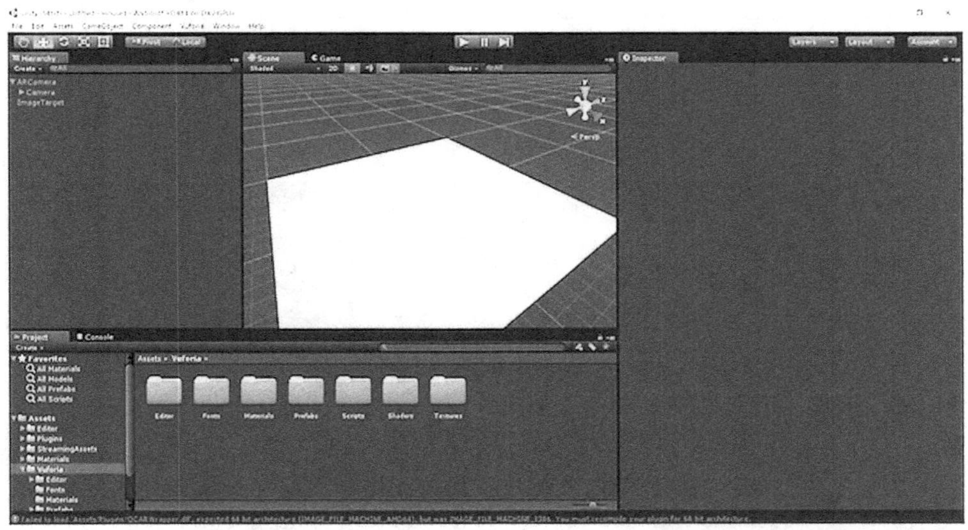

Step 10: In standard Unity practice, place your ".fbx" model above the tracker in the relative scale. Any animation, scripting, game logic, physics etc. can be added now. Most importantly, textures are added to the model. In this example, we will keep the application basic. Textures alone have been added. See Figure 12.

Step 11: Save the scene, and enter the build settings. Add current scene to the build settings. Choose Android as the publication platform. See Figure 13.

Step 12: Move on to Player Settings and make sure to choose the proper API. Name the application and define the proper settings. A few are shown below. See Figure 14.

Figure 10. The image shows the procedure to enter the license key from Vuforia™ Developer Portal and initialize the data

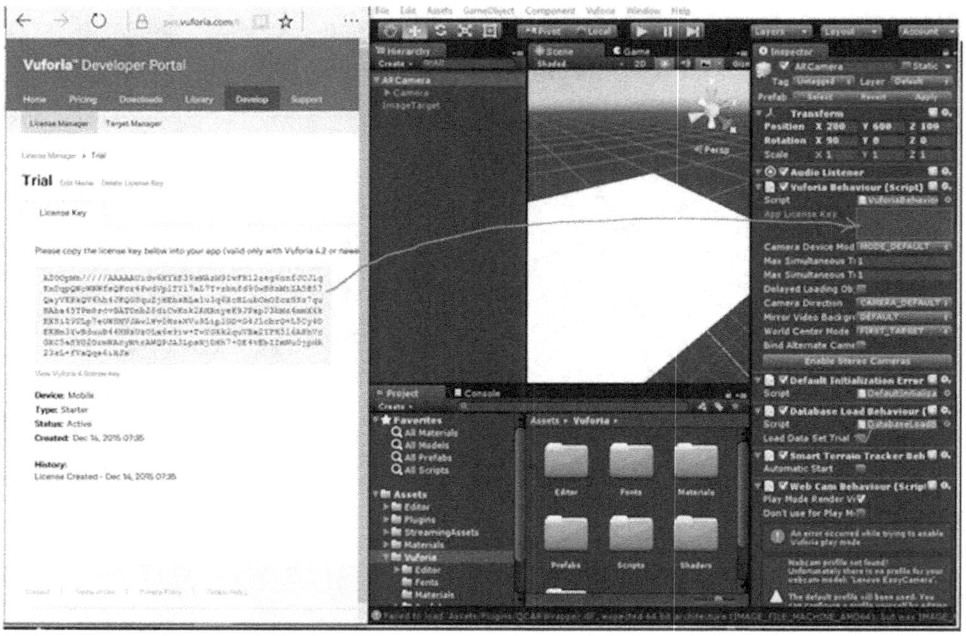

Figure 11. The target is initialized with the help of the ImageTarget prefab

Step 13: Compile and build the application. Push the application to your phone and test the build. See Figure 15.

Step 14: You are ready to publish the basic application. It is to be remembered much more customization and optimization is possible in this process. Various products we have created will showcase the power of this powerful backbone. This tutorial/methodology is just the tip of the iceberg. It is the easiest way to create an augmented reality app as per our knowledge.

Figure 12. A 3D apartment model has successfully been imported about the target into the current scene

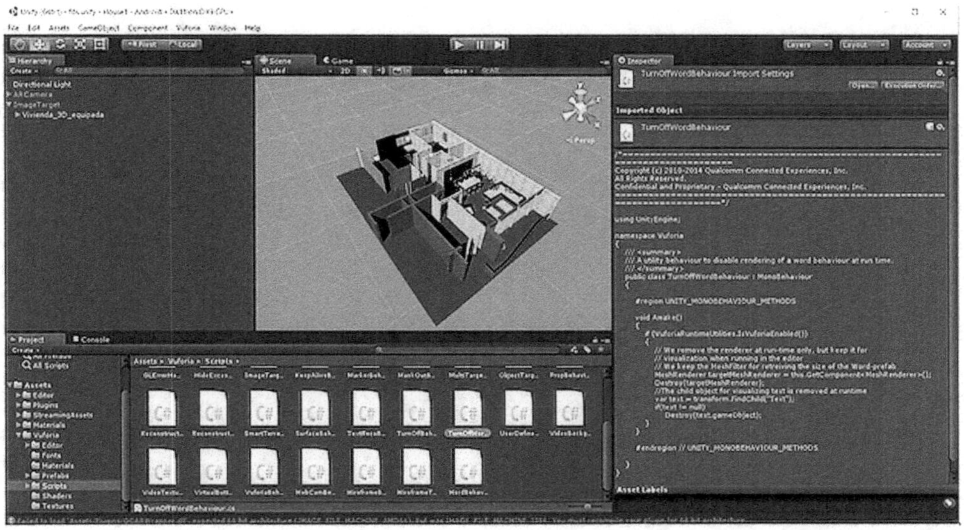

Figure 13. The scene is ready to be built into an application

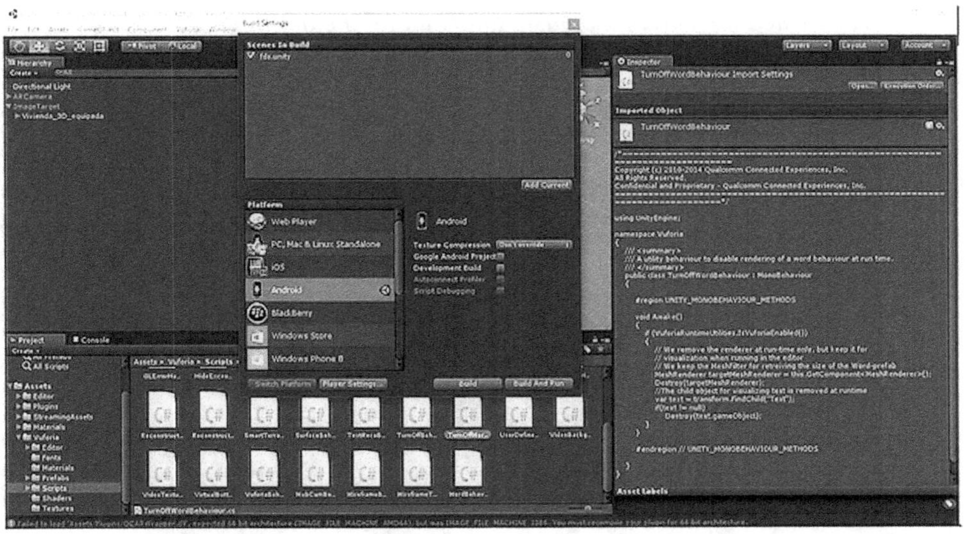

Virtual Reality

Virtual Reality takes augmented reality to the next level. It allows the end user to have a significant immersive and interactive experience. To create a virtual reality, two cameras are placed inches apart in the scene to mimic the stereo vison of human eyes. The user views the image from the left camera with the left eye, and similar for the right eye to create an illusion of depth. Using the gyroscope of the smartphone, we can track the direction the user is gazing. The main challenge of using a setup like Google Cardboard is that the application can only use two inputs, the gyroscope and the physical button provided. This limitation called for a drastic improvisation in the user interface.

Figure 14. Player Settings that can be used to begin the build process

Figure 15. We have reached the end of the basic tutorial and the application is being assembled

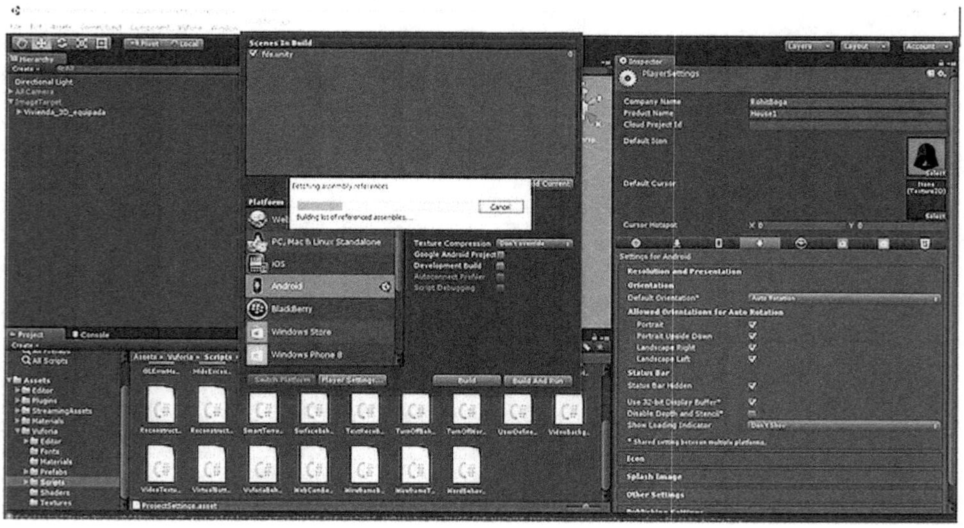

Unlike Oculus Rift, Google Cardboard does not natively utilize a controller. A controller would allow different actions to have their own physical buttons. A joystick would easily allow the person to explore the environment. An implementation like a joystick did not seem innate to what the Cardboard experience demanded. Instead, we used a technique where by tracking the user's gaze, when the person looks down in a direction, he walks in that direction. The physical clicker is used to interact with the scene. For example, a person could change the materials of different game objects by clicking the physical button.

Another significant problem that comes to mind is that of the camera not being a rigid object. It can hence pass through walls, doors, and other items. The challenge was to make the camera constrained like a real person so it is effected by gravity, collides with obstacles and other parameters. Necessity demanded innovation again. This time we applied a script that allowed a rigid capsule to follow the player's camera continuously as a stand in for the person's body. This capsule was given all the necessary properties.

Step 1: Without further ado, proceed to open Unity3D, create a new project with any assets you may require. It is advised you open a 3D project. See Figure 16.

Step 2: Stepping out of our comfort zone, instead of Vuforia™, we make use of the Cardboard SDK that is available free on Google (Google Developers, 2016). The current version of the SDK is v0.6. Import the SDK as shown. See Figure 17.

Step 3: In the prefabs folder of your recently imported package, drag and drop the CardboardMain prefab into the project hierarchy. Make sure that its location is at 0,0,0. Set the Unity Editor Emulation Set-

Figure 16. The Unity3D New Project Window is shown again

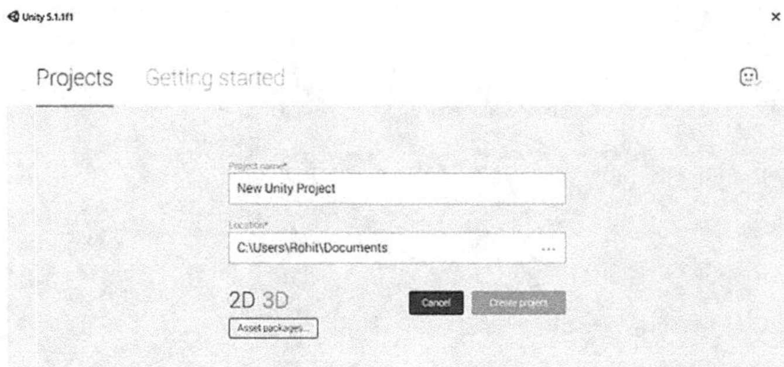

Figure 17. Visualization of the process to import custom packages to the scene

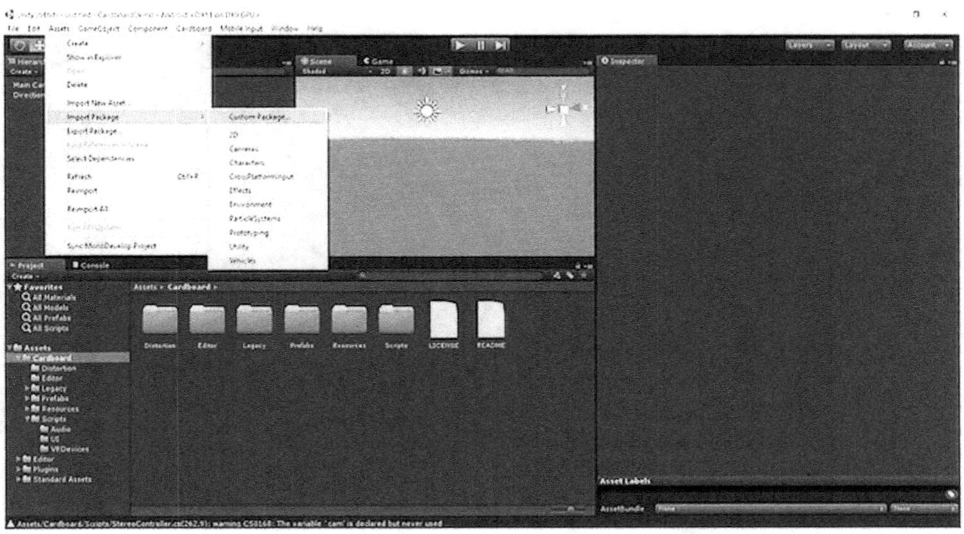

ting as Cardboard May 2015 or the model you are developing for. We add a physics raycaster to our CardboardMain camera so that we can detect hits against 3D objects with colliders. See Figure 18.

Step 4: Similar to AR, add the fbx model you wish to view, to the scene. Move the model to a location such that the room is surrounding the two cameras. Scale the model accordingly. An example is shown below. See Figure 19.

Step 5: To create an example of a click event, we will make a cube to manipulate. Place it in a location easily locatable in the scene for demonstrative purposes. Scale it accordingly. See Figure 20.

Step 6: Create a new C# script. To demonstrate the event, our example script will change the color of the cube. See Figure 21.

Figure 18. The Stereo Cameras are added at the global origin of the scene through the prefab

Figure 19. A texture-less model is added to the scene for demonstrative purposes

Figure 20. A cube is added to the scene to demonstrate an event trigger

Figure 21. A simple code that when executed changes the color of the game object it is attached to

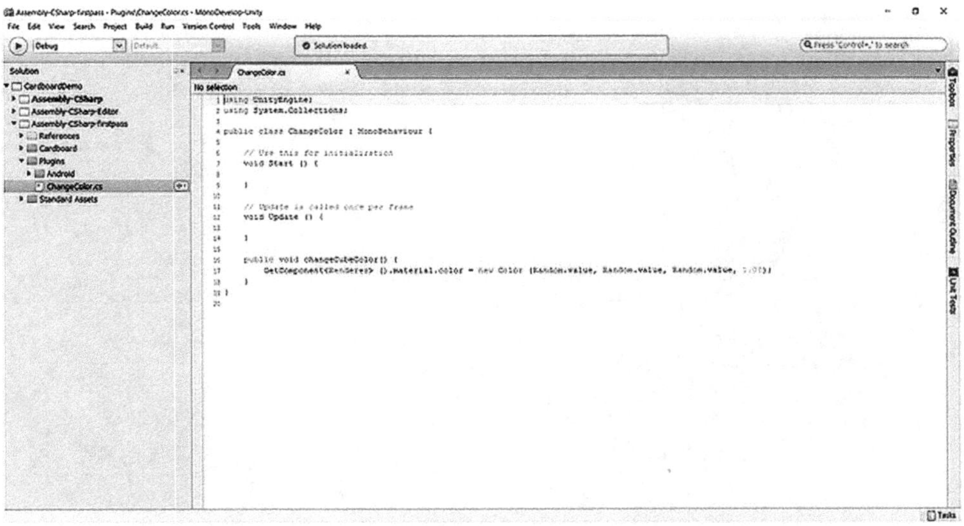

Step 7: Create an EventSystem to respond to various events triggers in the scene. Do this by right clicking on the hierarchy tab and navigating to the UI menu. Drag the GazeInputModule.cs from the scripts folder of the Cardboard SDK package. In the EventSystem, uncheck the Standalone Input Module and the Touch Input Module. You should end up with something similar to the screenshot shown below in Figure 22.

Step 8: Drag the newly created script for changing the color onto the cube. Add an EventTrigger on to the Cube gameobject. EventTrigger can be added using the Add Component button in the Inspector. See Figure 23.

Figure 22. Visualization that demonstrates how the GazeInputModule is added to the EventSystem

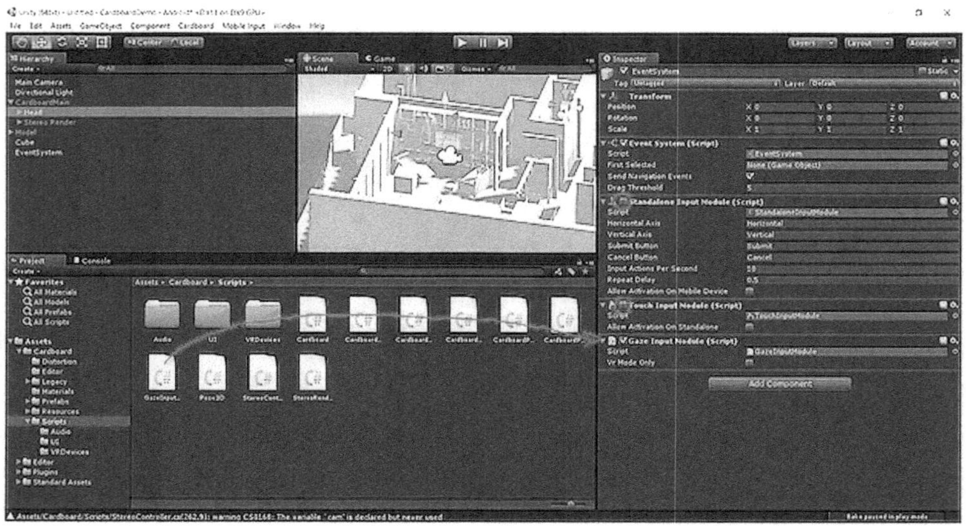

Figure 23. Visualization that details the technique to add the script to the game object

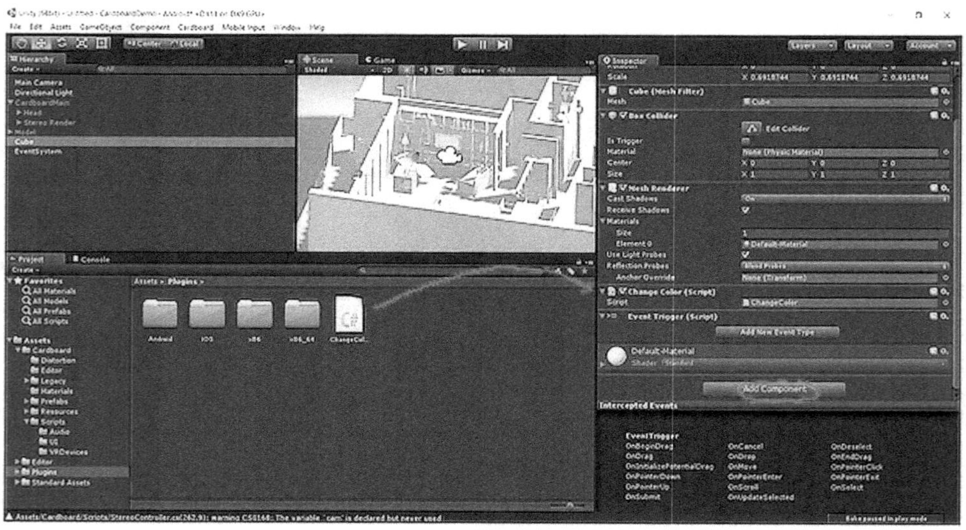

Step 9: Create a new Event type. For example, PointerClick is triggered when the Google Cardboard physical clicker is pressed. PointerExit will trigger an action when the person's gaze moves away from the game object. See Figure 24.

Step 10: Click the "+" symbol and drag the Cube game object to the Object box (below "Runtime Only"). This helps the EventTrigger to associate with the concerned object and different scripts attached to the object. Navigate through the Function menu and select the function, we created, to change the color of the cube. See Figure 25.

Step 11: The app is now ready to run. For emulation, delete main camera, and play the scene. To rotate your virtual head, hold alt for Horizontal/Vertical and ctrl for Tilt and move your mouse. Look at

Figure 24. The drop down menu with all the possible types of event triggers

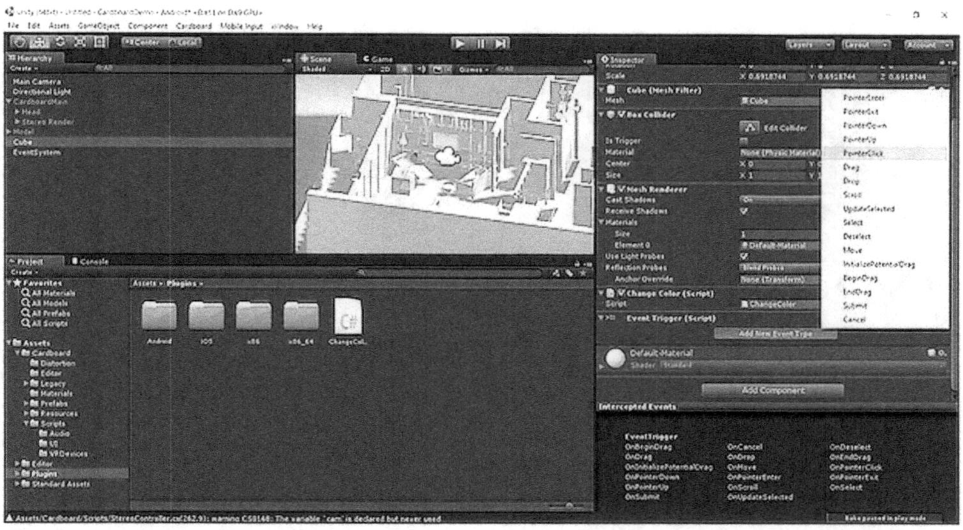

Figure 25. The function in the code created earlier is called by the event trigger

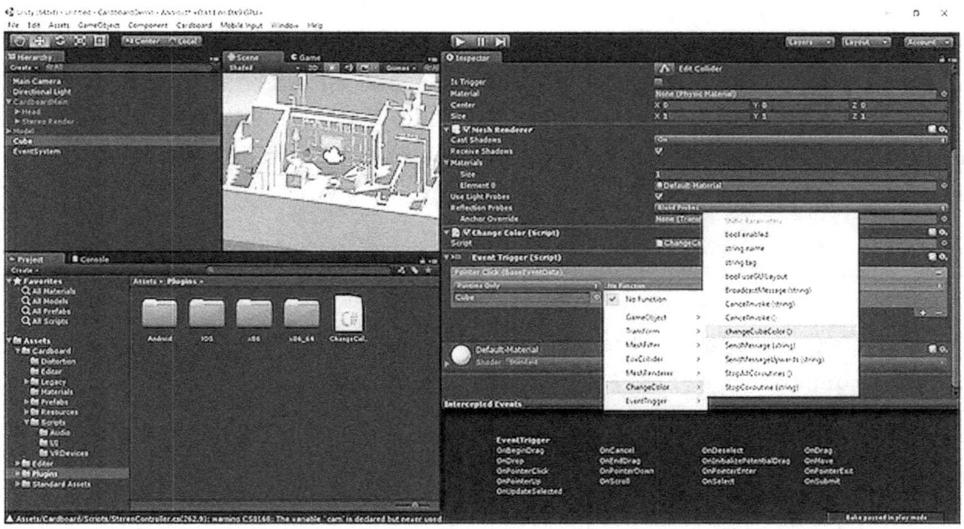

the cube and click with your mouse. The color of the cube will change. You may build the application for Android similar to the process followed in the AR methodology. See Figure 26.

DESCRIPTION OF SOFTWARE CODE/PROGRAM

Coding in Unity has been made simple for even an absolute beginner. Scripting is predominantly done in either Javascript or C#. Different properties of a scene like physics, graphics, materials, shaders, animation, sound, lighting, cameras etc can be altered by piecing together short lines of code. In this segment,

Figure 26. An emulation of the final working application is shown. On click, the cube changed color from the native white to a shade of green

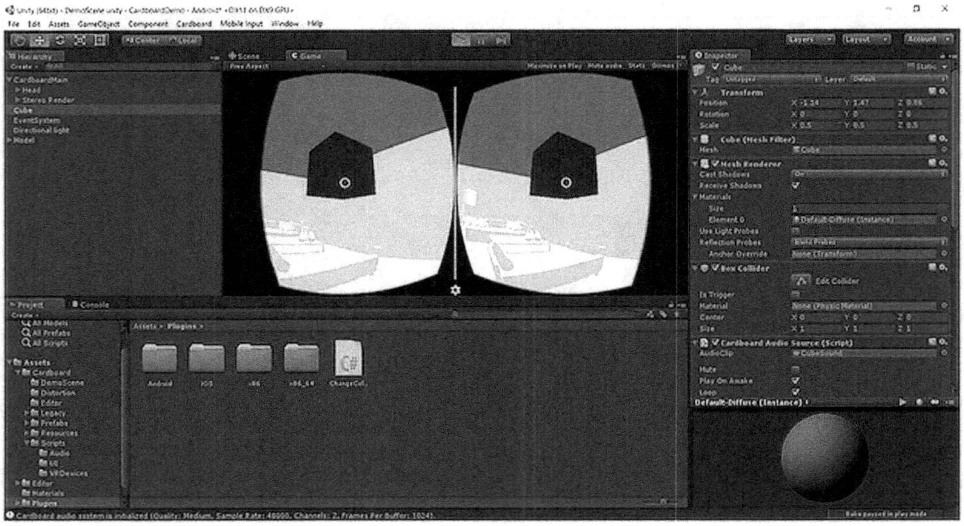

we will be discussing a few examples of how code can be integrated into a scene to add features and functionality to scenes we created.

In augmented reality, a majority of the script has to enable the end user to manipulate the model in the application. A C# function like below when attached to a button will allow the user to scale up the Model gameObject on button click. GameObject.FindWithTag ("Model") allows us to find and manipulate any gameObject with the "Model" tag attached to it.

```
public void ScaleUpButton ()
        {
                GameObject.FindWithTag ("Model").transform.localScale += new
Vector3 (scalingSpeed, scalingSpeed, scalingSpeed);
                }
```

For this script to work, a public float variable "scalingSpeed" is created and initialized with an arbitrary value of say 0.03f. Therefore, on button press, the function ScaleUpButton would execute which increments the localScale of the Model by a 3D vector of x,y, and z components equal to scalingSpeed. The scalingSpeed can be customized in the Inspector in the case that the variable is declared to be public.

```
        public void RotationRightButton ()
        {
                GameObject.FindWithTag ("Model").transform.Rotate (0, -rota-
tionSpeed * Time.deltaTime, 0);
        }
        public void RotationLeftButton ()
        {
```

```
                GameObject.FindWithTag ("Model").transform.Rotate (0, rota-
tionSpeed * Time.deltaTime, 0);
        }
```

Similarly, to rotate the Model gameObject, functions like RotationLeftButton and RotationRightButton are required. For these functions to work, a rotationSpeed float variable should be created and initialized with a value (for ex: 70.0f). Two buttons can be created on the GUI of the application. One for rotating the model in clockwise and the other for rotating the model counter-clockwise utilizing the respective methods above.

The SmoothFollow script can prove to be significantly useful while developing virtual reality applications. It allows gameObjects to follow another target gameObject flawlessly throughout the game environment.

```
using System.Collections;

public class SmoothFollowPlayer: MonoBehaviour {

        public GameObject Player;
        private Vector3 playerPosition;

        void LateUpdate () {
                playerPosition = Player.transform.position;
                Vector3 newPos = new Vector3(playerPosition.x, transform.
position.y, playerPosition.z); // not changing Y value
                transform.position = newPos;
        }
```

The script is applied to the object that is following a target. In the script, we declared a public GameObject instance named Player. In the inspector, we can drag and link the gameObject we wish to follow to this Player. Next, we create a 3D Vector and name it playerPosition. playerPosition will act as the place holder of the coordinates at which our current gameObject should be in order to follow the target gameObject. We define the 3D vector with the coordinates of the position at which our target gameObject that we are following is. The y coordinate is kept constant so that the current gameObject does not lose its height/vertical position in the scene and follows the gameObject at its original vertical plane. The last line assigns the current gameObject with the position it ought to be in.

Customization is a much sought after functionality in any application. Imagine a virtual reality application that helps the end user experience an apartment before it is constructed. It would be beneficial if the user could change the materials of different gameObjects by just clicking them. The materials will cycle through a few built-in materials for the user to visualize different color combinations. The code on the following page would help us do exactly that.

Materials that can be called through scripting must be placed separately in the Assets folder in a folder called Resources. The variable currentMat is initialized to help cycle through the various defined materials in the switch-case loop. In the example, an array is created for materials because the walls had more than one material attached to them. One material was wood and the other material was paint to make the walls more realistic. In each case, different materials are added to the array and applied to the wall. GetComponent<Renderer>().materials accesses the materials of the current gameObject.

```
using UnityEngine;
using System.Collections;

public class GenericWallChanger: MonoBehaviour {
        private int currentMat = 0;
        Material[] temp;
        void Start () {
                temp=GetComponent<Renderer>().materials;
        }
        public void changeWallMaterial() {
                switch (currentMat) {
                case 0:
                        temp[0] = Resources.Load("MAT_WOO_Wood_03",
typeof(Material)) as Material;
                        temp[1] = Resources.Load("MAT_PAI_Paint_04",
typeof(Material)) as Material;
                        //temp[1] = mat2;
                        GetComponent<Renderer>().materials = temp;
                        currentMat++;
                        break;
                case 1:
                        temp[0] = Resources.Load("MAT_WOO_Wood_03",
typeof(Material)) as Material;
                        temp[1] = Resources.Load("MAT_PAI_Paint_03",
typeof(Material)) as Material;
GetComponent<Renderer>().materials = temp;
                        currentMat++;
                        break;
case 1:
                        temp[0] = Resources.Load("MAT_WOO_Wood_03",
typeof(Material)) as Material;
                        temp[1] = Resources.Load("MAT_PAI_Paint_03",
typeof(Material)) as Material;
                        GetComponent<Renderer>().materials = temp;
                        currentMat++;
                        break;
                case 2:
                        temp[0] = Resources.Load("MAT_WOO_Wood_03",
typeof(Material)) as Material;
                        temp[1] = Resources.Load("MAT_PAI_Paint_02",
typeof(Material)) as Material;
```

```
                    GetComponent<Renderer>().materials = temp;
                    currentMat++;
                    break;
            case 3:
                    temp[0] = Resources.Load(“MAT_WOO_Wood_03”,
typeof(Material)) as Material;
                    temp[1] = Resources.Load(“MAT_PAI_Paint_05”,
typeof(Material)) as Material;
                    GetComponent<Renderer>().materials = temp;
                    currentMat = 0;
                    break;
            }
        }
}
```

CONCLUSION

Civil Engineering demands practical experience by default to completely appreciate the scale of the concept. However, such experiences are difficult to come-by. These experiences would demand the students to enter dangerous and distant environments. Unlike other branches of engineering, examples cannot be showcased and simulated in the classroom environment effectively. Augmented and virtual reality enters the picture to fill this gap.

In architecture classes, the teacher relies on the ability of the student's imagination to convert two dimensional figures in real time and simultaneously grasp the knowledge that is being assimilated. This can be extremely stressful for both parties. Taking the example of the Corinthian column in Figure 27, a flat drawing cannot even begin to describe the intricate details of the structure. Students are often left to fill in the holes themselves.

Figure 27. Typical representation of Corinthian column for architectural visualization

This application utilizes augmented reality to allow students to view the Corinthian column in its entire three dimensional glories leaving less room for imagination and more for education. Students can rotate the model, and move closer to observe finer details. The different parts of the column are labeled and provide a strong eidetic base for the young learners. The labels can be added within Unity utilizing pre-built gameObjects like 3D Text and cubes. The augmented reality can be integrated into current textbooks utilizing a companion smartphone application that utilize pictures embedded in the book as trackers.

First year civil engineering students may not be aware of the process in which a building is constructed. Visualization of a building construction in the classroom is again limited to photos and figures which can only describe so much. A 3D animated simulation of a building construction that happens on the student's desk is another league all together (Figure 28). An experience like that is unique and bound to leave an impression on the students that is much more impactful than a two dimensional slideshow. As the student presses a button, a building is constructed realistically in front of them. An untouched land is shown. Then, the land is excavated and prepared. The entire construction process is vividly showcased like never before.

Another profound application for this technology is in the real estate sector. Augmented and virtual reality can be utilized to provide a strong competitive advantage among real estate firms. Using the technology, a developer can sell his properties by showcasing the buildings to the customer before the construction is completed. The customer can get a life-like experience of the structure before any sig-

Figure 28. 3D simulation of entire construction process

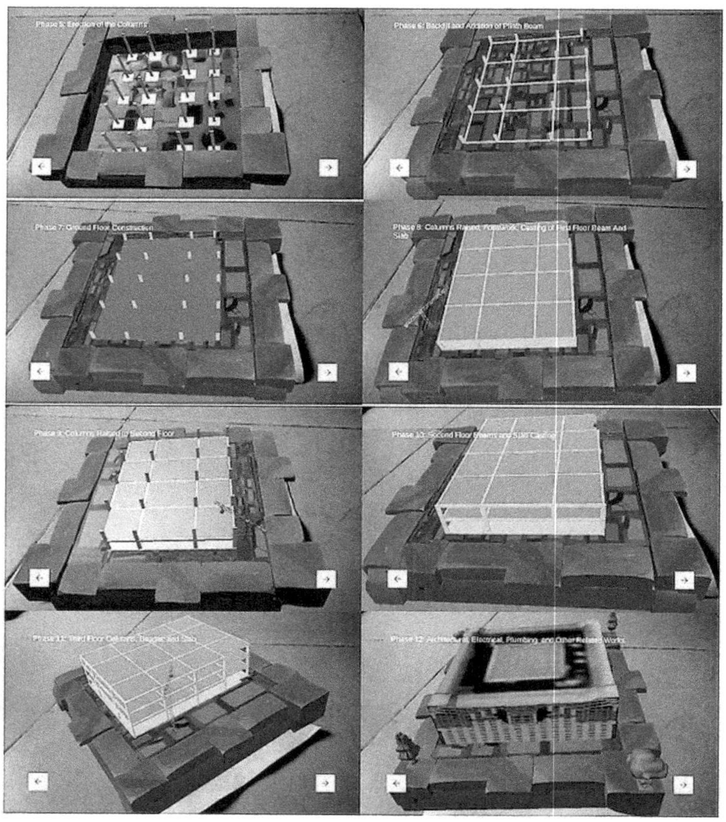

nificant investment is made. An example of the marketing approach is the development of a companion app for our university's brochure. When the user points the camera at the front page, our university's front gate pops up in the smartphone in three dimensions. Similarly, when the camera is pointed at the page describing our hostel, a three dimensional model of our hostel building is shown. See Figures 29, 30, 31, and 32.

To further improve the usefulness of a tie-in application of a magazine, it can provide the user with three dimensional models of the rooms within the building. This idea can also be extended to furniture. For example, the hostel rooms have bunker beds (Figure 33). People may not know what to expect. A simple AR application could help solve this problem.

Figure 29. 3D model of VIT Chennai Entrance Gate

Figure 30. Actual Photograph of VIT Chennai Entrance Gate

Figure 31. 3D model of VIT Chennai Hostel Building

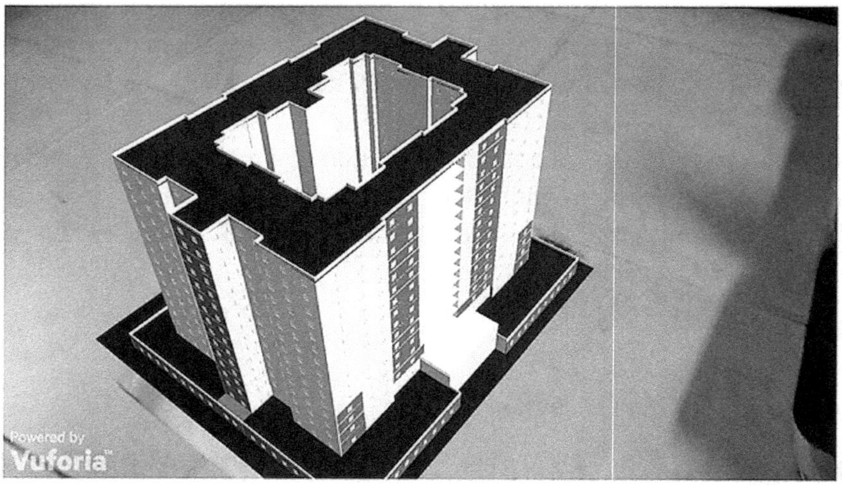

Figure 32. Actual Photograph of VIT Chennai Hostel Building

Small applications may be developed by real estate developers during pre-booking sales to help the customer visualize a house that is yet to be constructed. Apartments with furniture present in their respective places can be modeled in AutoCAD or Revit (Figure 34). The models can be imported into the Unity platform as described in the tutorial section of the chapter. A lightweight application can be developed and be used as a potent marketing tool. Instead of a plain line diagram that is usually showcased, a vivid 3D model is sure to grab the investors' attention.

Using more advanced practices and higher quality assets, we developed an immersive photo realistic apartment for a user to explore. In our application, the material of almost any wall or object can be changed with just a click. In the kitchen, any cabinet, fridge and stove can be opened and closed. The experience is interactive and immersive as promised. See Figures 35 and 36. This application is a commercial product that we intend to offer as a service. The user can move around freely in the apartment

Figure 33. Conceptual 3D model of Bunker Bed Model for a hostel room at VIT Chennai

Figure 34. Rendered 3D model of a furnished apartment for real-estate promotion

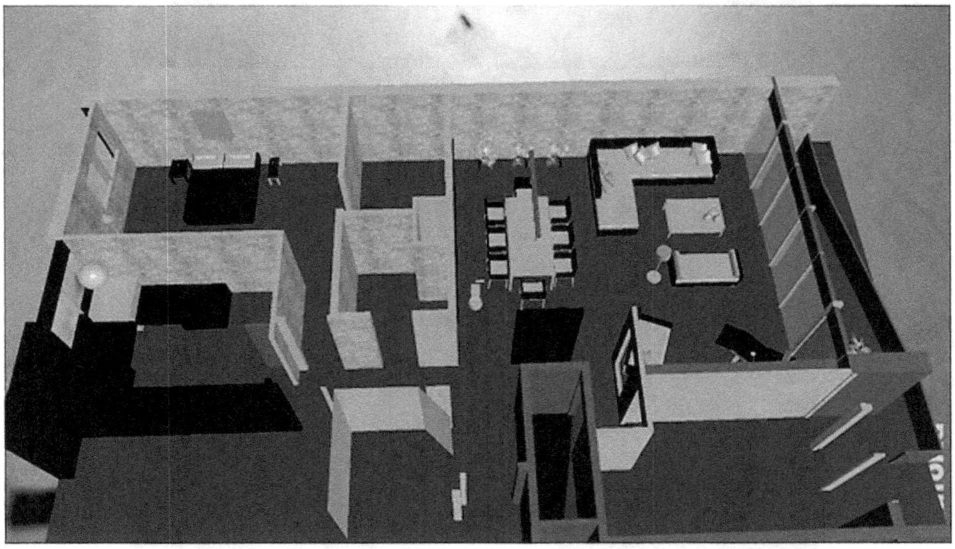

by just looking down in the direction the user intends to walk toward. This is accomplished predominantly using the open-source "autowalk" script. When the user wants to modify the texture/material of an object, the user gazes at the object and clicks the cardboard clicker to manipulate it. A sample of the said-application is shown below.

Figure 35. 3D visualization of the photo-realistic apartment model

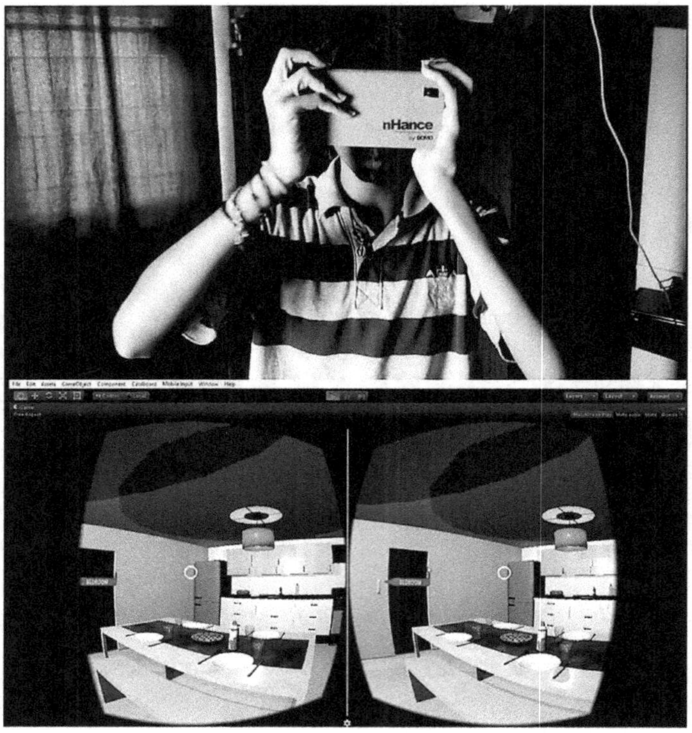

Figure 36. 3D visualization of the photo-realistic apartment model with parametric change characteristics

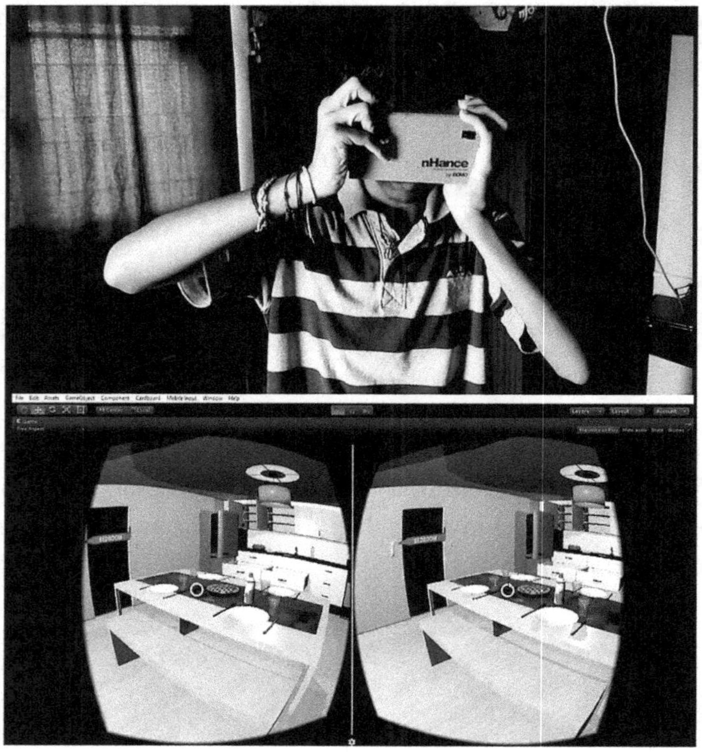

FURTHER RESEARCH

This chapter was inspired by the lack of adoption of AR and VR technology into the civil engineering branch. We wanted to tackle a real life problem in an innovative manner. Initially, we provided justification as to why the civil engineering branch must adapt to newer technologies as quickly as possible. We approached the problem statement and divided the workflow into the two major topics, AR and VR. Working with augmented reality, we developed a polished application that allowed the user to visualize multiple blueprints three dimensionally, complete with a user friendly interface. In the realm of Virtual Reality, we developed an advanced walkthrough of an apartment that provided an interactive and immersive end-user experience. Despite showcasing multiple applications of AR and VR in this chapter, we only just scratched the surface. The potential of this technology in Civil Engineering is immense. The technology can be extended into structural analysis where it can help students understand bending moment and shear force in structural members. It can be used as a visual aid in multiple courses. Taking the example of foundation engineering, it allows beginner students to understand the characteristics of different types of foundations. In a subject like Concrete Technology, an application can teach students how to use a complex machine like the UTM and perform various tests with the help of augmented animations. The technology can be implemented to help designers optimize space utilization and rectify errors in an early stage. Applications can also be developed to coordinate between the work being done at a site and the plan at headquarters. The translation of drawings into a building is not easy. Different structural elements must be identified and subsequently be constructed. Engineers can utilize augmented reality to check for any mistakes in the construction process. It can also be used as a visual aid to help explain workers exactly what work is to be done. The application of Virtual Reality can be extended to include an immersive training environment for construction workers.

REFERENCES

Ammari, K. E., & Hammad, A. (2014). Collaborative BIM-based Markerless Mixed Reality Framework for Facilities Maintenance. In *Proceedings of the 2014 International Conference on Computing in Civil and Building Engineering*. ASCE. doi:10.1061/9780784413616.082

Boeykens, S. (2011). *Using 3D Design software, BIM and game engines for architectural historical reconstruction*. Liege, Belgium: CAAD Futures.

Gaudiosi, J. (2016). *7 Ways AR and VR Will Change Tech in 2016*. Retrieved December 18, 2016, from http://fortune.com/2016/01/04/augmented-reality-vr-change-tech/

Google Developers. (2016). *Cardboard SDK for Unit*. Retrieved December 18, 2016, from https://developers.google.com/cardboard/unity

Harrison, L. T. (2003). *Introduction to 3D Game Engine Design Using DirectX 9 and C*. New York: Apress. doi:10.1007/978-1-4302-0821-1

Indraprastha, A. S., & Shinozaki, M. (2008). Constructing virtual urban environment using game technology. In *Proceedings of the Architecturein Computro, Integrating Methods and Techniques* (pp. 538–541). Antwerpen, Belgium: 26th eCAADe.

Lehtinen, S. (2002). Visualization and teaching with state-of-the-art 3Dgame technologies. In *Proceedings of the Connecting the Real and the Virtual-design E-ducation*. (pp. 538–541). Warsaw, Poland: 20th eCAADe.

McGrath, R. (2016). *The Pace of Technology Adoption Is Speeding Up*. Retrieved December 18, 2016, from https://hbr.org/2013/11/the-pace-of-technology-adoption-is-speeding-up

Reynes, J., & Yerli, C. (2007).The CRYENGINE technology applied to architecture. In *Imagina 2007*. Monaco: the 25th European Digital Content Creation.

Tulke, J., Nour, M., & Beucke, K. (2008). A Dynamic Framework for Construction Scheduling Based on BIM Using IFC. In *Proceedings of 17th IABSE Congress, Creating and Renewing Urban Structures – Tall Buildings, Bridges and Infrastructure*. Chicago: IABSE.

Uddin, M. S., & Yoon, S. (2002). Peter Eisenman's House X, Scheme G: 3D game engine for portable virtual representation of architecture. In *Proceedings of the Connecting the Real and the Virtual-design E-ducation*. (pp. 526–531). Warsaw, Poland: 20th eCAADe.

Wang, X., Love, P. E. D., Kim, M. J., Park, C.-S., Sing, C.-P., & Hou, L. (2013). A conceptual framework for integrating building information modeling with augmented reality. *Automation in Construction*, *34*, 37–44. doi:10.1016/j.autcon.2012.10.012

Wikipedia. (2016a). *Augmented reality*. Retrieved December 18, 2016, from https://en.wikipedia.org/wiki/Augmented_reality

Wikipedia. (2016b). *Virtual reality*. Retrieved December 18, 2016, from https://en.wikipedia.org/wiki/Virtual_reality

Yan, W., Culp, C., & Graf, R. (2011). Integrating BIM and gaming for real-time interactive architectural visualization. *Automation in Construction*, *20*(4), 446–458. doi:10.1016/j.autcon.2010.11.013

KEY TERMS AND DEFINITIONS

Android: It is the most popular mobile operating system developed by Google, based on the Linux kernel and designed primarily for touch screen mobile devices such as smartphones and tablets.

Android Application Package (APK): It is the package file format (essentially ZIP file formatted packages, based on the JAR file format) used by the Android operating system for distribution and installation of mobile apps and middleware.

Augmented Reality (AR): It is a technology that improves the person's optical sense by superimposing a computer-generated image over the user's view of the real world.

Building Information Modeling (BIM): It is a 3D model based digital representation that equips architecture, engineering, and construction professionals with the insight and tools to more efficiently plan, design, construct, and manage buildings and infrastructure providing a reliable basis for important decisions in the project's lifecycle.

Civil Engineering: It is a professional engineering discipline that deals with the design, construction, and maintenance of the physical and naturally built environment, including works like roads, bridges, canals, dams, and buildings.

Graphical User Interface (GUI): It is a type of user interface that enables communication/navigation/ and interaction with the electronic device through graphical icons and other visual indicators.

Real Estate: Real estate is a property comprised of land, including the natural resources of the land, including uncultivated flora and fauna, farmed crops and livestock, water and minerals as well as any physical property or improvements affixed to the land, including houses, buildings, landscaping, fencing, wells, etc.

Software Developer's Kit (SDK): A set of programs used by a computer programmer to write application programs for a particular platform. Typically, an SDK includes a visual screen builder, an editor, a compiler, a linker, and sometimes other facilities.

Unity 3D: It is a cross-platform software framework developed by Unity Technologies used for the creation and development of video games for PC, consoles, mobile devices and websites.

Virtual Reality (VR): It can be defined as the digital emulation of an environment that can be experienced by a person's multiple senses.

Vuforia: It is an Augmented Reality Software Development Kit (SDK) for mobile devices that enables the creation of Augmented Reality applications.

Chapter 13
Augmented Reality Implementations, Requirements, and Limitations in the Flipped-Learning Approach

Nilgun Tosun
Trakya University, Turkey

ABSTRACT

In the year 2016, flipped learning became an innovative learning approach. The students, in flipped learning, prepare to their courses watching the videos, listening to podcasts, and reading the articles prepared for themselves, and have a prior knowledge. In flipped learning approach, efficiency of learning materials is related to providing permanent learning, facilitating the learning and being interactive. For that reason, the students should be supported with stronger learning materials rather than the aforementioned materials. The most current one among these materials includes Augmented Reality implementations that have been prominent in recent years. In this chapter, flipped learning approach and augmented reality concept will be emphasized, augmented reality implementations possible to be used in flipped learning will be exemplified, and limitations will also be discussed.

INTRODUCTION

Education is one of the areas that the world leader, developed societies invest in. The most significant feature of societies which invest in education, which are also known as information societies is that they can successfully integrate training with new methods, approaches and technologies. In this way, it is possible to disseminate education services to a wide range of people and to ensure easy and permanent learning that is independent of time and place, to deliver updated information to people timely and to enable participatory education.

DOI: 10.4018/978-1-5225-2110-5.ch013

Together with the development of new learning models, technology got integrated to education, and this has always been an important field of research. The technological developments at social and global levels, people's desire for education in order to get prepared for their work life, their willingness to study without delaying their works or social responsibilities and their lifestyle are significant factors for the transformations and innovations in education.

Transformation and innovations in education have gained a different dimension with the beginning of the studentship of the generation which is called as 'digital natives'. Digital natives have features such as using information technologies, being active, impatient, energetic, and being interested in software that includes animation. Because of these features, the use of traditional methods, techniques and technologies in education will be insufficient for the learning needs of the digital natives and it will affect their academic success negatively. Furthermore, the integration of education with contemporary teaching technologies, teaching and learning models and approaches as a result of globalization, is a necessity that cannot be delayed.

The flipped learning model is at the forefront of integrations that have attracted attention and produced successful results in recent years. Flipped learning is considered to be an appropriate learning model for digital natives, for whom individuality is a distinctive feature.

In today's technology, 3D models, video, photo and audio content have the ability to overlay on the actual image, with augmented reality which has a wide range of application areas. The most important feature of the augmented reality applications is that; they are real-time, and they aim to enrich the real world with the virtual data developed through computer systems. The augmented reality, which combines the real world with virtual and facilitates the understanding of abstract concepts and events, enriches the learning environments thanks to these features, and it is considered important as it supports learning by doing and experiencing.

Considering the knowledge and technology skills, which are among the learning and renewal skills that 21st -century students should have, the use of augmented reality applications in the flipped learning model becomes important.

This section will focus on the concepts of flipped learning model and augmented reality, the importance and advantages of using augmented reality in education, examples of augmented reality, the necessity of using augmented reality in flipped learning, possible limitations in practice and suggestions for overcoming these limitations.

BACKGROUND

The concept of flipped learning was first introduced in 2007 by Bergmann and Sams from Woodland Park High School, when they sent the course videos they recorded to their students over the internet. According to Bergmann and Sams (2012); in the flipped learning model, students follow video lessons recorded by the teacher to complete their homework and classroom works and to perform laboratory applications. In this way; they are able to achieve much deeper learning compared to the traditional learning models.

One of the most accepted definitions and explanations about this model belongs to Eric Mazur. According to Mazur, the flipped learning model consists of five stages: (November & Full, 2012);

- Students watch the videos, listen to the podcasts and read the articles prepared for them to get prepared for the lesson and to get preliminary information. In the meantime, they determine the questions for the sections that they do not understand about the topics they are trying to learn.
- After access to the content, students are asked to communicate to the teacher the questions about the topics they were not able to understand.
- Students communicate these questions to the teacher using Facebook or another similar social tool.
- The teacher categorizes the questions communicate based on their priority. He/she prepares materials that will facilitate understanding of these questions and eliminate the difficulties in understanding. These materials are not instructional materials. They are application materials that will improve learning and ensure permanent learning. At this point, the teacher assumes that the students have come to class having studied the subject.
- The teacher creates an appropriate classroom environment so that the students can cooperate to answer and solve the questions and problems directed to them from the internet.
- Finally, the teacher uses a Socratic method in the classroom environment. This environment enables students to work together to solve questions and problems sent to the teacher by students. The teacher's role in the class is to listen to students and discussions and, when necessary, to deal privately with a student or student group.

Flipped learning model works contrary to traditional learning models. While in traditional education, the transfer of knowledge to the students is carried out by the teacher in the classroom environment, in flipped learning, a virtual environment on the internet is used for this purpose. Different kinds of learning materials are shared among students in this social network or virtual classroom environments. Whereas in traditional learning, homework assignments are given to have the students learn and internalize the information to achieve permanent knowledge; in flipped learning, highly interactive materials and activities are prepared to enable students to study and discuss with their friends in the classroom environment.

Brame (2013) points out some elements that will make flipped learning a success:

The first of these is to provide students with the opportunity to learn information before coming to class. Simple reading texts, videos, podcasts and screen displays can be used for this. Different type of learning materials is supportive for students with different learning styles, as well.

Another element is to encourage students to get prepared for class activities before coming to class. The most important and effective method used for this purpose is, giving marks. If class activities are evaluated with marks, more positive and successful results can be obtained from the model.

Another element Brame suggests for making the model a success is the creation of a mechanism to enable the students understand and internalize information. For example, quizzes or worksheets shared in the virtual environment also enable the teacher to gain preliminary information about the students and thus, it helps the teacher to organize the classroom activities by focusing on the significant topics and the topics the students have difficulty in learning. Wegerif (2013) also supports this element. According to him, it is an important step in the flipped learning model to predict where students will need assistance the most and to provide activities and experiences that meet with their difficulties and concerns in the class. In this way, students are in less need of the teacher's support in classroom activities and they can spend more time on activities such as analysis, discussion and synthesis.

The final element of focus is to prepare classroom activities that will encourage focusing on high-level cognitive activities. If students have acquired the basic level of knowledge outside the classroom, they will need an efficient time period to support deep learning in the classroom. What is important here is that; the activities to be carried out during this time period must be in parallel with the learning objectives and they shall be carried out in a well-disciplined way. There are two important issues that should not be overlooked at this point; students use new knowledge to deepen their learning and to improve their skills and they spend time in the classroom to achieve these.

Chen et al. (2014) also support Brame and define flipped learning as a model consisting of unique, independent and flexible learner-centered environments.

Just as the emergence of each new teaching model, the flipped learning model has been developed and used in order to overcome some shortcomings and problems of existing teaching models. One of the most important problems in current teaching models is the transfer of teacher-centered knowledge to technology-focused students, who are called as the 'digital native students'. These students, who are active, impatient and fond of technology, become passive learners in traditional teaching models and this is not their thing at all. For this reason, only the flipped learning model can encourage the learning of digital native students by enabling them use their technology skills and actively access to digital resources, make research and learn by doing. In this sense, flipped learning overlaps with structuralism. According to Norris and Solloway (2014), the flipped learning model which include social learning by making research and by doing and experiencing, is a training process that focuses on improving students' critical thinking, communication, collaboration and creativity skills.

Another problem in traditional teaching models is the disinterest of learners in classes. Traditional classroom environments, where a single type of learning material is used, where the teacher acts as a source transferring knowledge and the students only listen and take notes, do not appeal the students at all. Ordinary educational activities carried out in this way cannot go beyond being boring school activities for the students. Another reason for the disinterest of the students is that; the teachers do not attach importance to the attendance of the students to classes, which also decreases the interest and the motivation of the students. Factors such as improper design of learning environments, improper physical environment, heat and lighting, number of students, sound insulation, and missing learning materials also increase students' disinterest in the course. For these reasons, the flipped learning model, which supports home learning and research, is able to promote interest in the lesson by offering different and interactive learning environment materials. One of the important problems that contemporary digital native students face in traditional teaching approaches is the inability to communicate with the digital immigrant teachers and not to speak the same language. Digital immigrant teachers who are accustomed to traditional teaching methods and techniques, who instruct in a self-centered way and who cannot go beyond using data show, negatively affect the interest of the students. However; a teacher who uses the flipped learning model has to be tech-friendly and has to possess certain basic features such as selecting and preparing interactive learning material, sharing these on the internet, using e-measurement tool, sharing in social media or virtual class, using mobile information technologies. With respect to this, the flipped learning can be considered as a model that will increase the interest of students in lesson.

Especially because the internet is accessible and easy to use, many students can supply materials for the assignments, performance or project works by making use of the electronic media. This free-riding causes the students to prepare their homework without reading, questioning or analyzing. It is not difficult for students to copy such materials that are available on the electronic environment. In the flipped learning model, students do activities such as homework, problem solving, problem solving, application and laboratory work in the classroom environment, and so, the risk of copied homework is significantly eliminated.

Flipped learning also provided a solution to the group work problem experienced in traditional teaching. While it is often not possible to conduct group work for students for each course in traditional classroom environments with respect to materials, time and space, in flipped learning classes, it is easier to have the students perform group work and create a discussion environment among each other.

One of the problems experienced in classroom environments where traditional teaching models are used is the presence of students who have difficulty in understanding or who understand very quickly. Students who have a general difficulty in learning compared to other students in the class are often unable to ask teachers questions because of their shyness, or because of the crowded classes, the teacher cannot deal with these students particularly. On the other hand; the students who understand quickly compared to the other students in class can get bored while waiting for the others to understand things, because they are ahead of the class level ant thus, they may get demotivated. In both cases, the solution may be a flipped learning model. In this model, students can easily ask questions to their teachers from virtual classroom or social networking environments, or via e-mail, as they get prepared at home for the classroom activities. Teachers can present the topics by making use of classroom activities that include special materials for fast learner students or for students with learning difficulties.

How the students who cannot attend classes due to various reasons will learn and follow the topics is another problem in traditional classrooms. The flipped learning model is an important opportunity for the students who have missed classes, as the learning materials shared in the virtual classroom or social networks are constantly accessible everywhere.

An important reason to use the flipped learning model is the incredibly rapid development of technology. Rapid development of technology affects education as well as many areas, and by incorporating technology into educational environments, it becomes possible to use new methods and techniques in the learning process. Thus, many innovations emerge in the organization of learning environments (Usta & Korkmaz, 2010). Flipped learning can also be described as a major educational innovation of the modern time.

Flipped learning is a preferred model as it provides the students with the opportunity to participate more in group works and to take more responsibility thanks to the individual learning activities.

As well as the models applied, the technology and materials that are integrated into these models are also factors that directly affect the student's success. When distance education and flipped learning are considered together, the most remarkable technology/material for individual learners appears to be the 'augmented reality'.

The first known reference to augmented reality concept was provided in 1901 with the mentioning of the idea of overlaying metadata on top of people through the use of spectacles (Martinez et al., 2014).

Tom Caudell developed an eyewear that can be mounted on the head to support workers during the installation of electrical cables on aircraft and thus, became the first person to use the term 'augmented reality' in 1992 in its current meaning (Siltanen, 2012).

As a result of subsequent research and applications, different definitions have been made for the concept of augmented reality. Milgram and Kishino (1994) define augmented reality as the reality environment in which digital media products are used instead of real-world objects.

According to Azuma (1997), who has worked on augmented reality; augmented reality enables the real and virtual objects to co-exist in the same environment and interact in real time. Based on the work he has done, Azuma (2001) defines augmented reality as a system with three important characteristics: combination of the real and the virtual, real time interaction and real 3D recording (Siltanen, 2012).

Arvanitis et al. (2007) defines augmented reality as the systems that help students visualize complex spatial relationships and abstract concepts by combining virtual objects with real environments.

Chang, Morreale and Medicherla (2010) emphasize that; augmented reality acts as a perfect bridge between the real and the virtual. According to Özarslan's (2011) definition; augmented reality is the context where the real world meets the virtual world in real time and reaches the user in the same sensory area.

In augmented reality applications, first the image of the real world is taken with a camera. Materials prepared on the computer environment are connected from certain points to the pre-determined target points in the real world. The outcome is interpreted through software and the printed image is taken simultaneously. In simpler terms, augmented reality is the overlaying of virtual objects on real images by making use of technologies that have object recognition capabilities (Regis, 2013). Computer based materials can vary as 2 or 3 dimensional drawings, photographs, texts or sounds. In addition, combinations of these work together successfully (Augmented Reality Türkiye, 2010).

AUGMENTED REALITY PRACTICES IN EDUCATION

Augmented reality, which shows a rapid increase in usage in many areas such as health, tourism and transportation, has begun to be included in educational programs in recent years. A new era in education has been opened with the augmented reality practices developed for every level of learning, from pre-school to higher education. Particularly the diversification and acceleration of mobile technologies, developments in wearable technologies, programming and robotics are the factors accelerating the integration of augmented reality to education, just as other areas.

It is possible to use augmented reality applications for almost every subject, especially physics, chemistry, biology, mathematics, history, astronomy, music and medicine. Below in Figures 1 through 8 are some interfaces from some augmented reality software.

In many lessons, augmented reality is applied; as curriculum complementary and supportive elements, for guidance and promotional purposes, in educational games, on educational trips and in exercises (Çetinkaya & Akçay, 2013).

Augmented reality software can now be used in classroom activities in traditional face-to-face learning programs or in out-of-class activities such as in the flipped learning model, or in blended or distance education programs where the flipped learning model is applied. Because both in the flipped learning model and in blended and distance education, the basic philosophy is to have the students carry out their learning activities wherever, whenever and as much as they desire and often alone. In this case, it is a serious necessity for the teacher to make the learning facilitating technologies and materials available to the students. One of the most up-to-date technologies that can highly meet this necessity is the augmented reality applications. At this point, it is important to decide on which courses or topics to take advantage of augmented reality applications in distance education, taking into account the target audience and their competencies. In addition, the internet infrastructure, hardware, software, and support services needed to enable learners to take the best advantage of augmented reality applications should be fully prepared by distance education institutions.

Figure 1. Biology with LearnAR (Hernandez, 2013)

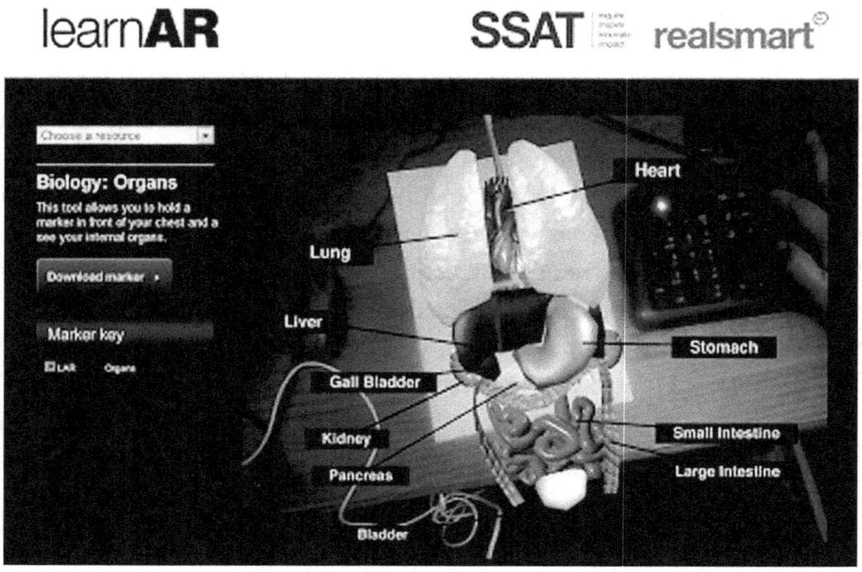

Figure 2. 3D Story-Maker (Jugaru, 2016)

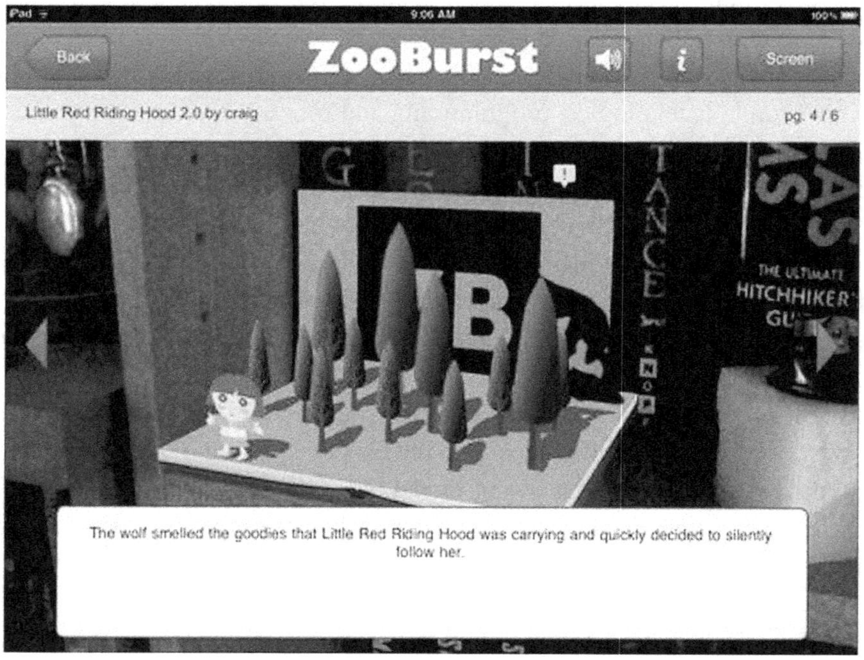

USE OF AUGMENTED REALITY IN FLIPPED LEARNING

Effective learning materials shared in the virtual environment and classroom activities come to the forefront in the success of the flipped learning model, thus, the students. The effectiveness of the learning materials shared in virtual environments is important in the flipped learning model. Because effec-

Figure 3. Space trip with Planet 3D (http://popartoys.com)

Figure 4. AR in architecture training (http://www.augment.com/augmented-reality-architecture/)

tive material provides permanent learning, facilitates learning and supports the students to learn in an enjoyable way without getting bored, as it encourages interaction. Activities such as problem solving, analyzing an event, practicing and product development are used to reinforce and assess learning both in the traditional classroom and in virtual classroom environments. For such activities, the teacher has to prepare interesting applications that are in line with the curriculum and appropriate to the level of the learners and their competencies. Considering that classroom, laboratory and workshop studies are carried out in blended education and virtual classroom studies are being conducted in distance education, distance education teachers are obliged to present to students the applications in which the effectiveness of classroom and virtual classroom activities is emphasized.

It may not always be possible for each teacher to prepare the learning materials shared in the virtual environment or the reinforcement and assessment materials to be used in the classroom in an effec-

Figure 5. Physics Laboratuary supported by AR (Radkowski, 2016)

Figure 6. Animals with ZooKazam (http://zookazam.com)

tive and professional way. This may be due to a number of reasons: The teacher may not have enough knowledge or skills about the software used to prepare the materials, may be foreign to the concept of e-material, may not have knowledge about the principles of e-material development or may not have sufficient experience about contemporary education technologies. These may all cause the students face learning difficulties in a distance education program where the flipped learning model is applied. This is because distance education students often conduct their learning alone. So it is necessary to apply technologies for these students to enable them to learn, especially by doing and living, without the need for additional guidance or a thought-provoking support. Such as in blended education classrooms, the use of realistic applications in distance learning virtual classrooms is also important to bring interaction to top level for the permanence and assessment of learning.

Figure 7. AR in teaching Turkish (Baş, 2015)

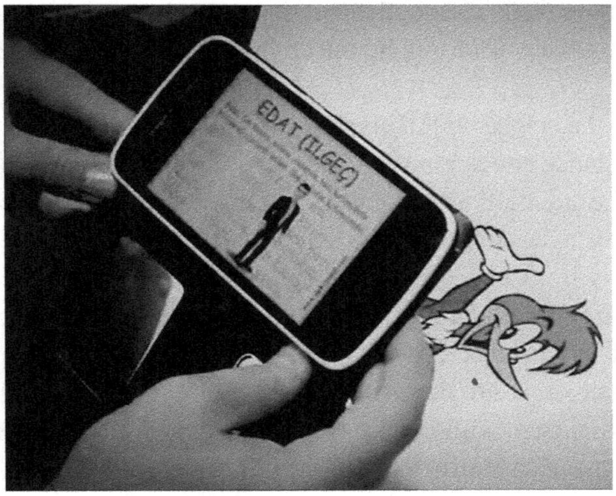

Figure 8. ARGE3D Geometry Coursebook (İbili & Şahin, 2013)

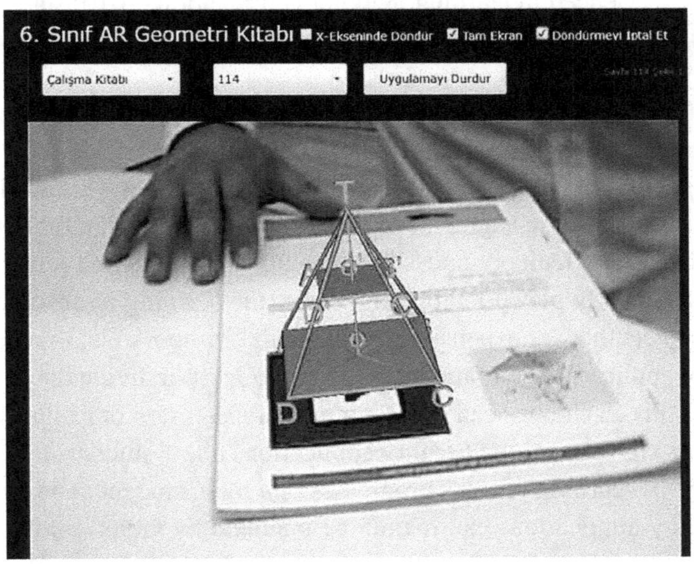

If distance education students are predominantly composed of digital natives; videos, presentations, discussions or paperwork on virtual media will be rather ordinary and inadequate for these students. Because the most significant features of digital native students are that; they prefer digital sources, they prioritize visuals and movements over plain text (Şahin, 2009) and they prefer smart phones and digital media applications such as e-mail (Bilgiç, Duman, & Seferoğlu, 2011). Static learning materials, such as plain text, presentations, podcasts and screen displays, can become boring after a while for digital native students. For this reason, if the flipped learning model is applied in distance education, students should be supported with highly powerful teaching materials and technologies. The most up-to-date and appealing of these materials are the Augmented Reality applications, which have been at the forefront recently in many areas, especially in education, and which are powerful enough for blended and distance learning students, as well. The augmented reality applications that are easily used with mobile devices have already found their place among the applications that are conveniently acquired and used by digital native students.

The lack of interest to the course and motivation are among the things the teachers mostly complain about the blended and distance education students, just as in the traditional face-to-face education. It is more difficult to attract the attention of students to the course due to intensive work life, responsibilities of family life, physical obstacles and trying to learn alone. In addition, the facts that certain subjects cannot be understood only by listening or watching, it takes some students a much shorter or longer time than the others to learn things and that, these students are not directed correctly, can cause motivation problems in blended and distance education programs where flipped learning model is applied. In distance education programs that are developed based on the flipped learning model, the motivation problem can mostly be eliminated with the use of interactive and multi-dimensional learning materials designed considering the theories of structuralist learning, independent study, equality and transactional distance. The key word at this point is the augmented reality applications. Augmented reality applications are effective, easy-to-use, user satisfactory and highly interactive learning materials that contribute to the learning process (Taşkıran et al., 2015; Lee, 2012). In this context, it is possible to minimize the motivation problems experienced in learning. Klopfer and Sheldon (2010) stated that, augmented reality would help the students acquire technology-based and immersive experiences by combining the real and virtual worlds.

Students also learn deeper with the tasks given and the contents and the sources of information they are provided with through the augmented reality applications. Moreover; students can associate previous information with deeper and more permanent links to new information (Kerawalla et al., 2006).

Another important feature of the augmented reality applications is that; they enable the three dimensional use of objects that are difficult or impossible to obtain in real world (Yuen et al., 2011). Augmented reality applications particularly play a facilitating role in the learning of abstract concepts, events and phenomena and the objects that are expensive and difficult or impossible, to obtain.

Augmented reality applications for students who try to learn individually at home or at work also act as teachers. And to this teacher we can load features to repeat, stop or restart the topics whenever we like. From this point of view, augmented reality applications allow students to learn at their own pace. So, it can be said that; the learning process progresses in a more student controlled way.

As augmented reality applications can mainly be operated by tablets and smartphones, which are widely used by the people today, students can continue to learn with the advantage of using a technology that they already know very well. Using new software in a technology they know very well will be interesting for students.

Given that many augmented reality applications are game-based, learning is more enjoyable for students. In game-based augmented reality applications, students can take roles and use their knowledge and skills to reach a certain goal. Thus, when the application is completed, an enjoyable learning activity is realized.

As the augmented reality applications can transfer the real world to the virtual world as it is, the knowledge and experiences gained with the augmented reality act as facilitating factors in analyzing the events and solving the problems faced in real life.

The students of flipped learning model based distance education program are supported with augmented reality applications Instead of being loaded with lots of unnecessary information, they will learn up-to-date, necessary and sufficient amount of information. In other words, students will not have to carry unnecessary cognitive load.

When the books, which are part of blended and distance education programs, are enriched with augmented reality applications, they will serve as an effective and powerful resource for students who are educated based on the flipped learning model, especially in environments that require individual study.

Because of the advantages and benefits addressed above, modern blended and distance learning programs based on the flipped learning model should integrate augmented reality applications. Thanks to constantly evolving and accelerating information technologies, augmented reality can be delivered to a greater number of distance education students.

REQUIREMENTS

In order to be able to use the augmented reality applications, the following hardware and software requirements shall be met.

Hardware Requirements

There are basically two types of augmented reality applications: Video-based and optical-based. Optical based systems are very expensive because they are based on imaging on the retina. Such applications are predominantly used on Android and I-phone platforms. On the other hand, video-based systems require a large processing volume because they process the pictures one by one on camera and present it to the user, and this can slow down or cause troubles in the application. Furthermore; augmented reality uses a complex positioning system to learn where the user is looking. First of all, the system determines the position of the user through GPS and then, it uses digital compass and accelerometer to find out the direction the user is looking at (Augmented Reality Türkiye, 2016).

Augmented reality technology is the combination of 4 different environmental units. These environmental units are; camera, computer infrastructure, marker and the real world. Increased reality can be described as three dimensional positioning of these four different units in the real world (Bilici, 2012).

All processes are created via the computer infrastructure in augmented reality. Generally, high configuration computers are required for real-time viewing of 3D models. However, given the fact that augmented reality applications are mostly designed to be used on mobile devices, personal computer, laptop computer, I-pad, I-phone and Android mobile devices are the examples to the computer infrastructure (Çakal & Eymirli, 2012).

In augmented reality, markers are used for the locational relations between the real world and the virtual world. While these markers used to be 2-bit figures in the first years of application, any colorful and even real object in life is now being used as location marker (Çakal & Eymirli, 2012). A marker based augmented reality system enables animation or 3D models display when a predefined visual is shown on a camera integrated device such as PC, I-phone or Android. If the object containing the visual is moved, the virtual media will also move in 3D (Mobil Artırılmış Gerçeklik, 2016). See Figures 9 through 12.

Figure 9. AR heading (Açık Bilim, 2012)

Figure 10. AR glasses (Atakul, 2016)

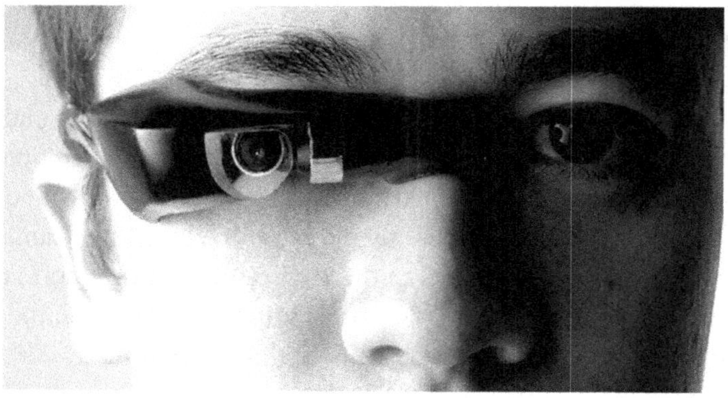

Figure 11. A marker sample (Çakal & Eymirli, 2012)

Figure 12. A marker sample (Çakal & Eymirli, 2012)

Software Requirements

For augmented reality applications, an interface that will interpret the virtual and real environment concurrently is required. This interface is usually sold as software packages registered by the software companies. Tools that facilitate the use of augmented reality applications are also designed together with the software. The software is generally known with a modeling tool, marker production tool, performance enhancing engine tool, a mobile application tool and a web interface tool (Çakal & Eymirli, 2012).

LIMITATIONS AND SUGGESTIONS

It is possible to encounter some problems when using augmented reality applications in blended or distance education programs based on the flipped learning model.

One of these problems is hardware or software based technical malfunctions that students may encounter during out-of- class work. For students who cannot overcome these problems on their own, this can lead to the disruption of studying or motivation problems. Students encountering technical malfunctions will need instant support for their resolution. This support can be given to students in the virtual environment through the frequently asked questions and intelligent teacher systems, or when necessary, by e-mail or telephone. In order to teach the students the ways to act in the event of a technical malfunction, students can be subjected to orientation training through a face-to-face training session to be performed at the beginning of the term or through video conference or an augmented reality application.

Special hardware and software are needed to in order to be able to use augmented reality in flipped learning. Augmented reality software development is a time-taking and tedious process carried out by a team of experts in the field. Therefore, this kind of software is expensive. Some hardware units needed to be able to use the augmented reality software may also be rather expensive from time to time. The lack of financial resources for students and distance education providers to fund the necessary software and hardware for augmented reality applications is one of the major obstacles in using these applications in distance education. If we view the situation at institutional level; finding a sponsor or developing projects will solve the problem to some extent. Thus; distance education institutions will both reduce their financial costs and provide a source of hardware and software for the students who need it.

To be able to use the hardware and software units necessary for augmented reality applications is important in flipped learning for teachers and students. Teachers who do not know how to use these units or who are not positive about using them constitute a problem for using the flipped learning model with augmented reality. It is also an important detail that; students are included in individual and group works after being informed and gaining experience about these special hardware and software. If augmented reality applications are to be used in the flipped learning model in a distance education institution, teachers and students should be informed about model and technology. The system will continue to work successfully when the teacher and the students become able to perform their tasks individually in this model. In particular, teachers involved in the digital immigrant class should be provided with up-to-date knowledge and skills in areas such as learning models, approaches and educational technologies. Participation in events such as in-service trainings, conferences and technology fairs should be supported in order to replace negative attitudes by the positive. Similarly, students who are not open to innovation or who are afraid of technology, who are not capable of learning alone can also be problematic in using augmented reality in distance education programs based on the flipped learning model. If possible, these students may be provided with face-to-face education, if this is not possible, they may be educated through the virtual environment.

A significant limitation for educational institutions for using the flipped learning model and the augmented reality technology is the managers who are unable to undertake the role of technology leadership. As a technology leader, managers should encourage teachers and students in this regard and be a pioneer in the preparation of appropriate environments.

It is inevitable for teachers to act as a guide and consultant in the use of augmented reality practices at home. It is the primary duty of the teachers to identify the software that covers the topics that are necessary to be learned, compatible with the curriculum, appropriate for the student level and that have language support, which are designed in consideration with design principles. Teachers should also be able to respond to the question whether augmented reality should be used for each course or topic in distance education. In blended or distance learning programs based on the flipped learning model, augmented reality should be preferred to teach abstract concepts or complex skills or that are difficult to understand by listening or watching. Software selections should also be made accordingly. Teachers should be knowledgeable about the choice of augmented reality software. For this reason, the programs of the teacher training institutions should include courses covering the design, development, selection and use of up-to-date training technologies.

The factor of balance shall not be ignored in learning materials shared in the virtual environment for the use of distance education students. Frequent and intensive use of augmented reality can lead to some problems in students. For example, using augmented reality application in an uncontrolled way can negatively affect students' reading habits. Moreover, students may regard these applications as games and become addicted to them. For this reason, it is important for teachers to share and use in materials that include reading, watching and practicing, in similar proportions.

Another problem that can be experienced in using augmented reality applications in distance learning programs based on the flipped learning model is slow internet connection in inadequate band-with and insufficient hardware. For these reasons, there may be synchronization problems such as delayed delivery of augmented reality images to students or late understanding of student instructions. In order to avoid such problems, it is necessary for both distance education institutions and for students to have fast internet infrastructure and appropriate hardware. So, at the beginning of the term, students should be asked to take necessary precautions in this regard.

CONCLUSION

The flipped learning model functions in a way that is different from the traditional learning models where the information is transferred by the teacher in the classroom environment and the learners try to learn by listening to the information. On the other hand; flipped learning uses a virtual environment on the Internet to transfer information. Different types of learning materials are shared with students in social networks or in virtual classroom environments. Whereas in traditional education, home assignments are given to students to have them learn and internalize information for permanent knowledge, in flipped learning, the students are provided with highly interactive materials and activities that enable them to study and discuss with their friends in the classroom environment.

The augmented reality, which can be defined as the environment where the real world meets the virtual world in real time and reach the user in the same sensory context, is one of the important technological developments that attract attention with successful applications in recent years. As is the case in many areas, it is ever increasingly being used in education.

Effective learning materials and classroom activities are significant for the flipped learning, and thus, for the students to achieve success. The effectiveness of learning materials shared in virtual environments is important in the flipped learning model. Because effective material enables permanent learning and facilitates enjoyable learning as it encourages interaction. Augmented reality applications are among the most effective and up-to-date materials.

Augmented reality applications can be used as learning materials is virtual sharing platforms in blended or distance learning programs based on the flipped learning model. They can be used as supplementary or assessment materials in classroom, laboratory or workshop studies in blended education, and in virtual classroom activities in distance education.

While the use of augmented reality applications in the flipped learning model has many benefits, there are some limitations, as well. Some of these are; inability to instantly solve the technical malfunctions that may arise during classroom or virtual, out-of-class individual studies, financial troubles, teachers' and managers' lack of knowledge and experience regarding the subject, their negative attitudes, applications that are not compatible with the curriculum, excessive and unintended use of augmented reality applications by the students. It is also necessary to take some measures in order to minimize these problems.

Supporting distance education based on the flipped learning model, which is one of the most suitable learning models for digital native students, with augmented reality applications is a significant issue that needs the investigation of combined use of concepts in distance education, as it enables creating rich and interactive learning environments, permits individual learning, overlaps with the structural approach, can be used with mobile devices and ensures enjoyable and permanent learning. It is considered that; carrying out researches on the combined use of the two concepts in different age groups and in different lessons and identifying the deficiencies and limitations with the data to be obtained will contribute to the improvement of the educational environments and elimination of the mistakes made in practice.

REFERENCES

Açık Bilim. (2012). *Zenginleştirilmiş Gerçeklik Artık Gözümüzde*. Retrieved from http://www.acikbilim.com/2012/06/dosyalar/zenginlestirilmis-gerceklik-artik-gozumuzde.html

Arvanitis, T. N., Petrou, A., Knight, J. F., Savas, S., Sotiriou, S., Gargalakos, M., & Gialouri, E. (2007). Human factors and qualitative pedagogical evaluation of a mobile augmented reality system for science education used by learners with physical disabilities. *Personal and Ubiquitous Computing*, *13*(3), 243–250. doi:10.1007/s00779-007-0187-7

Atakul, B. (2016). *Gelecek Vaat Eden 10 Artırılmış Gerçeklik Gözlüğü*. Retrieved from http://www.teknolo.com/10-artirilmis-gerceklik-gozlugu/

Augmented Reality for Architecture & Construction. http://www.augment.com/augmented-reality-architecture/

Augmented Reality Türkiye. (2010). Retrieved from http://augmentedrealitytr.blogspot.com.tr/

Azuma, R. (1997). A survey of augmented reality. *Presence (Cambridge, Mass.)*, *6*(4), 355–385. doi:10.1162/pres.1997.6.4.355

Baş, H. (2015). *Artırılmış Gerçeklik (Augmented reality) ve Türkçe Dersi*. Retrieved from https://www.youtube.com/watch?v=0zBmMwNlCvQ

Bergmann, J., & Sams, A. (2012). *Flip your classroom: Reach every student in every class every day*. International Society for Technology in Education.

Bilgiç, H. G., Duman, D., & Seferoğlu, S. S. (2011). Dijital Yerlilerin Özellikleri ve Çevrimiçi Ortamların Tasarlanmasındaki Etkileri. Akademik Bilişim 2011. İnönü Üniversitesi.

Bilici, F. (2012). *Artırılmış Gerçeklik Nedir?* Retrieved from http://artirilmisdunya.blogspot.com.tr/2015/12/artirilmis-gerceklik-nedir.html

Brame, C. (2013). *Flipping the classroom*. Vanderbilt University Center for Teaching. Retrieved from http://cft.vanderbilt.edu/guides-sub-pages/flipping-the-classroom/

Çakal, M.A. & Eymirli, E.B. (2012). *Artırılmış Gerçeklik Teknolojisi*. Kuzey Anadolu Kalkınma Ajansı, TRA1.

Çetinkaya, H. H., & Akçay, M. (2013). *Eğitim ortamlarında arttırılmış gerçeklik uygulamaları*. Retrieved from http://ab.org.tr/ab13/bildiri/114.pdf

Chang, G., Morreale, P., & Medicherla, P. (2010). Applications of augmented reality systems in education. *Proceedings of Society for Information Technology & Teacher Education International Conference 2010*, 1380-1385.

Chen, Y., Wang, Y., & Kinshuk Chen, N. S. (2014). Is FLIP enough? Or should we use the FLIPPED model instead? *Computers & Education*, *79*, 16–27. doi:10.1016/j.compedu.2014.07.004

Hernandez, J. (2013). *LearnAR. Herramienta de aprendizaje con Realidad Aumentada*. Retrieved from http://creaconlaura.blogspot.com.tr/2013_12_08_archive.html

İbili, E., & Şahin, S. (2013). Artırılmış gerçeklik ile interaktif 3d geometri kitabı yazılımın tasarımı ve geliştirilmesi: ARGE3D. *Afyon Kocatepe Üniversitesi Fen ve Mühendislik Bilimleri Dergisi*, *13*, 1–8.

Jugaru, G. (2016). *5 Top Augmented Reality Apps for Education*. Retrieved from http://www.hongkiat.com/blog/augmented-reality-apps-for-education/

Kerawalla, L., Luckin, R., Seljeflot, S., & Woolard, A. (2006). Making it real: Exploring the potential of augmented reality for teaching primary school science. *Virtual Reality (Waltham Cross)*, *10*(3-4), 163–174. doi:10.1007/s10055-006-0036-4

Klopfer, E., & Sheldon, J. (2010). Augmenting your own reality: Student authoring of science-based augmented reality games. *New Directions for Youth Development*, *28*(128), 85–94. doi:10.1002/yd.378 PMID:21240956

Lee, K. (2012). The future of learning and training in augmented reality. *InSight: A Journal of Scholarly Teaching*, *7*, 31-42.

Martinez, H., Skournetou, D., Hyppölä, J., Laukkanen, S. and Heikkilä, A. (2014). Drivers and Bottlenecks in the Adoption of Augmented Reality Applications. Journal of Multimedia Theory and Application 2(1), 2014, pp.27-44.

Milgram, P., & Kishino, F. (1994). A taxonomy of mixed reality visual displays. *IEICE Transactions on Information and Systems*, *77*(12), 1321–1329.

Mobil Artırılmış Gerçeklik. (2016). Retrieved from http://www.matbilisim.com/index.php/mobil-artirilmis-gerceklik-mobile-augmented-reality/, 14.09.2016).

Norris, C., & Soloway, E. (2014). *Personalized Learning, Flipped Classrooms, Video Watching: Last Gasps of the Old Education*. Retrieved from http://thejournal.com/articles/2014/03/31/broken-educational-model.aspx#6V6xhizZguxtx5hW.99

November, A., & Mull, B. (2012). *Flipped Learning: A Response To Five Common Criticisms*. Retrieved from http://novemberlearning.com/educational-resources-for-educators/teaching-and-learning-articles/flipped-learning-a-response-to-five-common-criticisms-article/

Özarslan, Y. (2011). Öğrenen içerik etkileşiminin genişletilmiş gerçeklik ile zenginleştirilmesi. *International Computer & Instructional Technologies Symposium (ICITS 2011)*. Fırat Üniversitesi.

PoparI. S. T. http://popartoys.com

Radkowski, R. (2016). *AR for Physics Education. The Augmented Reality Physics Experiments Toolbox*. Retrieved from http://arlab.me.iastate.edu/ar-physics.html

Regis. (2013). *Artırılmış Gerçeklik Nedir?* Retrieved from http://www.regisapp.com/augmented-reality-nedir

Şahin, M. C. (2009). Yeni bin yılın öğrencilerinin özellikleri. *Anadolu University Journal Of Social Sciences*, *9*(2), 155–172.

Siltanen, S. (2012). *Theory And Applications Of Marker-Based Augmented Reality*. Retrieved from http://www.vtt.fi/inf/pdf/science/2012/S3.pdf

Taşkıran, A., Koral, E., & Bozkurt, A. (2015). *Artırılmış Gerçeklik Uygulamasının Yabancı Dil Öğretiminde Kullanılması. Akademik Bilişim 2015*. Eskişehir: Anadolu Üniversitesi.

Usta, E., & Korkmaz, Ö. (2010). Öğretmen adaylarının bilgisayar yeterlikleri ve teknoloji kullanımına ilişkin algıları ile öğretmenlik mesleğine yönelik tutumları. *Uluslararası İnsan Bilimleri Dergisi*, *7*(1), 1335–1349.

Wegerif, R. (2013). *Dialogic: Education for the Internet Age*. London: Routledge.

Yuen, S., Yaoyuneyong, G., & Johnson, E. (2011). Augmented reality: An overview and five directions for AR in education. *Journal of Educational Technology Development and Exchange*, *4*(1), 119–140.

ZooKazam Magical Animals. http://zookazam.com

Chapter 14
Augmented Reality in K-12 Education

Lisabeth J. Leighton
Old Dominion University, USA

Helen Crompton
Old Dominion University, USA

ABSTRACT

Today the world revolves around technology in daily tasks. Society uses technology to communicate, trade, for business, politics, and to education. Through education, Augmented Reality (AR) takes on a new purpose. AR enables students to reach a higher level of education, and teachers to efficiently engage students to see real-world situations. The use of technology in classrooms is currently on the rise; but how are teachers using AR within their classrooms to engage and education children? This chapter discusses literature and research supporting the AR affordances in K-12 Education.

INTRODUCTION

In a society where our daily lives revolve around an ever-changing digital world, new methods of learning have been created using technology. We use technology daily to communicate, collaborate, entertain, invent, design, and teach (Kidd & Crompton, 2015). Over the last few years, the educational system has integrated various digital tools and technologies into classrooms for students and teachers in the United States at an accelerating rate (Bower, Howe, McCredie, Robinson, & Grover, 2013). Teachers are able to utilize different digital tools to help students not only think critically in the classroom, but also take information and apply it outside of the classroom (Baharti, 2014). Augmented Reality (AR) is one of the digital tools being integrated into classrooms. AR is defined as a technology applied to digital devices to link together the reality and virtual worlds by adding a virtual overlay to real world scenarios (Bower et al.,). These digital tools allow teachers and students to formulate, explain, learn, and let their creativity take their education to new levels. Through traditional teaching, this opportunity would not be possible for students. The use of AR in a classroom enables students to reach higher levels of thinking by connecting to real-world experiences (Bower et al., 2013). Although AR is available to many educators,

DOI: 10.4018/978-1-5225-2110-5.ch014

there are problems in using and applying it in the classroom (Demski, 2013). Teachers are still mainly relying on older methods of teaching to instruct students, as they lack training in how to use and integrate these digital tools into teaching and learning (Antonioli, Blake, & Sparks, 2014). Technology is often the tool selected to support didactic teaching in the use of slide presentations (Demski, 2013). Teachers also face obstacles, such as a lack of support from administration, a lack of personal experience with the technologies, and a fear of digital malfunctions (Spencer, 2012). There are a great many affordances in using AR in K-12 learning and with support these tools can be best used to support teaching and learning.

The purpose of this chapter is to articulate the affordances of using AR in K-12 classrooms. This is thematic review of the literature to understand what themes have developed from the published literature. This chapter begins with a definition of AR. This is followed by the unpacking of four trends in how AR can support teaching and learning. These affordances are that AR supports; authentic learning, contextualized learning, student-centered learning and enables students to better visualize subject content. Finally, the projected future path of AR is revealed to show where AR may benefit future students.

What is Augmented Reality?

Augmented Reality (AR) blends physical and digital worlds together in real time by using 3D technologies (Kidd & Crompton, 2015). Ionitescu and Radu (2015) articulated this further as they defined AR as "adding computer-generated content upon the real, physical objects in the world around us, by displaying overlays of information and digital content connected to physical objects and locations" (p. 105). This means that images on a piece of paper or our view of the environment can be enhanced with the addition of digital images, video, sound. This relatively new and still emerging technology can be applied to web-enabled devices such as tablets and iPads, smartphones, television, movies, desktop computers, and laptops. One common example of AR on television in the U.S. is when viewers watch sports, such as American football, and colored lines appear to explain what is happening in the game. Viewers see a blue line to symbolize the line of scrimmage, and a yellow line symbolizing the line for a first down. If you are present at the football game, there are no colored lines indicating the placements (Joan, 2015). These lines represent a basic AR technology by using an overlaid image on top of a preexisting real world object.

Two methods are used in AR, which are: Marker-based, which uses symbols or codes to create a digital image, and location-based, which uses GPS to pinpoint the user's location globally (Bharti, 2014). Virtual objects present in AR can include an assortment of texts, videos, audios, 3D models, animations, and images that are superimposed in the viewer's environment (Bower et al., 2013). Bower et al. listed basic hardware requirements for creating and viewing AR. These are a camera for video and image taking, computer storage space, a powerful computer processor, and a computer program to allow interaction between real and virtual objects. AR has progressively developed as a popular tool for use in real world applications.

AR Technologies and K-12 Education

AR is viewed via mobile devices such as tablets and mobile phones. There are various AR headsets available, such as the Oculus Rift, HTC Vive, VR One, Poppy3D, and Google Cardboard. Two of the most recent AR headsets that have advanced AR capabilities are the HoloLens and Meta 2.

Hololens is an advanced holographic computer providing the user a mixed reality experience that allows them to interact with the real world and digital content. Hololens has more computing power than the average laptop and as it does not have any wires, external cameras or a phone or PC connection needed, the user is completely untethered and able to move freely. The sensors in the Hololens capture the persons' activities as well as the surrounding environment as the high definition lenses with the optical projection system allows the user to view the holographic images projected over the real-world.

A hologram is a visual light created 2D or 3D object that is placed in the real world without any mass. These holographic images are placed on top of real world surroundings that only the individual wearing the HoloLens can see (Statt, 2015). For example, when using the HoloLens headset, the wearer could bring a holographic image of building plans on top of a clearing where the future building will be built. If several individuals are wearing the HoloLens, they are all able to see the same hologram when wirelessly connected to the same file or program. Unlike a virtual reality headset where the user can visit a location out of reach, the HoloLens allows users to overlay 3D images from anything, so long as it comes from a computer program (Statt, 2015). Statt (2015) also points out that while using the HoloLens, wearers are still able to move and communicate with people and their surrounding environment, unlike virtual headsets where the image hinders vision of the surrounding environment.

There is also a camera on the outside of the HoloLens that overlooks the area surrounding the operator. This feature allows for video chatting with others who can interact with your HoloLens by drawing images on their digital device that will appear in your vision (Statt, 2015). Statt (2015) was able to experiment with the HoloLens by rewiring a light switch while another individual coached him through the process by drawing images and arrows on a separate tablet that showed up on his HoloLens. In the future, HoloLens may be able to help businesses and companies interact and develop connections, help customers with numerous problems, create product designs, impact surgeries and the medical field, or assist in instructive settings.

In a university in North America, Microsoft HoloLens is using the device to see human anatomy in 3D verses 2D on a computer or in a book. Teachers and students are able to manipulate the human body for investigation verses using a cadaver. Simulations are created for students that enables them to fail and practice with the HoloLens, because in the real world medical students have little room for mistakes on human life. HoloLens transforms a student's learning environment to change how they see the world. The future of teaching and learning could be changed dramatically through HoloLens.

Similar to Holo Lens, Meta 2 is a wearable glasses set that allows the wearer to freely manipulate 3D objects projected in front of them from the AR lens. The lens projects 3D objects in front of the wearer who can manipulate the image through hand movements (Catanzariti, 2016). Meta creators are currently working on their second model to include a headset larger than Microsoft's HoloLens, which will the wearer to have an increased field of vision. An advantage to a wider lens decreases confusion and space limitations for the user. In addition of the wider lens, the Meta 2 will also feature a higher resolution display and can also be used with both PC and Mac systems. Unlike HoloLens, Meta glasses will have a heightened mapping space. This means that the location tracker within the headset will be to closely pinpoint the wearer's positioning within his or her environment. This heightened feature will be able to automatically add images to the user's environment based on where they are located. The Meta Vision AR lens will also be able to help decrease the billions of dollars and hundreds of lives lost as a result of human error (Meta, 2014). Meta glasses will allow medical students and doctors practice different scenarios to eliminate human error in future situations. Meta creators are also working towards creating interaction between wearers in different locations by using AR as means of communication

(Catanzariti, 2016). Meta Vision's AR glasses will drastically change how we communicate and interact with one another in the future.

Meta is currently working with the Royal Ontario Museum and Decimal Lab, a company working on bringing historical articles to life through AR glasses. The app will allow museum visitors to borrow a Meta headset and use AR in addition to the preexisting museum exhibits without any physical contact to the exhibits (Friberg, 2015). Teachers and their students would be able to take class field trips to museums with AR glasses to help students grasp different concepts and view historical battles or speeches as if they were present. Another Meta app, anARtomy, allows the operator to view the muscular and skeletal systems layered on top of their own body or another person in front of the AR headset. Meta's upcoming apps will allow students to manipulate different atoms and combine them to make new compounds and elements (Friberg, 2015). As Antonioli et al. (2014) stated in their research, students retain more material when they are doing, and Meta's glasses headset will provide those opportunities for students.

Affordances for Teaching and Learning

In a review of the literature, four trends were revealed in how AR can be used to extend and enhance teaching and learning. These affordances are the ability to support authentic learning, contextualized learning, student-centered learning visualize subject content. While these are highly interconnected, these have been separated to delineate each of these aspects.

Authentic Learning

Brown, Collins, and Duguid (1989) defined authentic activities as ''the ordinary practices of the culture'' (p. 34) AR technologies today are used to assist learning in a variety of different occupations. For authentic learning, Reeves, Herrington, and Oliver (2004) posited that the learning environments should include authentic tasks that:

- Have real-world relevance;
- Are ill-defined and include a number of sub-tasks;
- Are complex and require students to undertake complex investigations;
- Provide opportunities for students to investigate the tasks from different perspectives;
- Provide collaborative and reflective opportunities;
- Integrate across different subject areas;
- Include integrated assessment; and
- Allow competing answers or solutions.

AR is a tool that can be used to meet each of these criteria. It is versatile in that it can be used across subject areas and it has real-world relevance as it has the student directly connected with actual tasks that a person would do in the real world. For these reasons, AR has become a leading technology in various fields, such as surgery (Van Krevelen and Poelman, 2010). Doctors are able to utilize AR to work with patients to visualize the body system and discuss 3D imaging reports together (Yuen, Yaoyuneyong, & Johnson, 2011). AR aids surgeons with navigation and positioning during surgery (Yuen et al., 2011)). In the car industry, BMW and Volkswagen have already incorporated AR into their assembly lines (Van Krevelen and Poelman, 2010). Airlines use AR to superimpose information in real time to show the location of the plane and flight information for passengers through tablets on the plane (Bower et al., 2013) and the military uses AR that allows pilots to view instructions, maps, and enemy locations on a screen in front of them (Yuen et al., 2011).

AR allows teachers to enhance students' educational experience by allowing students to interact with 3D objects in their environment. Research findings show that students who use AR while learning content, were more likely to retain the information and construct real-world applications of the material, verses students who learned in a more traditional instruction method (Yuen et al., 2011). Seeing the added digital layer in their environment through the lens is relatively simple, but what incites the higher-level thinking is when the students are able to apply the knowledge of what they saw with the AR to the real world after the AR is removed. Through the technology, students are able to learn in a real-world situation, and develop 21st century skills required for a society that is technology enriched (Dede, 2008).

Student-Centered Learning

Student-centered learning is an educational philosophy that puts students in the center of the learning process. Using this approach, students play an active role in their learning using their own learning styles and strategies (Çubukcu, 2012). This strategy has become popular in the past 30 years as it improves skills such as critical thinking, problem-solving, and reflective thinking (Çubukcu, 2012). AR has the students seeing and experiencing what the subject content is rather than hearing about it described, often in a decontextualized manner, through a lecture approach. Pedagogical potentials are endless as a result of the multi-sensory experience permitted through AR (Kidd & Crompton, 2015).

When properly used in the classroom, the benefits are exceedingly constructive. Wu et al. (2013) found that AR helped students develop skills and knowledge, as well as facilitate skill acquisition. Teachers noticed a difference in student's abilities to retain and connect material to the real world after using AR in their lessons (Wu et al., 2013). Joan (2015) states, "the importance of AR is to implement novel technology to enhance learning approaches in education" (p. 11). As a result of AR technologies, students can become active participants in their education and have teachers incorporate more student-centered teaching methods (Kidd & Crompton, 2015). The potential in AR for learning to be engaging, motivating, collaborative, and authentic is remarkably high when used appropriately and accurately.

To reach the top of Bloom's Taxonomy hierarchy, students could learn to design and create their own AR programs to help them learn subject content. When simply using AR to reinforce or teach material, lower order-thinking skills, like application and comprehension, are supported (Bower et al., 2013). By having students designing AR programs, this will enable students to reach higher order thinking by analyzing, evaluating and creating (Bower et al.). Creation of AR programs supports the best practice of learning by doing. Programs, such as ZooBurst and MagicBook, allow students to create visual 3D interactive AR pop-up books using a handheld device (Kidd & Crompton, 2015). In a recent study conducted by Antonioli, Blake, and Sparks (2014), they found that AR had a positive impact on the way students were learning by increasing knowledge retention. Antonioli et al. (2014) discovered that the children had a high level of engagement while using the book apps on a handheld device. The use of AR allows students to reach both high level of engagement and creativity to increase their educational experience.

Researchers have conducted experiments to measure whether using AR in the classroom increases students' learning. Estapa and Nadolny (2015) conducted a study that measured the effects of student achievement and motivation through an AR enriched math lesson. They developed a lesson on dimensional analysis by having students calculate unit conversion and dimensional analysis with multi-step problems for a Mexico spring break trip. The marker-based AR used an interactive print document for students to manipulate with a mobile device. Another group of students completed the same project with their mobile devices, but did not have an interactive print document to use. In their results, Estapa and Nadolny (2015) found that students in both groups increased their overall achievement, but those with the AR interactive print document exhibited higher motivation throughout the study.

Two emerging practices with AR include educational games and lab experiments. Currently, AR game systems are already available such as the 'Nintendo 3Ds' and AR apps like 'Star Chart', which allows users to map out the stars at that time and location (Joan, 2015). These AR educational games allow students to organize and collect data within the game that help develop 21st century skills (Wu, Lee, Chang, & Liang, 2013). Two AR games include ARSoccer and Alien Contact. Alien Contact was developed to help students with math, language arts, and science literacy through a hand-held game that allowed students to collaborate with one another (Antonioli, 2014). Virtual soccer balls are controlled in ARSoccer through user kicking motions (Bower et al., 2013). Yuen et al. (2011), noticed that games involving AR exhibited stress-relieving capabilities when used for either educational or entertainment purposes. Manipulations of objects and symbols through these games and labs also allow students who struggle with certain phenomena to grasp material in a different method.

Contextualized Learning

Contextualized learning is learning that is directly connected to the real-world context in which is occurs. Baker, Hope, and Karandjeff (2009) postulate that context is important as it "helps students recognize the purpose and value of basic skills development to their academic or career advancement – enhancing the learning process and facilitating students' mastery of material" (p. 3). The contextualized approach stems from constructivist theory where learning is based on experience and individuals construct their own knowledge through these experiences (Rathburn, 2015). This philosophy has been developed over the years by a number of prominent scholars including John Dewey, Paulo Freire, Lev Vygotsky, Jean Piaget, and David Kolb.

It is important to note that authentic learning is one type of contextualized learning as it involves the student taking part in a real-world task. What makes contextualized learning broader in scope is that students do not have to be learning through an activity a person would undertake in a professional situation. It can just be learning while interacting or just connecting to a real-world phenomenon. Often, students are required to learn from decontextualized materials that provide little to no meaning to the students. For example, we ask students to learn about the concept of angle by showing them joined lines on a page. Students often struggle as they try to remember that these lines are supposed to have another meaning. By contextualizing those angles and pointing them out in a real world setting, students are more likely to remember as the concept of angle has been connected with a context they are familiar with. They can better understand why angles are important as they see them on buildings and playground materials. Students themselves also report a significant preference for problems that are relatable (Premadasa & Bhatia, 2013).

Chiang, Yang, and Hwang (2014), contextualized learning for one group of students as they measured the effects of AR being used in a Natural Science elementary class on student achievement. Students were separated into two different groups: AR based mobile learning, or inquiry-based mobile learning. The group of students in the AR learning approach classroom were encouraged to use GPS to pinpoint ecological locations and information. This is the contextualized group as students were able to understand where and the locations are geographically. The other group of students were placed in a classroom with a teacher using a traditional learning approach used with mobile devices. These students were not encouraged or permitted to use GPS and students only conducted independent research on the mobile device. The results showed that students in the AR based mobile learning group had increased skills such as attention, confidence, and retention beyond those who were grouped into the AR based mobile learning group that used inquiry-based mobile learning.

Visualization

Information visualization is to represent abstract information in a dynamic way that facilitates human interaction for exploring and understanding (Klerkx, Verbert, & Duval, 2014). Information visualized in this way makes use of the principles in Gestalt Theory in using human visual capacity as a powerful pattern-finding tool, to provide a vigorous means of making sense of the available data. Klerkx et al. (2014), describes how high quality learning materials such as texts, graphical illustrations, interactive demonstrations, tutorials, and audio and video presentations can be used for students to fully grasp and understand the meaning of a certain topic.

However, many representations provided in texts, tutorials and demonstrations depict prototypical examples that do not represent the diverse examples in the real-world (Crompton, Grant, & Shraim, in press). As AR is situated in the real-world the visual examples are varied and relevant. For example, in Crompton's (2015) study, fourth grade students learned about the mathematical classification of angle by using visualization supported by AR. In this empirical work, students looked for angles in the real world. Once an angle was identified, students used the Measure a Picture program on an iPad to view the real-world angle and use a dynamic protractor to measure that angle.

Programs for AR have been created to fit various subjects that allow the students to utilize visualization. LearnAR created curriculum activity packs for courses such as biology, chemistry, foreign languages, math, and English to support independent investigation for teachers and students in the classroom or at home (Yuen et al., 2011). Students learning a foreign language have the capacity to use AR simulation programs to help decipher tongue placement for correct pronunciation of words (Yuen et al., 2011). Teachers are even capable of assigning AR for homework by giving students an informational homework page that allows them to watch a video by scanning a page or QR code to use if assistance is required for completion (Bharti, 2015). Lee (2012) discusses the use of AR in biology classes to help students understand the human anatomy through computer-generated 3D models. Students are able to collaborate with one another on biological structures using these marker-based ARs (Lee, 2012). Other visual-based AR mobile apps, like AndAr and String, allow the user to take a photo of a tack on a bulletin board, and a 3D prism will appear on the screen to be manipulated by the user (Demski, 2013).

Using a handheld device with access to a camera, teachers can create mobile applications to be used alongside AR for students. For example, a quick response (QR) code application will help students in a biology class understand the human anatomy through computer-generated 3D models (Lee, 2012). Using the application, students can scan a QR code next to the definition of vascular arteries with their digital device camera that will enable them to see the inside of the body and locate vascular arteries that would not be visible without AR technology.

Future of Augmented Reality in Education

Although AR is not a new concept and is still in its infancy, it has immense potential for changing education (Billinghurst, 2002). Within the next few years, scholars postulate that AR will be widespread among college and university campuses (Yuen et al., 2011). Practices and activities involving AR are seemingly endless, and can be applied across subject areas. Currently in U.S. school systems, AR does not yet appear to be a popular tool among teachers. This is probably due to lack of funding or training on how to use AR effectively in the classroom and this will increase in the future. One explanation for Lee (2012) is the lack of financial support and awareness from both the government and academic community.

Dede (2008) suggests that due to the decrease in the cost of digital devices, AR will become more affordable for schools to purchase in the future and this could cause a rise in the use of AR. Furthermore, it is anticipated that further training will be available for teachers in the future. The increase in AR technology is not only expanding daily, but it is also being developed in various fields to benefit educational programs. For example, the cost of field trips each year prevents many students from hands-on experiences; however, schools can purchase AR technology that will allow students to experience a field trip without leaving the classroom. With technology continuously changing, it is extremely important for our educators to develop new methods of teaching to take advantages of the affordances of these devices for teaching and learning. Ionitescu and Radu (2015) posits that AR would benefit from companies producing easier to use AR programs and systems. It is exceedingly important to know how to use AR effectively in order for students to develop skills from the activity.

CONCLUSION

By expanding our teaching and learning environments, AR is a tool that can be used to support students in becoming 21st century thinkers and problem solvers. The increase in technology is daily life has dramatically increased in recent times; therefore, it is important to take advantage of these new tools and the affordances they provide. In this chapter, four overarching benefits have been identified of how AR can extend and enhance teaching and learning. These affordances are the ability to support; authentic learning, student-centered learning, contextualized learning, and visualize subject content.

Authentic learning is having students connecting to real-world tasks. Students can use a mobile device to become surgeons, pilots and many other important roles as learners get to act out this role in the safety of a digital environment. Learning with AR supports student-centered learning as it allows the students to see and experience what the subject content is. Contextualized learning is learning that is directly connected to the real-world context in which is occurs. AR is a tool that directly engages the student in looking at the world around them while providing additional supports such as prompts and tools to take the learner to a further depth of understanding than they would independently.

Information visualization is to represent abstract information in a dynamic way that facilitates human interaction for exploring and understanding (Klerkx et al., 2014). Graphical illustrations, interactive demonstrations, tutorials, and audio and video presentations can be used for students to fully grasp and understand the meaning of a certain topic. Empirical evidence show that AR can positively make a difference to how students learn and there is promise and potential for the future of AR.

REFERENCES

Antonioli, M., Blake, C., & Sparks, K. (2014). Augmented reality application in education. *Journal of Technology Studies*, *40*(2), 96–107. doi:10.21061/jots.v40i2.a.4

Baharti, P. (2014). How to use augmented reality in the classroom. *EdTech Review*. Retrieved from http://edtechreview.in/trends-insights/insights/1210-how-to-use-augmented-reality-in-the-classroom

Baker, E. D., Hope, L., & Karandjeff, K. (2009). *Contextualized Teaching and Learning Handout*. Retrieved from file:///D:/Downloads/CTL%20Brief.pdf

Billinghurst, M. (2002). Augmented reality in education. *New Horizons for Learning, 12*.

Bower, M., Howe, C., McCredie, N., Robinson, A., & Grover, D. (2013). Augmented reality in education-cases, places, and potentials. *Educational Media International*, *51*(1), 1–15. doi:10.1080/09523987.2014.889400

Brown, J. S., Collins, A., & Duguid, P. (1989). Situated cognition and the culture of learning. *Educational Researcher*, *18*(1), 32–42. doi:10.3102/0013189X018001032

Catanzariti, P. (2016, March 7). *What the meta 2 means for augmented reality developers*. Academic Press.

Chiang, T. C., Yang, S. H., & Hwang, G. (2014). An augmented reality-based mobile learning system to improve students' learning achievements and motivations in natural science inquiry activities. *Journal of Educational Technology & Society*, *17*(4), 352–365.

Crompton, H. (2015). Using context-aware ubiquitous learning to support students learning about angle. *Journal of Interactive Media in Education*, *1*(13), 1–11. doi:10.5334/jime.aq

Crompton, H., Grant, M. R., & Shraim, K. Y. H. (in press). Technologies to Enhance and Extend Children's Understanding of Geometry: A Configurative Thematic Synthesis of the Literature. *Journal of Educational Technology & Society*.

Çubukcu, Z. (2012). Teachers' evaluation of student-centered learning environments. *Education*, *133*(1), 49–66.

Dede, C. (Speaker). (2008). *Immersive interfaces for learning: Opportunities and perils* [motion picture]. Available from The President and Fellows of Harvard College: http://cyber.law.harvard.edu/interactive/events/luncheon/2008/12/dede

Demski, J. (2013). Jumping off the page. *T.H.E. Journal*, *40*(4), 19–22.

Estapa, A., & Nadolny, L. (2015). The effect of an augmented reality enhanced mathematic lesson on student achievement and motivation. *Journal of STEM Education: Innovation & Research*, *16*(3), 40–48.

Friberg, M. (2015, September 14). *Augmented reality in the enterprise*. Retrieved from http://blog.metavision.com/augmented-reality-in-the-enterprise/

Ionitescu, S., & Radu, C. (2015). Perspectives on implementing interactive elearning tools using augmented reality in education. *Elearning and Software for Education,* (1), 104-111. doi: 10.12753/2066-026X-15-015

Joan, R. (2015). Enhancing education through mobile augmented reality. *Journal of Educational Technology, 11*(4), 8–14.

Kidd, S., & Crompton, H. (2016). Augmented learning with augmented reality. In J. Lu, D. Churchill, & B. Fox (Eds.), *Mobile learning design theories and applications* (pp. 97–108). New York: Springer. doi:10.1007/978-981-10-0027-0_6

Klerkx, J., Verbert, K., & Duval, E. (2014). Enhancing learning with visualization techniques. In J. M. Spector, M. D. Merrill, J. Elen, & M. J. Bishop (Eds.), *Handbook of research on educational communication and technology* (pp. 791–807). New York: Springer. doi:10.1007/978-1-4614-3185-5_64

Lee, K. (2012). Augmented reality in education and training. *Tech Trends: Linking Research & Practice to Improve Learning, 56*(2), 13–21. doi:10.1007/s11528-012-0559-3

Meta. (2014, September 18). *Meta pioneers: Simx* [Video file]. Retrieved from https://www.metavision.com/meet-meta

Microsoft HoloLens. (2015, July 8). *Microsoft Hololens: Partner spotlight with Case Western Reserve University* [Video file]. Retrieved from https://www.microsoft.com/microsoft-hololens/en-us/commercial#fullbleed-volvo

Premadasa, K., & Bhatia, K. (2013). Real life applications in mathematics: What do students prefer? *International Journal for the Scholarship of Teaching and Learning, 7*(2), 20. doi:10.20429/ijsotl.2013.070220

Rathburn, M. K. (2015). Building connections through contextualized learning in an undergraduate course on scientific and mathematical literacy. *International Journal for the Scholarship of Teaching and Learning, 9*(1). Retrieved from http://digitalcommons.georgiasouthern.edu/ij-sotl/vol9/iss1/11

Reeves, T. C., Herrington, J., & Oliver, R. (2004). A development research agenda for online collaborative learning. *Educational Technology Research and Development, 52*(4), 53–65. doi:10.1007/BF02504718

Spencer, J. (2012, July 14). *11 reasons teachers aren't using technology*. Retrieved from http://www.spencerauthor.com/2012/07/11-reasons-teachers-arent-using.html/

Statt, N. (2015, January 24). Microsoft's Hololens explained: How it works and why it's different. *CNET*. Retrieved from http://www.cnet.com/news/microsoft-hololens-explained-how-it-works-and-why-its-different/

Van Krevelen, D. W. F., & Poelman, R. (2010). A survey of augmented reality technologies, applications and limitations. *The International Journal of Virtual Reality, 9*(2), 1–19.

Wu, H., Lee, S., Chang, H., & Liang, J. (2013). Current status, opportunities and challenges of augmented reality in education. *Computers & Education, 62*, 41–91. doi:10.1016/j.compedu.2012.10.024

Yuen, S., Yaoyuneyong, G., & Johnson, E. (2011). Augmented reality: An overview and five directions for AR in education. *Journal of Educational Technology Development and Exchange, 4*(1), 119–140.

Chapter 15
The Importance of Mobile Augmented Reality in Online Nursing Education

Belgin Boz Yuksekdag
Anadolu University, Turkey

ABSTRACT

The main topic of this chapter is the importance of mobile augmented reality in online nursing education. Firstly, imbalance nurse supply- demand will be discussed in the world and Turkey. Solution proposals of important institutions and organizations will be handled to increase the number and quality of nurses. One of these proposals is online nursing education. Secondly, the importance and use of mobile technology will be handled in online nursing education, and where mobile technology is used in nursing. Lastly, the use of mobile augmented reality will be tackled in theoretical and practical areas of online nursing education, and suggestions will be presented to online nursing education providers.

INTRODUCTION

In last century, the provision of health services has changed. The changing and developing healthcare system affects the basic health requirements of people, and also the nursing education. Some of these changes are very fast the increase in the world population, also increase the demand for health services, scientific advances, especially prevention of diseases, the development of diagnostic methods, the changing treatment methods, and technological developments. Because of these changes, the nurse educators have greater responsibilities. Nurses should have more comprehensive knowledge, enrich their care experiences, use technology and evaluate nursing purposes for support patients in complex system (American Association of Colleges of Nursing, 1998). Also, redesign of professional health education is necessary and timely, in view of the opportunities for mutual learning and joint solutions offered by global interdependence due to acceleration of flows of knowledge, technologies, and financing across borders, and the migration of both professionals and patients (Frenk et al., 2010). The purpose of this chapter is to address the situation of nursing in the world and in Turkey, to discuss solutions and to emphasize the use of mobile technologies, and the importance of mobile augmented reality practices in nursing education.

DOI: 10.4018/978-1-5225-2110-5.ch015

BACKGROUND

International Council of Nursing (ICN) (2006) stated that there is a link between the size of qualified nurses and health services to the community. There are the evidences on the increase of supply and demand imbalances of nurses, and major differences in nurse ratio in the world. ICN (2006) noted that the demands for both low and high- income countries have not been met.

In the United States, there are 140 thousand nurse shortages. It is expected to be 800 thousand nurses until 2020. According to the data of the World Health Organization, in 57 countries, it remains per thousand less than 2-3% nurse, midwife, and doctor. This ratio shows that some people in the world cannot receive even basic health care (Callister, 2012). According to Turkish Statistical Institute, there is only one nurse for every 549 people (TUIK, 2013). This number is very important to think situation of the healthcare services in Turkey.

Changes in the healthcare system, increasing number of the aware patients, and increase in the health care costs has also affected the nursing education. The focus is on specialized content knowledge and skills to prevent medical errors and ensure patient safety. On the other hand, nurses are expected to have high-level skills in working environments. These skills are access, analysis, synthesis, and evaluation. In this context, due to the increasing complex and vast amount of new data, nurse students have to know the ways of access to information. Thus, educators should pursue advances in technology and adopt effective teaching strategies to provide more effective learning. New tools are being developed to assist educators and nurse students to keep pace with the growing amount of new information. The work of educators is increasingly difficult. They have to gain students critical thinking and complex problem solving skills in many areas. In this context, in nursing education, technology can assist them to manage these expectations and make future health practitioners successful in their field (Ducut & Fontelo, 2008).

Tri-Council encourages for education on graduate, master and doctoral levels (TriCouncil, 2010). In health system, the higher-educated nurses are expected to work much more effectively (American Association of Colleges of Nursing, 1998). Aiken, Clarke, Cheung, Sloane, & Silber (2003) pointed out that nurses with Bachelor's and Master's degrees on average have better critical thinking skills than nurses with Associate's degrees. Because of this, hospitals recruit the higher educated nurses in order to reduce mortality rate. Also, authors add that hospitals strongly encourage their nurses to return to school to obtain higher degrees. In this context, online programs have a cost-effective option for both institutions and registered nurses who may not have access to traditional education to maintain a baccalaureate degree (McAlpine, Lockerbie, Ramsay, & Beaman, 2002; Ostrow & DiMaria-Ghalili, 2005). For this reason, the education that based on computer and Internet technology is seen as an alternative solution to increase the education levels of nursing (TriCouncil, 2010). In brief, the nurse shortage and the idea of increasing the level of vocational education has led to the growth of education programs that based on computer and Internet technologies (Schnetter, Lacy, Jones, Bakrim, Allen, & O'Neal, 2014).

According to American Associate of Colleges of Nursing (AACN) (2016) all nursing education programs delivered merely or in part via distance learning environment must meet the same academic program and learning support standards and accreditation criteria as programs provided in face-to-face formats, including the following:

- Student outcomes are consistent with the stated mission, goals, and objectives of the program
- The institution assumes the responsibility for establishing a means to assess student outcomes. This assessment includes overall program outcomes, in addition to specific course outcomes, and a process for using the results for continuous program improvement (AACN, 2016).

In the continuation of professional career, the continuous learning and the development of intellectual abilities are very important. The nursing education should gain problem solving ability in high level. Students should develop and use critical thinking skills. When the students graduate, they should interpret and use quantitative data. Additionally, they should use scientific knowledge and scientific process to develop, apply and evaluate nursing initiatives. In this context curriculum content and active learning strategies are the important issues in nursing education (American Association of Colleges of Nursing, 1998). According to American Association of Colleges of Nursing (1998) the high nursing education must be ready to handle the following priorities:

- In baccalaureate and graduate nursing programs, the development of skills in critical thinking and clinical judgment should be addressed as a priority in the curriculum
- Nurses should be prepared to practice across multiple settings, several of which will be nonconventional.
- Nursing curricula should increasingly emphasize primary health care, patient education, health promotion, rehabilitation, self-care, and alternative methods of healing.
- Programs must struggle for an ethnic and racial diversity among students and faculty that more closely mirrors society. This will require more aggressive and creative recruitment and retention programs.
- All curriculums should focus, at appropriate levels, on case management, research methods, quality indicators, health care policy and economics, outcome measures, financial management, legislative advocacy, trends toward privatization, and management of data and technology.

Advanced technology should be used to improve teaching methods at all program levels. In nursing education, creative paradigms are required for faculty development to meet the challenges of the future (American Association of Colleges of Nursing, 1998). Because of the increase and diversification of technology, educators have a crucial role on the recognition and effectively use of technology. Therefore, educators should follow technological developments, and use them properly in their fields (Akkoyunlu, 2002).

In order to adopt diverse learning methods to increase students' learning motivation, mobile devices, such as cell phones or Personal Digital Assistants (PDA), are also used in nursing activities or provided as teaching support (Wu, Hwang, Su, & Huang, 2012). Using mobile devices as a learning tool for nursing education can provide abundant clinical teaching materials; in addition, the learning process of using mobile devices can be used as a basis for evaluating students' learning achievement (Dearnley, Haigh, & Fairhall, 2008; Hung, Lin, & Hwang, 2010; McKinney & Karen, 2009). Mobile Learning is the use of wireless technology enabled mobile devices for learning at anytime, anyplace and anywhere (Asabere, 2012). Because of high performance and portable of mobile devices, augmented reality applications has shifted from computers towards mobile devices. Ortega et al. (2011) remarked that AR represents new mobile technologies that can support mobile learning (m-learning), where real-time views of real-world objects or environments may be enhanced by computer-generated media such as graphical, and video, sound or Web-based content. In terms of students' attention to the subject, augmented reality environments are more successful than regular desktop learning activities. At the same time, it was observed that increasing the interest of students to be three-dimensional the visualizing objects (Winn, Windshcitl, Fruland, & Lee, 2002). Hsiao & Rashvand (2011) stated that AR learning environments create the depth learning by student interest and attention to subjects, especially the most difficult classes. Boz Yüksekdağ

(2013) declared that Anatomy is the hardest lesson for nurses, and they want the visualization to better understand the subjects. AR environments can help the better learning of this lesson. In another study of author, the nurses wanted to have the visual anatomy and physiology courses supported with videos (Boz, 2008). In this context, AR applications are very useful in nursing education.

THE USE OF MOBILE DEVICES IN ONLINE NURSING EDUCATION

In today's learning environments, mobile devices should enhance the motility of learners, the mobility of learning and allow the learner to develop new skills in information communication technology (ICT). Mobile learning has the potential to supplement ICT, online learning and the traditional teaching and learning methods to educate practitioners in the clinical practice area (Clay, 2011). Ally & Prieto-Blázquez (2014) noted that mobile learning is not about the technology, it is about the learner. The learner is mobile and is at the Centre of the learning, and the technology allows the learner to learn in any context (Zhong, Xue, Liu, & Yuan, 2010).

When generally summarized, mobile learning is portable, flexible but more expensive for users. According to Guo, Chong, & Chang, (2007), in spite of the demand for healthcare and nursing professionals have increased most of the countries in the world, the supply of the nurses could not keep up with the demand. How to promote the quality and effectiveness of nursing education has become an important issue. Facing this issue, researchers must take into account the design of teaching materials, tools and modes, and how to efficiently and effectively share nursing knowledge and practical experience to equip students with better abilities to deal with unexpected clinical situations (Guo, Chong, & Chang, 2007). In order to adopt diverse learning methods to increase students' learning motivation, in addition to online teaching, mobile devices, such as cell phones or Personal Digital Assistants (PDA), are also used in nursing activities (Garrett & Jackson, 2006). The mobile technologies are used among nurses for such purposes as tracking medication, supporting preference-based care planning and research (Miller, Shaw-Kokot, Arnold, Boggin, Crowell, Allegri, Blue, & Berrier, 2005). Embi (2012) stated that for more than a decade now, handheld computers or personal digital assistants (PDA) have become standard equipment for medical students and clinicians to get a handle on the increasingly complex and large amount of information. The role of PDAs in improving practice by reducing medical errors is a common theme in the literature (Garret & Jackson, 2006). Thompson (2005) noted that with the significant growth in handheld technology, expanding healthcare applications and ease of use the PDA is becoming an integral part of healthcare and the practice of professional nurses. He asserted that about the potential of handheld computers and they will transform nursing practice, making it more efficient, safer and of a higher quality (Thompson, 2005). The handheld computers have significant advantages such as small size, mobility, flexibility, and ease of use. On the other hand using the handheld computer to organize data, access information, and share data enhances point of care capabilities (Tooey & Mayo, 2004). Given the advantages of handheld technologies, the exponential growth of its use in health care and the computing and data management capabilities of the PDA, it would seem a logical and powerful tool to support the mobile use of portfolios in the clinical learning environment (Garrett & Johnson, 2006). Kuiper (2008) noted that the PDA supports clinical reasoning during reflective activities. But the researcher adds more investigations are needed to carry out an evaluation of clinical reasoning at the point of care (Kuiper, 2008).

Using mobile devices as a learning tool for nursing education can provide plentiful clinical teaching materials; in addition, the learning process of using mobile devices can be used as a basis for evaluating students' learning achievement (Dearnley, Haigh, & Fairhall, 2008; McKinney & Karen, 2009). From related research, it has been found that mobile devices are often used as a knowledge acquisition tool in clinical nursing. Such a learning mode provides feasibility, but lacks functions of interaction and exercise. Therefore, Wu, Hwang, Su, & Huang (2012) noted that in the nursing programs that emphasize actual practice, how to establish a learning environment that provides personalized guidance and feedback for students to practice skills and apply knowledge in clinical situations is worth exploring.

Clay (2011) explored that if mobile devices could be used in pedagogically effective ways to support and enhance the learning and acquisition of clinical skills in the clinic environments. Participants in the study each received a handheld mobile device (iPod) that had been loaded with several Reusable Learning Objects (RLO) outlining each aspect of the physical examination to be performed. At the end of the module, each participant completed an evaluation questionnaire. Participants confirmed that mobile learning afforded flexibility in time and place of learning and captured their interest in the learning material. The results of this study show that the use of mobile technology for skill acquisition is creative and innovative for learners.

Medley & Horne (2005) emphasize the significance of simulation technology for nursing education. They said with simulation technology, nurse students can gain and improve skills in a safe, non-threatening, experiential environment that also provides opportunities for decision making, critical thinking, and team building. On the other hand, with the help of augmented reality projections of bones, muscles, nerves and other internal body parts, learners can practice procedures on mannequins in a somewhat real-world game of Operation. Applications can also help to teach learners about anatomy by allowing them to peer under the skin and reveal the inner workings of human bodies (Cameron, 2010).

Altmann & Brady (2005) conducted a survey of 190 student nurses and found that the software program most used by students in the clinical setting as a drug guide, followed by a laboratory guide. The medication administration is a skill. The absence or inadequacy of this skill causes high anxiety among students, and is the source of the greatest number of clinical errors. According to the researchers a PDA drug guide can help reduce medication errors (Altmann & Brady, 2005). Tooey & Mayo (2004) stated that the use of PDAs helps nurses to find appropriate medications easily, check for drug interactions, calculate dose and consult other clinical references all at the bedside. George, et al. (2010) said that nurse students indicate that PDA use improved their efficiency, using PDAs as a student would contribute to their future use of handheld technology. Also, students found PDAs to be an effective educational tool. PDAs are proving to be effective educational tools that result in improved efficiency, decreased medical error, and increased professional confidence (George, et al., 2010). The most significant benefit identified was the ability to quickly access current drug databases and nursing reference books. Additional benefits included the ability to manage patient and procedure information, bedside data entry, patient data collection and health management, and improved health care team communication (Davenport, 2004).

Hand-held technology is emerging as an effective clinical tool to aid evidence-based practice and support the educational needs of clinical staff (Honeybourne, Sutton, & Ward, 2006). Loyd (2012) aims to design a prototype mobile virtual patient. In this study, the researcher covers the design and development of a prototype mobile virtual patient system and evaluates the effects of screen size and mobility. The results of study show that mobile devices are sufficient to incorporate dynamic interaction of a conversational agent, and mobile virtual patients would be beneficial to nursing students. Also, mobile virtual patient co-presence is higher when using a tablet, mobile ease of use was highest on the tablet, and tablet provided the highest results for interacting in a realistic manner. Researcher asserted that all devices mobile virtual would be beneficial to nursing students and thus could be used as a learning device. The tablet provided participants with the best experience with a large screen size, highest co-presence and enjoyable to use. Thus, the tablet would be the ideal device for implementing a mobile virtual patient for training nursing students. Participants did not rate the smartphone as high as the other device however, it was never rated negatively.

Dickens (2016) asked six nurse educators to describe their experiences with the integration of mobile technology as an active learning tool. Participants had used mobile technology for at least one year, and were teaching in the first semester of an ADN (Associate Degree Nursing) nursing program. The main emphasis for nurse educators is the need for ADN programs to devise an innovative curriculum that will include mobile technology as an active learning tool. Kenny, Van Neste-Kenny, Burton, Park, & Qayyum (2012) declared that a high level of ownership and use of mobile devices among the participants of study. Also, both nurse faculty and nurse students were highly confident in their use of mobile technologies and prepared to engage in mobile learning. In short, when studies examined, mobile devices are useful both in clinical education and pre-clinical education for nurse students.

MOBILE AUGMENTED REALITY AND ONLINE NURSING EDUCATION

Zhu, Hadadgar, Masiello and Zary (2014) asserted that AR has the potential to provide powerful, contextual, and situated learning experiences. Students can use AR to construct new understanding based upon their interactions with virtual objects, which bring underlying data to life. There is a need to increase effective pain management education in undergraduate nursing programs. Evans & Mixon (2015) stress the importance of augmented reality scenarios in clinical education. They add simulation is an effective educational methodology to provide clinical situations to undergraduate nursing students in a safe environment that does not affect patient care.

Garrett et al. (2015) emphasize that the advent of new information technologies and clinical simulation in Professional health education has resulted in many new opportunities for lab-based clinical skills teaching and learning. But, although the development of clinical simulation technologies, the techniques used to familiarize students with new equipment and clinical procedures have remained fundamentally unchanged. They did a study on if new mobile augmented reality (AR) technologies have the potential to enhance the learning of clinical skills in the lab. The results of study show that students clearly expressed a comfort with the technology, and both students and faculty identified the ability to access resources to support self-directed learning and review of skills as positive attributes of using AR. However, technical issues such as slow response times and incompatible smartphones interfered with resource access and frustrated some students, potentially having a negative impact on their learning. Students gave positive feedback regarding the value of mobile access and having AR resources available "at the bedside" where they were practicing (Garrett et al., 2015). Wu, Lee, Chang, & Liang, (2013) remarked that AR systems and environments could help learners develop skills and knowledge in a more effective way. Additionally, they asserted that AR environments can facilitate skill acquisition.

Carlson & Gagnon (2016) conducted a Project. The ARISE project merges the concepts of simulation with augmented reality and game-based situated learning theory within 150 scenarios for health care students. Results from a preliminary trial in nursing shows ARISE may enhance health care simulation, learning, and critical thinking through providing a more authentic, engaging experience. Frenk et al. (2010), emphasized that AR provides rich contextual learning for students to aid in achieving core competencies, such as decision making, effective teamwork and creative adaptation of global resources towards addressing local priorities.

AR is usually used in learning of the anatomy in medicine. There are studies on this subject. Kiourexidou et al. (2015) said that augmented reality is increasingly applied in medical education mainly because educators can share knowledge through virtual objects. Their study describes the development of a web application. It enhances the medical knowledge of users with regards to the anatomy of the human heart by means of augmented reality. The results of study showed that it is understood that the design and the implementation of the educational application using AR and incorporating the three dimensional model of the heart of the human body, is a relatively new tool for the education of young physicians. On the other hand, the researchers asserted that it is a difficult process altogether. However, it is feasible and doable. In another study, Juan, Beatrice, & Cano (2008) presented that an AR system for learning the interior of the human body. Participants like this system, and said that it is useful for learning of the human body and other subjects. Lee (2012) explained AR can be used to study the anatomy and structure of the body in biology. Teachers could use AR technology to show what organs of human beings consist of and how they look by watching 3D computer-generated models in the real classrooms. Moreover, students may be able to study humans' organs independently with their camera-embedded laptops and AR markers that connect PCs with AR information about biological structures of the human body (Lee, 2012).

Augmented reality (AR) as an interactive technology has increasingly earned public relevance during the last few years (Olsson, Kärkkäinen, Lagerstam, & Ventä-Olkkonen, 2012). The use of tools supported by technology to visualize threads, it helps students to understand real-world situations and problems. Because data generated at the computer cannot be used effective and desired manner in the real environment, three-dimensional visualization even is inadequate in the teaching of subjects that difficult to understand for some students, it is reported that a need for different environments that will contribute to the reduction of such problems. This situation has helped the rapid development of application fields such as augmented and virtual environments (Kirkley & Kirkley, 2005). Augmented reality and virtual reality environments are different from each other, despite the fact that the concepts often to be expressed in the same category (Bound, Haniff, Beber, & Steiner, 1999).

Augmented Reality (AR) is a technology which allows computer generated virtual imagery to exactly overlay physical objects in real time. Unlike virtual reality (VR), where the user is completely immersed in a virtual environment, AR allows the user to interact with the virtual images using real objects in a seamless way (Zhou, Duh, & Billinghurst, 2008). Thomas, John and Delieu (2010) declared that AR allows a user to interact with virtual content in 3D space. AR is a development of Virtual Reality (VR) that allows the user to see both 3D computer-generated content and the real world concurrently in a composite image on a computer display. In AR the user is able to interact with both the real and virtual elements of the environment as though they were both situated in the real world (Thomas, John, & Delieu, 2010).

Höllerer & Feiner (2004) defined that augmented reality (AR) is related to the concept of virtual reality (VR). They add that AR brings about an interactive experience. AR aims to supplement the real world, rather than creating an entirely artificial environment. VR attempts to create an artificial world that a person can experience and explore interactively, predominantly through his or her sense of vision, but also via audio, tactile, and other forms of feedback. The physical objects in the individual's surroundings become the backdrop and target items for computer-generated annotations (Höllerer & Feiner, 2004). In AR, the environment is real, but extended with information and imagery from the system. In other words,

AR bridges the gap between the real and the virtual in a seamless way (Chang, Morreale, & Medicherla, 2010). Another definition comes from Milgram & Kishino (1994) AR is in reality environment instead of real-world objects used a digital media products.

Augmented Reality aims at simplifying the user's life by bringing virtual information not only to his immediate surroundings, but also to any indirect view of the real-world environment, such as live-video stream. AR enhances the user's perception of and interaction with the real world. While Virtual Reality (VR) technology, or Virtual Environment as called by Milgram, completely immerses users in a synthetic world without seeing the real world, AR technology augments the sense of reality by superimposing virtual objects and cues upon the real world in real time (Carmigniani, Furht, Anisetti, Ceravolo, Damiani, & Ivkovic, 2011). When summarized that an AR system combines real and virtual objects in a real environment, registers (aligns) real and virtual objects with each other; runs interactively, in three dimensions, and in real time (Azuma, 1997; Azuma, Baillot, Behringer, Feiner, Julier, & MacIntyre 2001).

Krevelen and Poelman, (2010) emphasized that the technological demands for AR are much higher than for virtual environments or VR, which is why the field of AR took longer to mature than that of VR. However, the key components needed to build an AR system have remained the same since 1960s. Displays, trackers, and graphics computers and software remain essential in many AR experiences (Krevelen & Poelman, 2010). Carmigniani et al. (2011) said that the main devices for augmented reality are displays, input devices, tracking, and computers. According to them there are three major types of displays used in Augmented Reality. These are head mounted displays (HMD), handheld displays and spatial displays. HMD is a display device worn on the head or as part of a helmet and that places both images of the real and virtual environment over the user's view of the world. Handheld displays employ small computing devices with a display that the user can hold in their hands. There are currently three distinct classes of commercially available handheld displays that are being used for augmented reality system: smart-phones, PDAs and Tablet PCs (Carmigniani, et al., 2011). Spatial Augmented Reality (SAR) the user's physical environment is augmented with images that are integrated directly in the user's environment, not simply in their visual field. For example, the images could be projected onto real objects using digital light projectors, or embedded directly in the environment with flat panel displays (Raskar, Welch, & Fuchs, 1998).

Mobile AR is particularly applicable whenever people require informational support for a task while needing to stay focused on that task. It has the potential to allow people to interact with computer-supported information (which might come from databases or as a live feed from a remote expert), without getting distracted from the real world around them. This is a very important feature for the mobile worker, or for anybody who needs or wants to use their hands, and some of their attention, for something other than controlling a computer (Höllerer & Feiner, 2004). The use of mobile technologies such as PDAs, computer tablets, Smart phones, Ultra Mobile PCs (UMPC) and other devices that can be easily transported and provide easy access to information by either direct storage on the device or Internet access is rapidly increasing (Miller, Shaw-Kokot., Arnold, Boggin, Crowell, Allegri, Blue, & Berrier, 2005).Henrysson (2007) asserted that realizing AR on mobile phones would make this technology available to millions of users and, in addition, provide a rapidly developing research platform. For educational purposes, AR based concepts are often integrated into solutions for elementary and high school settings. Applications targeting higher education often deal with technical subjects (Gu, Li, & Duh, 2011). Only few applications for medical training make use of the technology; often, they are still designed for a stationary setting (Yeo, Ungi, Thainual, & Lasso, 2011). Due to the ubiquitous availability of mobile phones and tablets and their ever-increasing computing power, it seems feasible to implement the concept of augmented reality on mobile devices. An Augmented Reality powered learning environment for mobile phones based this concept. Currently, content for a basic course in legal medicine is available; a preliminary evaluation has been performed to evaluate the concept regarding its effect on the learning experience (Jan, Noll, Behrends, & Albrecht, 2012). Looked at separately, neither mobile learning nor AR is really new to education. Through mobile learning approaches, users generally benefit from the provided location and time independent learning setting, while augmented reality learning environments have the advantage of delivering repeatable, standardized content in an interactive manner (Billinghurst & Dünser, 2012).

SOLUTION AND RECOMMENDATION

The shortage of well-educated nurse can be eliminated by online nursing programs. But these programs should provide to nurses a good nursing knowledge and skills such as critical thinking skills, decision making. To gain these skills, visualization is important. Many difficult subjects can be learned with a three-dimensional visualization. Nurse students and nurses can better understand with mobile AR environments. Moreover, whenever they want, and where they want. Especially, they are not afraid of some difficult lessons like anatomy, physiology. On the other hand, nurses practice in a safe environment without causing damage to patients. Specially, nursing interventions such as intramuscular, subcutaneous, and intravenous injections, or to insert catheter like Foley catheter can be learned before going the clinic. So, learners will have more self-confidence.

CONCLUSION

In this chapter, the importance of mobile augmented reality was discussed in online nursing education. Firstly, the imbalance nurse supply- demand was discussed, and it is seen that the number of nurses is not enough in the world and Turkey. The important organizations and institutions such as ICN, AACN, TriCouncil declared that nursing education programs based on computer and Internet technologies are one of the alternatives to increase the number and quality of nurses. Secondly, the importance and use of mobile technology was handled in online nursing education. Mobil technology can use as an active learning and evaluating tool in nursing education. Lastly, the use of mobile augmented reality was tackled in theoretical and practical areas of online nursing education. AR creates an interactive experience for users. AR applications are very important to increase the number of well-educated nurses. AR is very important especially in applied and intensive theoretical courses for learners.

REFERENCES

Aiken, L. H., Clarke, S. P., & Cheung, R. B. (2003). Educational levels of hospital nurses and surgical patient mortality. *Journal of the American Medical Association*, *290*(12), 1617–1623. doi:10.1001/jama.290.12.1617 PMID:14506121

Akkoyunlu, B. (2002). Educational technology in Turkey: Past, present and future. *Educational Media International*, *39*(2), 165–174. doi:10.1080/09523980210155352

Ally, M., & Prieto-Blázquez, J. (2014). What is the future of mobile learning in education? Mobile Learning Applications in Higher Education. *Revista de Universidad y Sociedad del Conocimiento, 11*(1), 142–151. doi:10.7238/rusc.v11i1.2033

Altmann, T. K., & Brady, D. (2005). PDAs Bring Information Competence to the Point-of-Care. *International Journal of Nursing Education Scholarship, 2*(1). doi:10.2202/1548-923X.1127 PMID:16646904

American Associate of Colleges of Nursing. (2016). *Alliance for nursing accreditation statement distance education policies*. Retrieved August 8, 2016, from http://www.aacn.nche.edu/publications/position/distance-ed-statement

American Association of Colleges of Nursing. (1998). *A Vision of Baccalaureate and Graduate Nursing Education: The Next Decade. Journal of Professional Nursing, 15(1),*59–65. doi:10.1016/S8755-7223(99)80027-1

Asabere, N. Y. (2012). Towards a Perspective of Information and Communication Technology (ICT) in Education: Migrating From Electronic Learning (E-Learning) to Mobile Learning (M-Learning). *International Journal of Information and Communication Technology Research, 2*(8), 646-649. Retrieved August 2, 2016 http://citeseerx.ist.psu.edu/viewdoc/download?doi=10.1.1.301.3049&rep=rep1&type=pdf

Azuma, R. T. (1997). A survey of augmented reality. *Presence (Cambridge, Mass.), 6*(4), 355–385. doi:10.1162/pres.1997.6.4.355

Azuma, R. T., Baillot, Y., Behringer, R., Feiner, S. K., Julier, S., & MacIntyre, B. (2001, November/December). Recent advances in augmented reality. *IEEE Computer Graphics and Applications, 21*(6), 34–47. doi:10.1109/38.963459

Billinghurst, M., & Dünser, A. (2012). Augmented Reality in the Classroom. *Computer, 45*(7), 56–63. doi:10.1109/MC.2012.111

Bound, A. C., Haniff, D. J., Baber, C., & Steiner, S. J. (1999). Virtual Reality and Augmented Reality as a Training Tool for Assembly Tasks. *Information Visualization, IEEE*, 32-36. Retrieved June 26, 2016 http://ieeexplore.ieee.org/stamp/stamp.jsp?tp=&arnumber=781532

Boz, B. (2008). *Determining education needs in online learning and learners' opinions: A study of medical workers* (Master dissertation). Anadolu University, Eskisehir, Turkey.

Boz Yüksekdağ, B. (2013). *The attitudes of psychiatry nurses towards the distance nursing education* (Doctoral dissertation). Marmara University, Istanbul, Turkey.

Callister, L. C. (2012). Issues in global nursing education. *American Journal of Maternal Child Nursing, 37*(6), 403. Retrieved June 16, 2016, from http://journals.lww.com/mcnjournal/Citation/2012/11000/Issues_in_Global_NursingEducation.13.aspx.

Cameron, C. (2010). *How augmented reality helps doctors save lives – read write*. Retrieved August 9, 2016, from http://readwrite.com/2010/06/02/how_augmented_reality_helps_doctors_save_lives/

Carlson, K. J., & Gagnon, D. J. (2016). Augmented Reality Integrated Simulation Education in Health Care. *Clinical Simulation in Nursing, 12*(4), 123–127. doi:10.1016/j.ecns.2015.12.005

Carmigniani, J., Furht, B., Anisetti, M., Ceravolo, P., Damiani, E., & Ivkovic, M. (2011). Augmented reality technologies, systems and applications. *Multimedia Tools and Applications*, *51*(1), 341–377. doi:10.1007/s11042-010-0660-6

Chang, G., Morreale, P., & Medicherla, P. (2010). Applications of augmented reality systems in education. In D. Gibson & B. Dodge (Eds.), *Proceedings of Society for Information Technology & Teacher Education International Conference 2010*, (pp. 1380- 1385). Chesapeake, VA: AACE.

Clay, C. A. (2011). Exploring the use of mobile technologies for the acquisition of clinical skills. *Nurse Education Today*, *31*(6), 582–586. doi:10.1016/j.nedt.2010.10.011 PMID:21112132

Davenport, C. (2004). *Analysis of PDAs in Nursing: Benefits and barriers.* Retrieved from http://www.pdacortex.com/Analysis_PDAs_Nursing.htm

Dearnley, C., Haigh, J., & Fairhall, J. (2008). Using mobile technologies for assessment and learning in practice settings: A case study. *Nurse Education in Practice*, *8*(3), 197–204. doi:10.1016/j.nepr.2007.07.003 PMID:17765015

Dickens, M. R. (2016). *Exploration of the use of mobile technology in an ADN program* (Doctoral dissertation). Available from ProQuest Dissertations and Theses database. (UMI No. 10017631)

Ducut, E. & Fontelo, P. (2008). Mobile Devices in Health Education: Current Use and Practice. *Journal of Computing in Higher Education, 20*(2), 59–68. DOI: 008-9003-210.1007/s12528

Embi, P. J. (2001). Information at hand: Using handheld computers in medicine. *Cleveland Clinical Journal of Medicine, 68*(10), 840–842. Retrieved August 10, 2016, from https://www.researchgate.net/profile/Peter_Embi/publication/11752751_Information_at_hand_Using_handheld_computers_in_medicine/links/0deec528f91773f605000000.pdf

Evans, C. B., & Mixon, D. K. (2015). Students Knowledge of Post-op Pain Management after Participation in Simulation. *Pain Management Nursing*, *16*(6), 930–937. doi:10.1016/j.pmn.2015.07.006 PMID:26697818

Frenk, J., Chen, L., Bhutta, Z. A., Cohen, J., Crisp, N., Evans, T., & Zurayk, H. et al. (2010). Health professionals for a new: Transforming education to strengthen health systems in an interdependent world. *Lancet*, *376*(9756), 1923–1958. doi:10.1016/S0140-6736(10)61854-5 PMID:21112623

Garrett, B. M., & Jackson, C. (2006). A mobile clinical e-portfolio for nursing and medical students, using wireless personal digital assistants (PDAs). *Nurse Education Today*, *26*(8), 647–654. doi:10.1016/j.nedt.2006.07.020 PMID:17011674

Garrett, B. M., Jackson, C., & Wilson, B. (2015). Augmented reality m-learning to enhance nursing skills acquisition in the clinical skills laboratory. *Interactive Technology and Smart Education*, *12*(4), 298–314. doi:10.1108/ITSE-05-2015-0013

George, L. E., Davidson, L. J., Serapiglia, C. P., Barla, S., & Thotakura, A. (2010). Technology in Nursing Education: A Study of PDA Use by Students. *Journal of Professional Nursing*, *26*(6), 371–376. doi:10.1016/j.profnurs.2010.08.001 PMID:21078507

Gu, J., Li, N., & Duh, H. B.-L. (2011). A Remote Mobile Collaborative AR System for Learning in Physics. *IEEE Virtual Reality Conference 2011,* 257-258. Retrieved April 25, 2016, from http://ieeexplore.ieee.org/stamp/stamp.jsp?tp=&arnumber=5759496

Guo, S. H. M., Chong, P. P., & Chang, H. K. (2007). Mobile learning in nursing practical training: An applicability analysis. *International Journal of Mobile Learning and Organisation, 1*(4), 342–354. doi:10.1504/IJMLO.2007.016175

Henrysson, A. (2007). *Bringing Augmented Reality to Mobile Phones Anders*. Department of Science and Technology Linköpings Universitet.

Höllerer, T. H., & Feiner, S. K. (2004). Mobile Augmented Reality. In Telegeoinformatics: Location-Based Computing and Services. Taylor & Francis Books Ltd.

Honeybourne, C., Sutton, S., & Ward, L. (2006). Knowledge in the Palm of your hands: PDAs in the clinical setting. *Health Information and Libraries Journal, 23*(1), 51–59. doi:10.1111/j.1471-1842.2006.00621.x PMID:16466499

Hsiao, K. F., & Rashvand, H. F. (2011). Body Language and Augmented Reality Learning Environment. *International Conference on Multimedia and Ubiquitous Engineering*. doi:10.1109/MUE.2011.51

Hung, P. H., Lin, Y. F., & Hwang, G. J. (2010). Formative assessment design for PDA integrated ecology observation. *Journal of Educational Technology & Society, 13*(3), 33–42.

International Council of Nursing. (2006). *The Global Nursıng Review Initiative. The Global Nursing Shortage: Priority Areas for Intervention*. Retrieved August 30, 2016, from http://www.icn.ch/images/stories/documents/publications/GNRI/The_Global_Nursing_Shortage-Priority_Areas_for_Intervention.pdf

Jan, U., Noll, C., Behrends, M., & Albrecht, U. (2012). mARble - augmented reality in medical education. *Biomedizinische Technik. Biomedical Engineering, 18*, 67–7. doi:10.1515/bmt-2012-4252

Juan, C., Beatrice, F., & Cano, J. (2008). *An Augmented Reality System for learning the interior of the human body*. Retrieved August 29, 2016, from http://users.dsic.upv.es/~mcarmen/docs/HumanBody.pdf

Kenny, R. F., Van Neste-Kenny, J. M. C., Burton, P., & Qayyum, A. (2012). Using Self-Efficacy to Assess the Readiness of Nursing Educators and Students for Mobile Learning. *IRRODL, 13*(3), 277-296. Retrieved June 25, 2016, from http://www.irrodl.org/index.php/irrodl/article/view/1221/2261

Kiourexidou, M., Natsis, K., Bamidis, P., Antonopoulos, N., Papathanasiou, E., Sgantzos, M., & Veglis, A. (2015). Augmented Reality for the Study of Human Heart Anatomy. *International Journal of Electronics Communication and Computer Engineering, 6*(6), 658-663. Retrieved August 19, 2016, from http://ijecce.com/administrator/components/com_jresearch/files/publications/IJECCE_3629_Final.pdf

Kirkley, S., & Kirkley, J. (2005). Creating Next Generation Blended Learning Environments Using Mixed Reality, video Games and Simulation. *TechTrends, 49*(3), 42–53. doi:10.1007/BF02763646

Kuiper, R. (2008). Use of Personal Digital Assistants to Support Clinical Reasoning in Undergraduate Baccalaureate Nursing Students. CIN. *Computers, Informatics, Nursing, 26*(2), 90–98. doi:10.1097/01.NCN.0000304776.40531.bc PMID:18317260

Lee, K. (2012). Augmented Reality in education and training. *TechTrends*, *56*(2), 13–21. doi:10.1007/s11528-012-0559-3

Loyd, J. W. (2012). *Mobile virtual patients for nurse training* (Master dissertation). Available from ProQuest Dissertations and Theses database. (UMI No. 1531108)

McAlpine, H., Lockerbie, L. Ramsay, D, & Beaman, S. (2002). Evaluating a Web based Graduate Level Nursing Ethics Course: Thumbs Up or Thumbs Down? *The Journal of Continuing Education in Nursing*, *33*(1), 12-18. DOI: 20020101-0410.3928/0022-0124

McKinney, A. A., & Karen, P. (2009). Podcasts and videostreaming: Useful tools to facilitate learning of pathophysiology in undergraduate nurse education? *Nurse Education in Practice*, *9*(6), 372–376. doi:10.1016/j.nepr.2008.11.003 PMID:19124275

Medley, C. F., & Horne, C. (2005). Using Simulation Technology for Undergraduate Nursing Education. *The Journal of Nursing Education*, *44*(1), 31–34. PMID:15673172

Milgram, P., & Kishino, F. (1994). A taxonomy of mixed reality visual displays. *IEICE Transactions on Information and Systems*, *77*(12), 1321–1329.

Miller, J., Shaw-Kokot, J. R., Arnold, M. S., Boggin, T., Crowell, K. E., Allegri, F., & Berrier, S. B. et al. (2005). A study of personal digital assistants to enhance undergraduate clinical nursing education. *The Journal of Nursing Education*, *44*, 119–126. PMID:15673170

Olssona, T., Kärkkäinena, T., Lagerstama, E., & Ventä-Olkkonen, L. (2012). User evaluation of mobile augmented reality scenarios. *Journal of Ambient Intelligence and Smart Environments*, *4*, 29–47. doi:10.3233/AIS-2011-0127

Ortega, L. D. M., Barachino, P. R., Rodriguez, J., Lourdes, M., Gonzalez, H., Ramon, J., & Salvador, O. T. et al. (2011). Using m-learning on nursing courses to improve learning, CIN. *Computers, Informatics, Nursing*, *29*(5), 311–317. doi:10.1097/NCN.0b013e3181fcbddb PMID:21084973

Ostrow, L., & DiMaria-Ghalili, R. A. (2005). Distance education for graduate nursing: One state school's experience. *The Journal of Nursing Education*, *44*, 5–10. PMID:15673168

Raskar, R., Welch, G., & Fuchs, H. (1998). Spatially Augmented Reality. *First International Workshop on Augmented Reality*. Retrieved June 12, 2016, from http://citeseerx.ist.psu.edu/viewdoc/download?doi=10.1.1.17.4260&rep=rep1&type=df

Schnetter, V. A., Lacy, D., Jones, M. M., Bakrim, K., Allen, P. E., & O'Neal, C. (2014). Course development for web-based nursing education programs. *Nurse Education in Practice*, 1–6. doi:10.1016/j.nepr.2014.06.007 PMID:25154660

Thomas, R. G., John, N. W., & Delieu, J. M. (2010). Augmented Reality for Anatomical Education. *Journal of Visual Communication in Medicine*, *33*(1), 6–15. doi:10.3109/17453050903557359 PMID:20297908

Thompson, B. (2005). The transforming effect of handheld computers on nursing practice. *Nursing Administration Quarterly*, *29*(4), 308–324. doi:10.1097/00006216-200510000-00004 PMID:16260994

Tooey, M. J., & Mayo, A. (2004). *Handheld technologies in a clinical setting: State of the technology and resources.* Critical Care Nurse. Retrieved May 25, 2016, from http://web.b.ebscohost.com/ehost/pdfviewer/pdfviewer?vid=3&sid=ae11ccab-4ef24341-9045-3f2e4d66eb80%40sessionmgr102&hid=116

Tri-Council for Nursing. (2010). *Tri-Council for Nursing Issues New Consensus Policy Statement on the Educational Advancement of Registered Nurses.* Retrieved January 25, 2016, from http://www.aacn.nche.edu/Educationresources/TricouncilEdStatement.pdf

TUIK. (2013). *Turkish Statistical Institute.* Retrieved January 2, 2016, from http://www.tuik.gov.tr/PreTablo.do?alt_id=1095

van Krevelen, D. W. F., & Poelman, R. (2010). A Survey of Augmented Reality Technologies, Applications and Limitations. *The International Journal of Virtual Reality*, *9*(2), 1–20. doi:10.1007/s11528-012-0559-3

Winn, W. D., Windschitl, M., Fruland, R., & Lee, Y. L. (2002). When does immersion in a virtual environment help students construct understanding? *Proceedings of the International Conference of the Learning Science.* Mahwah, NJ: Erlbaum. Retrieved April 25, 2016, from http://www.hitl.washington.edu/people/tfurness/courses/inde543/READINGS-03/WINN/winnpaper1.pdf

Wu, H.-K., Lee, S. W.-Y., Chang, H.-Y., & Liang, J.-C. (2013). Current status, opportunities and challenges of augmented reality in education. *Computers & Education*, *62*, 41–49. doi:10.1016/j.compedu.2012.10.024

Wu, P.-H., Hwang, G.-J., Su, L.-H., & Huang, Y.-M. (2012). A Context-Aware Mobile Learning System for Supporting Cognitive Apprenticeships in Nursing Skills Training. *Journal of Educational Technology & Society*, *15*(1), 223–236.

Yeo, T. C., Ungi, T., Thainual, U.-P., & Lasso, A. (2011). The Effect of Augmented Reality Training on Percutaneous Needle Placement in Spinal Facet Joint Injections. *IEEE Transactions on Bio-Medical Engineering*, *58*(7), 2031–2037. doi:10.1109/TBME.2011.2132131 PMID:21435970

Zhong, L., Xue, Q., Liu, H., & Yuan, J. (2010) Research on 3GMobileLearning Based on Cloud Service. *International Conference on E-Product E-Service and E-Entertainment (ICEEE)*, 1-4. Doi:10.1109/ICEEE.2010.5660179

Zhou, F., Duh, H. B., & Billinghurst, M. (2008). Trends in augmented reality tracking, interaction and display: A review of ten years of ISMAR. *2008 7th IEEEACM International Symposium on Mixed and Augmented Reality, 2*(4), 193–202. DOI doi:10.1109/ISMAR.2008.4637362

Zhu, E., Hadadgar, A., Masiello, I., & Zary, N. (2014). *Augmented reality in healthcare education: An integrative review.* Doi:10.7717/peerj.469

KEY TERMS AND DEFINITIONS

Augmented Reality: Interaction of the user's with real environment and virtual objects in real time.

Mobile Learning: Learning is with mobile devices without depending on time and location.

Nurse: A professional who has to learn lifelong on nursing knowledge.

Nursing: A practical discipline.

Online Nursing Education: Nursing education based on computer technologies.

Three-Dimensional Visualization: There is the width, depth, and height of a picture or image.

Virtual Reality: Virtual reality is to imitate the real-life environment through computers.

Chapter 16
Design, Development, and Marketing Process of Video Games

Devkan Kaleci
İnönü University, Turkey

Tansel Tepe
Kilis 7 Aralık University, Turkey

ABSTRACT

In this study, general information about video game and video game sector, process of steps when designing, developing, and marketing a video game are presented in detail. Since game designers attempt to create their own type and develop new game type, there is no certain acceptance about suitable categorization for video games and certain number of video games types. For this reason, it is firstly mentioned about video game concept as well as systematically categorizing all game types in this study. Afterwards, each of these game categories and game types are described in details. Furthermore, it is aimed to raise awareness of researchers interested in this area, and to help in removing false or incomplete information about video games in the process of video game development.

INTRODUCTION

Game has an important role in life of people of all genders and ages, particularly in children lives. The word game etymologically stem from "jump, bound, prance, leap" in Turkish. But, although the term game has not a well accepted definition, it can be difined in two basic ways; the committed behaviours unconsciously without considering results or the all of committed actions freely, knowingly, a rule that certain in a specific time and place. The main purpose of game is to Improve phsical, cognitive and sensorial features, have fun, let off steam, relax, develop imagination, motivate, learn or teach something educationally.

Games, programmed with several technologies and played by crerating a visual user interface with a video renamed as video games. Video games include activities between player-player or player-artificial

DOI: 10.4018/978-1-5225-2110-5.ch016

intelligence. Players control the game with a keyboard, mouse, joystich, motion and sound sensing devices. To know why people play viedo games are the basic for video games investigations and ther are some studies trying to explain why people tend to play video games in literature. Sherry and Lucas (2001) state reasons of playing a game as challenge, rivalry, fantastic environment, exhilarant effect and variety. In this work, participations mention the purposes of playing a game as relaxing, keeping away from stress, spending time, can't find something to do. The other work in this subject is made by Kirriemuir (2002). In this work, purposes of playing a game are specified as fantastic environments, curiosity, rivalry. According to Rouse, players have some anticipations from video games. The anticipations are shown as consistency of video games, understanding limits of world of video games, working out rational solutions in video game, directing players while playing games, progression in achieving the tasks step by step, overtaken itself to game, experiencing temporary failures, players know chance of success, players don't have to repeat itself in a continuingly, getting stuck in game desperately, players want to be a part of game.

In this work, the general information about video games are presented and mentioned about historical development process of video games kinds of video games are categorized, necessary processing steps for designing are explained, and the developing and marketing process of video games are presented. There main purposes of the work is to eliminate information that are presented as inaccurate or incomplete about video games which we can encounter these in many websites, present the steps in creating video games in detail and raise the awareness in rapid increasing numbers of curious researchers about industry of video games.

VIDEO GAME INDUSTRY AND ITS HISTORICAL DEVELOPMENT PROCESS

The origins of first video game are based upon video game which is called Nimatron and later called Nim by Edward Cordon in 1940, ten thousands of people played Nimatron but artificial intelligence would win ninety percent probability (Wikipedia, 2015). Between 1956-1961, Mouse in the Maze (a mouse that searches cheese in a labyrinth), HAX (in the game, various visual forms and sounds can be made with two keys on the consoles), Tic-Tac-Toe (players can write X and O with a light pen in the game), works graphical in computers are called TX-0, were developed in MIT. According to Friedman (1995), the beginning of video game history is accepted as "Spacewar" that was designed by enginners of MIT in 1962. The game was prepared in Digital PDP-1 master computer. Digital PDP-1 master computer pass into the history by developing the first interactive war gameIn the progress of time it become popular took place in all of computers. Ralph Bea made a game "Corndog", running on standard television in 1966 and then he developed games like ping pong and shooting target in 1968. In 1969, Ken Thompson designed a game "Space Travel", it worked in MULTICS operating system. Space travel simulates motions of solar system and it is about taking spacecraft to the surface. Later, Ken Thompson and Dennis Ritchie designed the game in PDP-7 assembler language for bringing the costs down. In this period, the first UNIX version of "Space Travel" was found in development process of UNIX operating system to learn software development for machine (Ritchie, 2007).

In 1970s, television started to come into prominence as a visual output unit. In 1971, "Galaxy Game" was designed in Stanford University that includes vector graphic terminal and also which is the first pay game. In the same year, Don Doglow from Pamona University designed the first baseball game, it is played as by one by matches or complete season matches. Also in 1971, text-based version of "Star

Trek" game was designed again by Mike Mayfield in Sigma 7 computer. Until 1972, games was being played just in laboratory of universities. In 1972 games started to reach the large masses and it began with "Pong" game that was made by Atari Company. In 1972 the first video game console, "Magnavox Odyssey" was put on the market in USA. There were plugin circuit cards on the console rather than slots for game and cartridge. Also in 1972, Hunt the Wumpus, an adventure game that was text-based and created with the style of hidden- searched, was designed for PDP-10 machines by Gregory Yob. "Maze War", was developed in NASA's Ames Research Center in California, "Spasim", was developed in Plato, formed a basis for today's first person shooter games (FPS). Also in 1974, Brad Fortner and his friends developed a flight simulator for education. The first written with Fortran language text-based games "Colossal Cave Adventure" was served by Will Crowther in 1975. In the same year, Daglow designed the first acting game (RPG) "Dungeon". Also in this year, "DND" is based on Dungeons and Dragons was designed in CDC computer at the system of Plato. In 1976, Atari Company made sales of 26 million dollars and leaded up in game industry. In 1977 Kelton Flinn and John Taylor developed the first version of text-based air war game "AIR". In the same year, Dave Lebling, Marc Blank, Tim Anderson and Bruce Daniels designed "Zork" game. In 1977, there was a sectoral crisis due to the widespread use of "Pong" machines and this crisis affected lots of game companies in a negative way. But, the cirisis was resolved with "Space Invaders", made by Atari Company in 1978. In 1979, Atari Company provided playing video games on TV.

In 1980, Michael Toy, Glenn Wichman and Ken Arnold found "Rogue" game as a result of two years study. Nintendo Company marketed "Donkey- Kong" in 1981, Namco Company marketed "Pac-Man". In 1981, sales figures of game industry got six billion dollars. In 1982, Commodore 64 broke sales record. In 1982, "Pole Position" and "Pitfall" were marketed and in 1983, "Mario Bros" and "Dragon's Lair" were marketed. There was a big cirisis as a result of lots of unqualified games to be marketed and a lot of company bankrupted. Video game industry had fallen to less than 800 million dollars. When NES has been marketed by Nintendo Company, ındustry slipped from USA to Japan and put an end to process of crisis. In 1984, "King Quest" and "Elite" games showed up. In 1985 Russia contributed to game world with "Tetris". In the same year, Nintendo Company introduced the new system of entertainment and Atari Company returned with 16-bit 520ST. In 1986, "The Legend of Zelda" and "Metroid" were designed. In the same year, Sega Company included Sega Master System game console to game market, Electronic Arts Company introduced "Baseball" game that was an important step for sport games. Again in the same year, "Mega Man", "Final Fantasy" and "Metal Gear" were introduced. In 1989, Nintendo Company marketed "Gameboy" device in the same year, "Popoulos" and "Prince of Persia" games were written. As Amiga computers were sold for a million, the game world developed (Overmars, 2012).

Within 1990s, PCs became cheaper and popular so PC started play an important role in popularizeting of computer games and video games. In the same years, CD and 3D were viewed in games and video games and the industry continued to grow rapidly. Besides Amiga and Atari ST started to die out and computers, consoles became basic platforms for games. In 1991, Nintendo Company marketed 16-Bit Super-NES. In 1992, although computer gamess' sales felt, Nintendo Company reached a sales figure of seven million dollars. In 1993, ID Software developed a technology with FPS (first-person shooter) games that gamers can watch the game with player's eyes of the game. At the beginning of 1990, Sega and Nintendo's consoles had %80 of game markets. FPS games became more popular with "Doom (1994)" and "Quake (1996)" games. In the same year Atari Company began to appear with Jaguar 32 bit. In 1996, Nintendo Company put onto the market ULTRA 64. In this year, multiplayer games started to making commercial. In 1997, artificial intelligence applications developed and a computer won chess champion of the period for the first time. In the year, three-dimensional graphics acceleration was standardized as 3D-FX. In 1998, Playstation began to dominate consoles.

In 2000s, games began to be played in social networks and mobile devices. In this year, game industry precluded theatre, cinema, music industries. Development movements in games leaded to consoles from computers. “Diablo 2” game was sold a million in one week. “SIMS” game was sold 2.3 million in one week at all platforms. In 2001, Nintendo Company marketed Gamecube as for that Microsoft put on the market XBOX. In 2003, turnover of the EA Company reached 2.5 billion dollars. In 2004, game sales industry reached about 7.3 billion dollars. Nintendo Companymarketed DS while Sony company marketed PSP in portable games. In 2005, numbers of subscribes of World of Warcraft overcame four million dollars and NFL 2006 of EA Company was sold 1.7 million dollars for the first week (Laird, 2005). In 2005, Xbox 360 marketed game console furthermore in 2006, Playstation 3 put on the market game console. Nowadays, game industry is cleaned up. Moreover some games carried interests single-handedly surpassed other industries. Game industry, one of the few sectors not affected by the economic crisis in the last of 2000, continued to grow rapidly with 2010. While marketed “GTA 5” in 2013 overcame 800 million for the first day, “Call of Duty: Black Ops” reached 650 million dollars for the five days. If the conditions became different in mobile games market and games have been integrated into world-wide about 1.1 million mobile phones that have became ready to be released (Paavilainen, 2004). The number of those who choose to massive multiplayer online role-playing game has frown steadily and the type of game that emerged with increasing use of social networks and games, has been having quite a large userbase. It is guessed that the size of the video games industry will overcome 110 billion dollars. With the last of 2015 game industry is developing in countries producer of game consoles such as firstly Japan, USA and then South Korea and China. Video games has became an interesting sector not only in some countries but also in countries like Turkey. A clear market analysis cannot be made in game industry for evaluation of Turkey but it is guessed that profits derived from the game industry is about 150-200 million dollars. About 20 company have personnels more than some companies’s personnels that have 10 people groups. Game industry has been developed each day. Especially in last days, games in social networks are preferred in Turkey and a lot of company that produces games in this area, are opened because of the fact that the cost of games is less. While these companies are supported by foreign capital, some of them are formed from small groups. Digital gaming federation of Turkey (TUDOF) was established firsty in the world within the scope of General Directorate of Youth and Sport in order to raise the awareness of public, provide that the partner get know each other, provide that the public benefit from the games but the closed up aftersome time.

People’s meeting age with games are decreasing everyday. The amount of the game users in Turkey is not known exactly but it is increasing day by day. The use of widespreading social networks and to be able to playing games in this social networks are important factors for increasing numbers of gamers. According to data of ESA (Essential Facts Association) in 2015, viedo game player under 18 years generate %26 of the population, player at the age 18-35 generate %30 of the population, player at the age 26-49 generate %17 of the population and players at the age fifty and over fifty generate %27 of the population. Men constitute %56 while women constitute %44 of video game players. Once again to ESA data in 2015 the best selling video games are %28.2 Action games, %21.7 Shooter games, %13.3 Sport games and %9.5 Role-Playing games.

TYPES OF VIDEO GAMES

Fast growing and diferentiating video games everyday are difficult to categorize into a certain systematic structure. And there isn't a certain and accepted work in this subject. In the context of the type and structure of the new video games have begun to take more interwined that wherrets the situation. The work tries to make a specific system with the styles of games and the work unlike previous studies categorise video games as three basic category which are according to scenario features, playedat platforms and used graphics.

According to Scenario of Video Games

The number and diversity of video games are increasing everyday in parallel with fast growing of video games industry. There isn't a certain judgement about the styles of video games because game designers are in an effort for creating their own kinds and developing them. From simple to complex video games have a scenario like a film or a novel. Prensky (2001) collected video games in eight categories; adventure, action, fight, puzzle (crossword), role-playing, sport, strategy and simulations.

- **Action Games:** It is a type of video game that physical challenges like eye-hand coordination and reaction time come into prominence. Sometimes it is described as a first person shooter game. The best knownn sub-types of action games are fight, shooter, and platform games. Arcade and video console games like "Super Mario", "Sonic", "Hedgehog", "Labyrinth" games like "PacMan", duty and fight games like "Counter Strike" and "Medal of Honor" and race games like "Need For Speed" are some type of action games.
- **Sport Games:** Generally the real life sports transfered to electronic environment. Gamers can direct both a character and a team. Sports in this games are branched according to types of sports. Talent comes into prominence in these games. Form of playing game simulates the truth and increase people's wishes and likes. "Fifa", "Pes", "Footbal Manager" series in football, "NBA" series in basketball, "NHL" series in ice hockey are exemples to sport game.
- **Role-Playing Games:** The aim of these games is to manage players in the game by gamer put himself instead of a player of game and adopt to the player in the game. Gamer who replaces the player takes all decisions instead of the player furthermore gamer has physical, social and spiritual features of the players. So gamer resolves a problem by the help of knowledge and features of player. "Dota", "World of Warcraft", "Star Wars: Old Republic", "Final Fantasy 4", "Metin 2", "Never Winter Nights", "Fallout", "Fable", "Dragon Age" series are exemples to these games.
- **Adventure Games:** The aim of these games is to manage players in the game by gamer put himself instead of a player of game and adopt to the player in the game. Gamer who replaces the player takes all decisions instead of the player furthermore gamer has physical, social and spiritual features of the players. So gamer resolves a problem by the help of knowledge and features of player. "Dota", "World of Warcraft", "Star Wars: Old Republic", "Final Fantasy 4", "Metin 2", "Never Winter Nights", "Fallout", "Fable", "Dragon Age" series are exemples to these games.
- **Fight Games:** In this type, real fighting games transfer to video games. Chosen player by the gamer should win the fight with other fighters or opponent in a specific time. In these games, gamer must replace one of the options presented in the game or a specific player. "Street Fighter", "Mortal Combat", "Marvel vs Capcom", "Tekken" are exemplified to this type.

- **Strategy Games:** The type is divided into two; real time strategy in which gamers don't wait for the turn of each other and game continues to play mutually or turn-based strategy games in which gamer must wait the turn of the other for lunging. In this type, the aim can be developed starting over from scratch by managing an army or a civilization with using different strategies. These games have spcific rules. "Civilization", "Roller Coaster Tycoon" and "Red Alert" are in real time strategy and games such as chess are in turn- based strategy.
- **Puzzle Games:** These games create a platform where develops visual literacy of people. Moreover people solve problems virsually and stories of games are simlified. Generally these games are fictionalised over people's desire to know "Tetris", "Devil Dice", and "Portal" is in this type.
- **Simulation Games:** In this type, the primary condition is to ensure environment based on reality. Generally these are used educational for reducing the cost and risk factors. "Sim City", "The Sims" and flying and driving simulations are in the type.

According to Used Platforms

- **Console Games:** Console of game is an entertainment device that can be connected to the TV outside of the computer for playing game. Firstly Nintendo and Atari Companies developed cartridge game consoles and these are the beginning of console market. Nowadays it becomes devices, utilizer the latest hardware units. Games designed for these platforms have the latest and advanced graphics features in parallel with hardware unit. A large part of the game market is formed with games designed for these platforms.
- **PC Games:** PC games are the type of games in these platform that is designed for PCs and its derivatives. These are games in which gamer can use available technologies and new generation of control mechanisms may be used conventional control devices like keyboard and mouse. A lot of game, are played in consoles, markets for electronic environment.
- **Handheld Games:** Video games transfer to small and portable equipment for playing interactive electronic games. Playstation Vita, Nintendo 3ds, Pandora, etc. are exemplified for these games.
- **Mobile Games:** Mobile games can be used in smart devices such as smartphones, tablets and TVs. These games have a wide range of gamers of all ages and genders. Mobile games can be played with SMS or WAP, hardware units of the device, mobile networks or wireless.
- **Web Browser Games:** These games are played via web browser. These are worked generally by used plugins developed for standard web technologies or web browsers. Social network games and web browser games are in this type. These are downloaded by utilizing serviced web browsers in the market service and this is in a type of plugin of web browser.
- **Social Network Games:** These can be played via a social network, gamers can play with their friends or gamers can be friend with other players of the game. After social networks became popular, these have came up. "Farmville", "Candy Crush", "Second Life", "Salem and Roblo" are in this type.
- **Arcade Games:** These are played at game arcade machines with jettons. Gamer's style of playing game is very important for playing longer because these machines have not recording and retrieval features. Arcade games were a single judge of game market in 1970s and 1980s and these have been played by gamers in game arcades. Today these are played with various emulators. Gamers, who love nostalgia, prefer this type.

According to Graphics Type

- **2D Games:** They are formed with two dimensional graphics. In 2D games, it is worked with X and Y axes on the coordinate system. The third axis, Z, isn't in it. Design 2D games is easy so it is located very much in the game market. Designers, who first began designing game, begin with 2D games because it is easy to play. If the designer has informations about codes, "Game Maker" provides convenience to designers or if the designer doesn't have informations about codes, "Eclipse Evoluation" provides convenience to designers.
- **3D Games:** The term of 3D is formed with the term of three dimensional. 3D objects are gotten by gaining depth to a two dimensional drawing. Having three dimensions of an object means that game has height, depth and width. The type of games that is developed by using three dimensional models and environments, is named as 3D games. A lot of new games are in the type. 3D games can be designed by the help of game engines like Blender, Unity 3D, Unreal. Although these engine have internal modeling capabilities, advanced 3D applications software are preferred for a good 3D game design.
- **Text-Based Games and Sound Games:** These can be played entirely with texts and include simple graphics. It has been designed in 1970s. In text-based games expression is at the forefront more than visual quality. In the games, game is visualised by depicting writing instead of graphics. Gamers must make some decisions in the game. According to the decisions, the course of the game changes. "Atlantis Remixed" is an example to this type. Sound games are electronic games that give audible and tactile feedback rather than visuality. Generally, the game is designed for sight-disabled people and people with visual distress. "Grail To The Thief" is an example.

According to Types of Player

- **Single Player Games:** The game includes just one player and the whole process is managed by the player. The game is controlled by a gamer and there isn't an external intervention and all players in the game accompany to the player who controls game. Almost all FPS games are in this group. Also these includes online and offline version in parallel with developing networks and internet technologies.
- **Multi-Player Games:** Multiplayer games can be played on the internet or a network and a computer. These became almost the most popular game with increasing the mass of gamers. Besides new generation game consoles support to multi-player games. "World of Warcraft", is developed by Bilizard Company has online user more than 10 million. From 2004 to today, numbers of gamers have surpassed 100 million. Also, "League of Legends" and "Dota 2" are examples to this group.

Stages of Preparation a Video Game

Although curious designers who want to design video games don't know exactly the idea of making a video game, this is a complicated and long process. When it is realised challenges of the process, it will be understand that prepare a quality game literally is so difficult. Although video games may be formed with 2D graphics by a designer, video games may require a team workk for 3D graphics, animations, films, artificial intelligence components and voiceover. Nowadays when it is asked to design a video game by pursue gamer's satisfiction, entertainment, motivation, interaction and also an educational goal, to be realised of this game by a single designer is almost impossible stage. Some specific process steps for designing an effective video game is prepared, process steps like deciding the type of game, developing the scenario, preparing graphics, coding, testing, marketing. These are shown as follows in Figure 1.

Deciding the Type of Game

The first step of process of preparing a video game is deciding to type of game. Before designers start to design a video game, they must decide to game will have which features, game will be designed for whom, game will be designed at which, game will have which graphic features. If designers start to design a video game by considering these, these provide precluding some problems. While designers decide to type of game, they must evaluate ideas of gamers, analysis to successful games, know features that gamers like game for these. Before designers prepare a video game, they must decide game will be 2D or 3D, game will be prepared at which (computer, console, mobile, etc.), game will be which type in this direction they must utilise technologies that are convenient to games's type and platform.

Determining of the Scenario

The designer should form primarily a game focus in this chapter. The game focus is a description chapter that designer of the game tells shortly the basic features. The that description chapter is told which part of the game being entertaining, compulsive and matchles, and the aim of the game. Elements that gaining identity to the game and the most important in the game explaine. It is guide in the process of the game's development. A game focus is the most important idea about designer of the game. Focus is the necessary designer of the designing game to forming in supposed the budget, and designing the form and wanted the time. According to selected the game type, it is possible identify the environment, events and characters during the scenario. The roles of character who are in the play should be determined. The designers should be set well the limitations of the environment and features of the characters should be detect around the these limitations. Also, in the progress of secenario's determining can be explained in the grade of level design that determining the difficult degree of game and explaining what it needs in the passing which between the chapters.

The level designer should be take advantages from the engine of the game, from aspect of artistic and mechanic of the game in the level designing because the levels can be built. Therefore it is, the level designing is a place where being appeared highest the problems in the game. A stage in a race game can be a track available in the game or can be different stadium from a different sport game. Also, after the duties fulfilled in the adventure and duty games, passing to a new map indicates passing of the next level. Especially, in daily games during the process of the game to see a level pass screen is not at least as boring as to see a loading game screen. A good designer specify the areas where is storage abridge-

Figure 1. Process steps for designing an effective video game

ment and to create stage appropriately, it situate in the available places to make it enjoyable. The line of the level is quiet important to flow of the game. The game can be seem as different from two games. If all of strategic level firstly were established and after these, all of adventure level were put in order. The line of level should be shown balance the equal as expect of the gamers. For example; there are 3D levels and six monters in a game, it will be logical if these monsters can be established balanced to 5th, 10th, 15th, 20th, 25th and 30th levels. The inter of levels can change one-two level or small difference can be shown. The first levels can be relatively more simple, the next levels can be more compelling. A level of compenent was formed by adventure, research, discovery, solved crossword (puzzle), telling story, aestetic and stabilization of entire (Rouse, 2005). By not moving away from the focus of the game, it is necessary to specify the importants part of each level. While the refuging and protecting were being purpose in a level, other level of purpose is fighting and bumping. Adventure is the most distinctive component in a level. Nearly, entire of the game have adventure component expect for the some adventure and the crossword games. For example; killing the demons by a trigger man in the "Doom", exploding the mushroom in the "Super Mario 64", killing the mutants in "The Fallout", killing the guardian in "The Metal Gear Solid" and as "The Need For Speed" is a race game hat have adventure the more make speed toward rivals. Instead of justly providing the bridge between two adventure set, if the research/ the discovery can be designed correctly, it can be quiet enjoyable to the gamers. The level designers may demand the gamers to discover some chapters of the game on their own. The gamers should solve some the crosswords, because they can be success in the game. There are hidden guns and the gamers should find these bacause of attaining a result. İt is possible that there is laser ray and it may be turn another direction to contining the game. It is in this case, the gamer should think simply including the physics rules.

For example; the gamer can find a mirror, it was placed correctly in front of ray and the gamer can send the correct corner. The game is story that is quite important the appropriate component of story and levels. The levels are one of the part of component that telling the story of the game in that games. The gamers can require getting in contact or conversing with characters who are in the levels. Much as the level designer has not to quiet move away from the story, the story designer should find sufficiently the independence because the best level order can be constituted by level designers. After entire of other component were finished, the quene is aestetic of vision and sound of level. The middle- level that is suitable for require of the story of the game, looks good, play good and and goes on fastly, exists all the time. The game should be charming with vision and sound of game, and with its visuality. The chapter of aestetical regulation takes time considerably. The level structure should produce where the gamer should been, where the gamer can come or not, entire of these components should be ideally compatible and well balanced each other because a good and qualified game level can be formed. The video games have two level flows; the linear levels (action/adventure, role playing and race games) and the nonlinear levels (strategy, sport and multi-player death-match games). If designer is enough creative and try to effort, the designer is not sharply maket he destinction between the two levels and the designer can design together because the game mechanic can do more differented and interesting. Rouse (2005) states that a good level design have this features.

- The gamer should not jam a apparent environment.
- The lower target should be understandable, clear and these should contribute to the main target.
- The signal should be direct understandably to the gamers for the correct parts, if the target is not create the confusion.

- The critic roads is clearly and definitly, this is a good method to preventing the gamers of confusing. So, the critic roads should be clear to go.
- The gamers should be give the limited returns in the level.
- The first success is too important for the gamer, because of firstly giving unsuccess sense and forming the dissapointment to the gamer, it can be reason to giving up.
- The roads that can be travellable and passable environment, should be state clearly.
- The right of election should given to the gamers, the options should be given.

Design of the Models, the Graphs and the Envirenments

This grade contains forming and designing of the entire visual compenents that were appeared in the scenario grade, by graph designer who are in the crew of video game project. The entire of visual compenents do with two and tree-dimensional visualization and modeling programs. The internal passing films and necessary animation for video game were formed. Also, the difficulty degree of game and artificial intelligence design that forming the logic of the game should analyze in this grade.

The scenary of the majority of game look like the each other when it thinks to the main meaning. Generally, there are a hero and difficulty that were overcomed by the hero. Consequently, the hero overcomes the whole difficulty and finishs the game. Finishing of a game is not as importantas how and which levels were got ower For this reason. The most important difference is environment of the playing the game, the characters and animations because of seperating the other games. Good designing visual elements come from primarily the most important step to design a good game.

Coding

The game engine that will be used in the making of the game in many of the functions, in the object, in the code-class. Visuallibraries and also two and third-dimensional visual game where the audience is located, is the game development environment. The game developers can easily generate the gones by writing less code with game enginesand with pre-written code (Gelibolu, 2013; Tuğtekin & Kaleci, 2011). There are many fee (Unity,Unreal Engine,Cry Engine,ets) and free engine game (Ogre, Belender 3d, Delta 3d, etc.) on the market (Kaleci, Kıran & Dinçer, 2012). The graphics processinglibrary known to be used in the coding grade, which is advantageous it. The most popular familiar graphic processing libraries are Microsoft Direct X Library that contained within direct 3D and open-sourse Open GL that developed by Khronos Grap.

Testing

The testing process is not merely a debug. While the debugging is the more programming based, participants who are involved in te testing process, examine the game and try to specify the game is fun or there is the error in the game mechanics. It is quite important to determine who will test the game. Many testers can tell what they liked or did not liked the game. However, the minority of them will explain according to having own level how to resolve this problem and why they do not like it. One of the main points of working with testers have recognized their mindset and prejudies. If tester's past experiensies, prejudies and personal features is not known, it will lower the chances of trust the donuts. It is important to follow the people who will be selected for the test team to the test the game. Testing process is one of the important element cycle that should be included from the playable stage of the game until offered up for sale. After the controller is placed exactly, the game is opened the door to first impression of the testers. The traditional game testers is entered, after the game almostly is got over and cored the details of game, anymore. When the testers recognized the error that is about the game or coding in the testing process, this error is reformed project team, after that, the game is again taken to the testing process. In the event that the error does not find at the end of testing process, the game's output can take suitable as wanting.

Forming of the Design Document

The design document includes detailed written descriptions of the game that is from the stage idea until market launch grade and it suggests about the vision of the game. People who is in the project team someone who wanted to research can scan and attain easily thanks to the reference. A design document have the basic content as the table of contents, the summary overview, the mechanics of the game, the artificial intelligence, the game elements, the game summary, the game progress, the system menus (Rouse, 2005). The table of the contents is the first palace to look in the document. If the table of contents is as detailed as, people can reach much more quickly and reliable it. Generally, the new users and new team members maybe a good start with a short executive summary to understanding the game. A large of the game is explaned the readers in this grade. It helps to explain the essence of the game. The design document is the most important part of the game mechanics. The game mechanics is the place where yield is given, subject of the game and the design of the basic game-play mechanic. This section may be cited as department of playing the game. This part is described how the game is played and which players have been allowed to play in the game. This part of the game is described what kind of action entire gamers can do and what kind of movements they can do. For example; how he crossword works, how to change, how to play the game to do the redirect and how to create the section that specifies in this part of the game. All kinds of action as hitting the objects, secret passegeways and falling to the grand, opening and closing doors, etc, are explained in the section of game mechanic.

When the gamers play the game, what the gamers will be in the game, that should be described. This is the desription of what kind of camera to use the gamers how they see the world, gamers can effect how the position of the camera. The artifical intelligence part of the game describes how they reacts to the movements of the gamers. İt explain the situation that is as components of how to treat the gamers faced in the game, and they will do what the situation is in, and what to do when the game did not do anyting. The game elements are divided the three section; characters, items an mechanisms. Characters are all different types of artifical inteligence agents and all the personalities whih can be faced in the game. The items are all of the goods or entity that the gamers in the game can get. Also, mechanisms can be expressed such as game system in the game. If the story of game is explained easily and clearly to the readers, it assists the readers for understanding most of the game. The progress of the game can be the longest part in the design document. It assists about what kind of backgrounds needed to create for both the art team and level designers. The items which are for the smooth progress of the game, which object wil be and where the objects will be, these should be written here. It should be mentioned that a gamer would be faced with difficulty. The system menu is described in the game menu and the gamers is introduced in the section where they face at different times of the options on the screen. In the game, what kind of tools to be used must be specified such as is based moused with click method? Is enter or with the keys of the keyboard? Or both of them to use in the system menu. The design documents can be difference from game to game. The reason of being difference from project to project, the game is not in the standart design form. The design document should be merely filled with the useful information, it should not be inflated with unneccesary information.

Marketing

Nowadays, the important situation is not only process of the design and development of the games but also the marketing process of the games. The number exceeds twenty in the world created the playing field of business models. There are four basic business models; these are “Play the full version and completely free”, “Play the full version with fee”, “Play the free to the trial version but play paying the money for extra features” and “Play with the advertisement in the game”, and the competition in the market is incresing day by day. As a result of this competition, the companies have developed different models on the marketing of games. The steam is the most prefered between developed marketing models, it is developed by the Valve Corporation. The Steam services in the digital distribution and the digital rights protection. Many games are sold through steam and program in the industry. Nowadays, the marketing of games began to the used intensive as Steam digital environments. The Steam interface is given in Figure 2.

Before the games bunch to the market, it should be tried to be followed by the right advertising and right of publicity strategies. The nature of the game should be featured by advertising the more effective and more reasonable instead of unnecessary and expensive promotions. To be disgussed by right companies or right person will speed up marketing process of the game during the promotion.

Figure 2. Steam interface

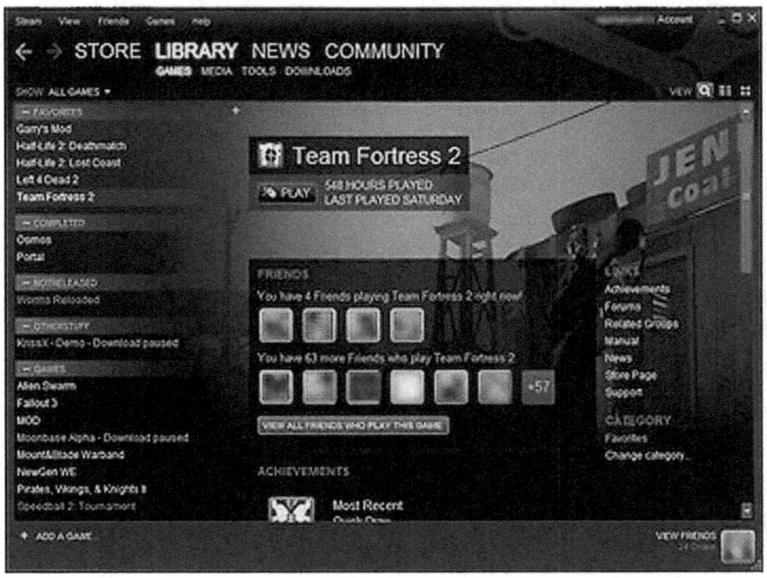

CONCLUSION AND SUGGESTIONS

First of all, general information about video games and video games with details of the process of the historical development is presented. After that, purposed for the new types of games categorized are described by the authors. Finally, the game design steps are given. At the beginning of the design grade of the gamer that grade was stated that the user had to be decided to the type and mass, and in which platform created. Then in this grade given the importance of the creation of a scenario according to selecting game type, in this process, depending on the type of game, environments, actions and characters of the selection process, tasks, determination levels and limitation of the game are disgussed. The next grade is based on determination appropriate graphics, the progress of design to environments and characters, creating of animation and the design of artificial intelligence. At on other grade is told which coding language is appropriate, determination of limitation of coding and the coding process was done for video games.

The next grade explains, after the design process of the game comes from the testing process of the game in which should be specified the errors, and then the errors are solved, retesting the game and after the process of testing, the errors are not happen in the game, the output of the game can get by suitable. Finally, it was stated that it is important that identification of the distribution of the video games when video games are released to the environment, the appropriate advertising of video games and also discussed with suitable the companies and promotion be made to video games. Also, while the video games prepare, the design document is importing. Those who want to design a video game during the process of designing of the game should follow some offers below:

- **It Should Be a Good Scenario:** Every game has either a simple or complex scenario. A good game very succesfull with the fiction of games can be designed.
- **Graphical Components are İmportant:** Actually, how realistic the games are the better. For this reason, the gamer fascinates from the models selected for the games and affect the quality of the graphics, these factors are one of the most important factors that affects the gamer. The game does not only have high quality graphics but also the game's graphics must be harmonies and impressive.
- **A Video Game Should be Primarily İnteresting:** During the game, the gamer should be feel like as if the gamer was in the game itself.
- **The Game Should Contain Elements of Curiosity:** During the game and in the pass beetween sections, if the gamers wonder that they faces with, they will be rush into the game. The game should be encouraged be to gamer.
- **When a Video Game is Designed, İt Must Contain Elements of the Fighting:** The problem situations are given to the gamer for the purpose of the game. The gamer overcomes the difficulties in this problem the situation, however, will be able to come as we struggle with challenges.
- **The Control Function is an İmportant Factor is Designing a Video Game in the Game:** The gamers should be able to control freely the characters as he/she wants. The gamer should be able to flexible on how he/she wants to set the key controls.
- **The Fantasy Elements Should be Present in the Game While Desinging the Video Games:** The behaviours that will not be possible in real life, with fantasy elements in the game will have the opportunity to perform. Thanks to fantasy elements in the game, the gamer will be able to reach the peak of satisfaction.

- **The Gamer Should be Treated Carefully on Elimination of Technical Problems, Because the Gamer is Not İnserted in the Game:** If the gamer is not go to the certain part of the game and he/she has the technical problems, the gamer probably will leave from the game.
- **The Video Games Should be Designed to the Flezibility Because the Gamer can Recognize:** The gamer should be able to save the game anywhere and leave it, the gamer should be given the sufficient time to make moves.
- **The Gamer Should be Prepared with Explicit Directions for Simplicity in the Game:** The game should be included character who can interact the more in the game instead of NFCs (Non-player character) who are more unneccessary characters.
- **The Sound and Animation Should be Considered to be Consistent in the Video Games:** For example; the sound that comes from before and after the movement of the characters of the game can cause the gamer to fall from his/ her eyes.
- **The Multiplayers Can Demand to Play the Same Game Together, This Should be Taken into Consideration:** The players may be wish to play into the same game, the players demand that are simultaneously both socializing and living different experiences.

REFERENCES

ESA. (2015). *Essential facts about the computer and video game ındustry*. Retrieved November 16, 2015, from http://www.theesa.com/wp-content/uploads/2015/04/ESA-Essential-Facts-2015.pdf

Friedman, T. (1995). *Making sense of software: Computer games and interactive textuality*. CyberSociety. *Sage (Atlanta, Ga.)*, 73–89.

Gelibolu, F. (2013). Eğitsel dijital oyunlar: Kuram, tasarım ve uygulama. Pegem Akademi.

Kaleci, D., Kıran, H., & Dinçer, S. (2012). Açık kaynak kodlu 3D oyun motorları. *Akademik Bilişim 2012 Konferansı Bildirileri, Uşak Üniversitesi, Uşak*. Retrieved November 16, 2015, from http://ab.org.tr/ab12/bildiri/139.pdf

Kirriemuir, J. (2002). Video gaming, education and digital learning technologies. *D-Lib Magazine*, *8*(2). doi:10.1045/february2002-kirriemuir

Laird, J. (2005). *History of computer games*. Retrieved October 1, 2015, from http://www.emunix.emich.edu/~evett/GameProgramming/History.pdf

Overmars, M. (2012). *A brief history of computer games*. Retrieved September 9, 2015, from http://www.cs.uu.nl/docs/vakken/b2go/literature/history_of_games.pdf

Paavilainen, J. (2004). *Mobile games: Creating business with N-GAGE Nokia*. New Riders.

Prensky, M. (2001). *Fun, play and games: What makes games engaging. From digital game-based learning*. Retrieved September 9, 2015, from http://www.marcprensky.com/writing/Prensky%20-%20Digital%20Game-Based%20Learning-Ch5.pdf

Ritchie, D. (2007). *Yes, a video game contributed to Unix Development*. Retrieved September 9, 2015, from http://sites.fas.harvard.edu/~lib215/reference/history/spacetravel.html

Rouse, R. (2005). *Game Design Theory & Practice* (2nd ed.). Wordware Publishing.

Sherry, J. L., & Lucas, K. (2001). *Video game uses and gratifications as predictors of use and game preference*. Retrieved September 22, 2015, from http://icagames.comm.msu.edu/vgu%26g.pdf

Tuğtekin, U., & Kaleci, D. (2011). 3D modelleme tekniği kullanılarak bilgisayar oyunu tasarımı. *Akademik Bilişim 2011 Konferansı Bildirileri, İnönü Üniversitesi, Malatya*. Retrieved November 16, 2015, from http://ab.org.tr/ab11/kitap/tugtekin_kaleci_AB11.pdf

Wikipedia. (2015). *History of video games*. Retrieved September 29, 2015, from https://en.wikipedia.org/wiki/History_of_video_games

KEY TERMS AND DEFINITIONS

Designing Steps of Video Games: These are the steps required during the design of video games.

Game Industry: It is a economical sector about video game development, marketing and sales.

Historical Process: The change process of video games from their emergence to now.

Video Game Design: It is about the stages of a video game before it was sold.

Video Game Preferences: Is is a preferance of video game players according to the types and characteristics of a video game.

Video Game Types: It is a method of classifying video games in different categories.

Video game: It is a type of game played using a visual user interface (television, monitor, etc.) with a video that sends an image signal, such as a computer or game console.

Compilation of References

Ab Hamid, S. H., & Yu Fung, L. (2007). Learn Programming by Using Mobile Edutainment Game Approach. In *Proceedings of DIGITEL, 1st IEEE International Workshop*. IEEE. doi:10.1109/DIGITEL.2007.31

Abdoli-Sejzi, A. (2015). Augmented reality and virtual learning environment. *Journal of Applied Sciences Research, 11*(8), 1–5.

Abdüsselam, M. S. (2014). Fizik öğretiminde artırılmış gerçeklik ortamlarının kullanımlarına ilişkin öğretmen ve öğrenci görüşleri: 11. Sınıf manyetizma konusu örneği. *Pegem Eğitim ve Öğretim Dergisi, 4*(1), 59–74.

Abdüsselam, M. S., & Karal, H. (2012). Fizik öğretiminde artırılmış gerçeklik ortamlarının öğrenci akademik başarısı üzerine etkisi: 11. Sınıf manyetizma konusu örneği. *Eğitim ve Öğretim Araştırmaları Dergisi, 1*(4), 170–181.

Abt, C. C. (1987). *Serious games*. Lanham, MD: University Press of America.

Açık Bilim. (2012). *Zenginleştirilmiş Gerçeklik Artık Gözümüzde*. Retrieved from http://www.acikbilim.com/2012/06/dosyalar/zenginlestirilmis-gerceklik-artik-gozumuzde.html

Ahn, S., Ko, H., & Yoo, B. (2014). Webizing mobile augmented reality content. *New Review of Hypermedia and Multimedia, 20*(1), 79–100. doi:10.1080/13614568.2013.857727

Aiken, L. H., Clarke, S. P., & Cheung, R. B. (2003). Educational levels of hospital nurses and surgical patient mortality. *Journal of the American Medical Association, 290*(12), 1617–1623. doi:10.1001/jama.290.12.1617 PMID:14506121

Akkoyunlu, B. (2016). *Son Zamanların Çılgınlığı "Pokemon Go" Oyunu Hakkında Neler Biliyoruz?* Retrieved 3 August, 2016 from http://www.bilimiletisimi.com/show.php?id=5103

Akkoyunlu, B. (2002). Educational technology in Turkey: Past, present and future. *Educational Media International, 39*(2), 165–174. doi:10.1080/09523980210155352

Albrecht, U. V., Folta-Schoofs, K., Behrends, M., & von Jan, U. (2013). Effects of mobile augmented reality learning compared to textbook learning on medical students: Randomized controlled pilot study. *Journal of Medical Internet Research, 15*(8), e182. doi:10.2196/jmir.2497 PMID:23963306

Albrecht, U.-V., Jan, U., von, , Kuebler, J., Zoeller, C., Lacher, M., & Muensterer, O. J. (2014). Google Glass for documentation of medical findings: Evaluation in forensic medicine. *Journal of Medical Internet Research, 16*(2), 53. doi:10.2196/jmir.3225 PMID:24521935

Aleven, V. (2013). Help seeking and intelligent tutoring systems: Theoretical perspectives and a step towards theoretical integration. In *International Handbook of Metacognition and Learning Technologies* (pp. 311–335). Springer New York. doi:10.1007/978-1-4419-5546-3_21

Alexander, B. (2004). Going nomadic: Mobile learning in higher education. *EDUCAUSE Review, 39*(5), 29–35.

Ally, M. (2009). Introduction. In Mobile learning: Transforming the delivery of education and training (pp. 1-8). Athabasca: Athabasca University Press.

Ally, M., & Prieto-Blázquez, J. (2014). What is the future of mobile learning in education? Mobile Learning Applications in Higher Education. *Revista de Universidad y Sociedad del Conocimiento*, *11*(1), 142–151. doi:10.7238/rusc.v11i1.2033

Ally, M., Schafer, S., Cheung, B., McGreal, R., & Tin, T. (2007). Use of mobile learning technology to train ESL adults. *Proceedings of the 6th Annual International Conference on Mobile Learning.*

Altmann, T. K., & Brady, D. (2005). PDAs Bring Information Competence to the Point-of-Care. *International Journal of Nursing Education Scholarship*, *2*(1). doi:10.2202/1548-923X.1127 PMID:16646904

Alvear, D., Abreu, O., Cuesta, A., & Alonso, V. (2013). Decision support system for emergency management: Road tunnels. *Tunnelling and Underground Space Technology*, *34*, 13–21. doi:10.1016/j.tust.2012.10.005

American Associate of Colleges of Nursing. (2016). *Alliance for nursing accreditation statement distance education policies*. Retrieved August 8, 2016, from http://www.aacn.nche.edu/publications/position/distance-ed-statement

American Association of Colleges of Nursing. (1998). *A Vision of Baccalaureate and Graduate Nursing Education: The Next Decade. Journal of Professional Nursing, 15(1),*59–65. doi:10.1016/S8755-7223(99)80027-1

Ammari, K. E., & Hammad, A. (2014). Collaborative BIM-based Markerless Mixed Reality Framework for Facilities Maintenance. In *Proceedings of the 2014 International Conference on Computing in Civil and Building Engineering.* ASCE. doi:10.1061/9780784413616.082

Antonioli, M., Blake, C., & Sparks, K. (2014). Augmented reality application in education. *Journal of Technology Studies*, *40*(2), 96–107. doi:10.21061/jots.v40i2.a.4

Apple ITunes. (2016). *AR Liver ViewerApp*. Retrieved from https://itunes.apple.com/us/app/ar-liver-viewer, 2016

Apple ITunes. (2016). *ZooBurswt App*. Retrieved from https://itunes.apple.com/us/app/zooburst

Ardito, C., Buono, P., Costabile, M. F., Lanzilotti, R., & Piccinno, A. (2009). Enabling interactive exploration of cultural heritage: An experience of designing systems for mobile devices. *Knowledge, Technology & Policy*, *22*(1), 79–86. doi:10.1007/s12130-009-9079-7

Arkin, I. E. (2013). *English-Medium Instruction in Higher Education: A Case Study in a Turkish University Context* (Unpublished doctoral dissertation). Eastern Mediterranean University.

Armitage, R. (2002). *To CCTV or not to CCTV? A review of current research into the effectiveness of CCTV systems in reducing crime*. Retrieved from https://epic.org/privacy/surveillance/spotlight/0505/nacro02.pdf

Arslan, R. S., & Saka, C. K. (2010). Teaching English to science students via theme-based model of content-based instruction. *Journal of Turkish Science Education*, *7*(4), 26.

Arth, C., Grasset, R., Gruber, L., & Wagner, D. (2015). *The History of Mobile Augmented Reality*. arXiv preprint arXiv:1505.01319

Arvanitis, T. N., Petrou, A., Knight, J. F., Savas, S., Sotiriou, S., Gargalakos, M., & Gialouri, E. (2007). Human factors and qualitative pedagogical evaluation of a mobile augmented reality system for science education used by learners with physical disabilities. *Personal and Ubiquitous Computing*, *13*(3), 243–250. doi:10.1007/s00779-007-0187-7

Asabere, N. Y. (2012). Towards a Perspective of Information and Communication Technology (ICT) in Education: Migrating From Electronic Learning (E-Learning) to Mobile Learning (M-Learning). *International Journal of Information and Communication Technology Research, 2*(8), 646-649. Retrieved August 2, 2016 http://citeseerx.ist.psu.edu/viewdoc/download?doi=10.1.1.301.3049&rep=rep1&type=pdf

Atakul, B. (2016). *Gelecek Vaat Eden 10 Artırılmış Gerçeklik Gözlüğü.* Retrieved from http://www.teknolo.com/10-artirilmis-gerceklik-gozlugu/

Attewell, J., & Webster, T. (2005). Engaging and supporting mobile learners. In *Proceedings of M-Learning 2004: Mobile learning anytime everywhere (A book of papers from M-LEARN 2004, Italy).* Learning and Skills Development Agency.

augment.com. (2016). *5 reasons to use Augmented Reality in Education.* Retrieved 23 April, 2016 from http://www.augment.com/blog/5-reasons-use-augmented-reality-education/

Augmented Reality for Architecture & Construction. http://www.augment.com/augmented-reality-architecture/

Augmented Reality Türkiye. (2010). Retrieved from http://augmentedrealitytr.blogspot.com.tr/

Azuma, R. T. (1997). *A Survey of Augmented Reality.* Teleoperators and Virtual Environments.

Azuma, R. T. (1997). A survey of augmented reality. *Presence (Cambridge, Mass.), 6*(4), 355–385. doi:10.1162/pres.1997.6.4.355

Azuma, R. T. (2004). Overview of augmented reality.*Proceeding of SIGGRAPH '04.*

Azuma, R., Baillot, Y., Behringer, R., Feiner, S., Julier, S., & MacIntyre, B. (2001). Recent advances in augmented reality. *IEEE Computer Graphics and Applications, 21*(6), 34–47. doi:10.1109/38.963459

Baharti, P. (2014). How to use augmented reality in the classroom. *EdTech Review.* Retrieved from http://edtechreview.in/trends-insights/insights/1210-how-to-use-augmented-reality-in-the-classroom

Bai, Z., Blackwell, F., & Coulouris, G. (2013). Th rough the looking glass: Pretend play for children with autism.*IEEE International Symposium on Mixed and Augmented Reality 2013* (pp. 49–58). Adelaide, Australia: IEEE. doi:10.1109/ISMAR.2013.6671763

Baker, E. D., Hope, L., & Karandjeff, K. (2009). *Contextualized Teaching and Learning Handout.* Retrieved from file:///D:/Downloads/CTL%20Brief.pdf

Bandura, A. (1977). *Social learning theory. Englewood Cliffs*, NJ: Prentice-Hall.

Baş, H. (2015). *Artırılmış Gerçeklik (Augmented reality) ve Türkçe Dersi.* Retrieved from https://www.youtube.com/watch?v=0zBmMwNlCvQ

Bayirtepe, E., & Tuzun, H. (2007). The effects of game-based learning environments on students' achievement and self-efficacy in a computer course. *Hacettepe University Journal of Education, 33*, 41–54.

Beatty, K. (2003). *Teaching and Researching Computer-Assisted Language Learning.* Essex, UK: Pearson Education Limited.

Beauchemin, R. W. (2016). Augmenting education: Using augmented reality technologies. In D. Mentor (Ed.), *Handbook of Research on Mobile Learning in Contemporary Classrooms* (pp. 160–180). Hershey, PA: IGI Global. doi:10.4018/978-1-5225-0251-7.ch008

Belbin, M. (2003). *Management Teams: Why They Succeed or Fail.* Oxford, UK: Butterworth-Heinemann.

Bergmann, J., & Sams, A. (2012). *Flip your classroom: Reach every student in every class every day*. International Society for Technology in Education.

Berque, D. A., & Newman, J. T. (2015). GlassClass: Exploring the Design, Implementation, and Acceptance of Google Glass in the Classroom. In R. Shumaker & S. Lackey (Eds.), Lecture Notes in Computer Science: Vol. 9179. *Virtual, Augmented and Mixed Reality* (pp. 243–250). Cham: Springer International Publishing. doi:10.1007/978-3-319-21067-4_25

Bijker, W. E., & Law, J. (1992). *Shaping technology/building society: Studies in sociotechnical change. Inside technology*. Cambridge, MA: MIT Press.

Bilgiç, H. G., Duman, D., & Seferoğlu, S. S. (2011). Dijital Yerlilerin Özellikleri ve Çevrimiçi Ortamların Tasarlanmasındaki Etkileri. Akademik Bilişim 2011. İnönü Üniversitesi.

Bilici, F. (2012). *Artırılmış Gerçeklik Nedir?* Retrieved from http://artirilmisdunya.blogspot.com.tr/2015/12/artirilmis-gerceklik-nedir.html

Billinghurst, M. (2002). Augmented reality in education. *New Horizons for Learning, 12*.

Billinghurst, M. (2002). Augmented Reality in Education. In *New Horizons for Learning*. Seattle, WA: Technology in Education.

Billinghurst, M., & Dünser, A. (2012). Augmented Reality in the Classroom. *Computer*, *45*(7), 56–63. doi:10.1109/MC.2012.111

Bimber, O., & Raskar, R. (2005). *Spatial augmented reality: Merging real and virtual worlds*. Retrieved 11 July, 2016 from http://web.media.mit.edu/~raskar/SAR.pdf

Blikstein, P., & Krannich, D. (2013). The makers' movement and FabLabs in education. In J. P. Hourcade, N. Sawhney, & E. Reardon (Eds.), *Proceedings of the 12th International Conference on Interaction Design and Children* (p. 613). Academic Press. doi:10.1145/2485760.2485884

Bodine, K., & Gemperle, F. (2003). Effects of functionality on perceived comfort of wearables.*Seventh IEEE International Symposium on Wearable Computers*, 57–60. doi:10.1109/ISWC.2003.1241394

Boeykens, S. (2011). *Using 3D Design software, BIM and game engines for architectural historical reconstruction*. Liege, Belgium: CAAD Futures.

Bokyung, K. (2009). Investigation on the relationships among media characteristics, presence, flow, and learning effects in augmented reality based learning. In P. A. Bruck (Ed.), *Multimedia and E-Content Trends* (pp. 21–37). Wiesbaden: Vieweg Teubner; doi:10.1007/978-3-8348-9313-0_3

Boletsis, C., & McCallum, S. (2013). The table mystery: An augmented reality collaborative game for chemistry education. In M. Ma, M.F. Oliveria, S. Petersen & J.B. Hauge (Eds.), Serious games development and applications (pp. 86-95). Springer Berlin Heidelberg.

Bonk, C. J., Lee, M. M., Reeves, T. C., & Reynolds, T. H. (Eds.). (2015). *MOOCs and open education around the world*. Routledge.

Botden, S. M., & Jakimowicz, J. J. (2009). What is going on in augmented reality simulation in laparoscopic surgery? *Surgical Endoscopy*, *23*(8), 1693–1700. doi:10.1007/s00464-008-0144-1 PMID:18813987

Bottino, R. M., Ferlino, L., Ott, M., & Travella, M. (2007). Developing strategic and reasoning abilities with computer games at primary school level. *Computers & Education*, *49*(4), 1272–1286. doi:10.1016/j.compedu.2006.02.003

Bound, A. C., Haniff, D. J., Baber, C., & Steiner, S. J. (1999). Virtual Reality and Augmented Reality as a Training Tool for Assembly Tasks. *Information Visualization, IEEE*, 32-36. Retrieved June 26, 2016 http://ieeexplore.ieee.org/stamp/stamp.jsp?tp=&arnumber=781532

Bower, M., Howe, C., McCredie, N., Robinson, A., & Grover, D. (2013) Augmented reality in education – cases, places, and potentials. In *Proceedings of the IEEE 63rd Annual Conference International Council for Educational Media* (ICEM). doi:10.1109/CICEM.2013.6820176

Bower, M., Howe, C., McCredie, N., Robinson, A., & Grover, D. (2013). Augmented reality in education-cases, places, and potentials. *Educational Media International*, *51*(1), 1–15. doi:10.1080/09523987.2014.889400

Boz Yüksekdağ, B. (2013). *The attitudes of psychiatry nurses towards the distance nursing education* (Doctoral dissertation). Marmara University, Istanbul, Turkey.

Boz, B. (2008). *Determining education needs in online learning and learners' opinions: A study of medical workers* (Master dissertation). Anadolu University, Eskisehir, Turkey.

Bozdogan, D. (2015). MALL Revisited: Current Trends and Pedagogical Implications. *Procedia: Social and Behavioral Sciences*, *195*, 932–939. doi:10.1016/j.sbspro.2015.06.373

Brame, C. (2013). *Flipping the classroom.* Vanderbilt University Center for Teaching. Retrieved from http://cft.vanderbilt.edu/guides-sub-pages/flipping-the-classroom/

Brandt, H., Nielsen, B. L., Georgsen, M., & Swensen, H. (2015). Augmented Reality for Science Education. Lecture Notes in Computer Science.

Brenner, W., Karagiannis, D., Kolbe, L., Krüger, J., Leifer, L., Lamberti, H.-J., & Zarnekow, R. (2014). User, Use & Utility Research. *Business & Information Systems Engineering*, *6*(1), 55–61. doi:10.1007/s12599-013-0302-4

Bressler, D. (2006). Mobile Phones: A New Way to Engage Teenagers. In J. Trant & D. Bearman (Eds.), *Informal Science Learning. Proceedings of Museums and the Web 2006*. Toronto: Academic Press.

Bressler, D. M., & Bodzin, A. M. (2013). A mixed methods assessment of students flow experiences during a mobile augmented reality science game. *Journal of Computer Assisted Learning*, *29*(6), 505–517. doi:10.1111/jcal.12008

Breuer, J. (2011). *Spielend lernen? Eine Bestandsaufnahme zum (Digital) Game-Based Learning.* Landesanstalt für Medien NRW. Retrieved May 4, 2016, from http://lfmpublikationen.lfm-nrw.de/index.php?view=product_detail&product_id=190

British Council. (2015). *The State of English in Higher Education in Turkey*. British Council Turkey. Retrieved from http://www.britishcouncil.org.tr/sites/default/files/turkiyede_yuksekogretim_kurumlarindaki_ingilizce_egitimi.pdf

Bronack, S. C. (2011). The role of immersive media in online education. *Journal of Continuing Higher Education*, *59*(2), 113–117. doi:10.1080/07377363.2011.583186

Brown, J. S., Collins, A., & Duguid, P. (1989). Situated cognition and the culture of learning. *Educational Researcher*, *18*(1), 32–42. doi:10.3102/0013189X018001032

Brown, R. W. (1995). Autorating: Getting individual marks from team marks and enhancing teamwork. In *Proceedings of Frontiers in Education Conference* (vol. 2, pp. 15-18). Atlanta, GA: IEEE Press. doi:10.1109/FIE.1995.483140

Bruner, G. C. II, & Kumar, A. (2007). Attitude toward Location-based Advertising. *Journal of Interactive Advertising*, *7*(2), 3–15. doi:10.1080/15252019.2007.10722127

Bruner, J. S. (1966). *Toward a theory of instruction.* Cambridge, MA: Belknap.

Bucki, J. (2016). *What is mobile computing?* Retrieved 14 May, 2016 from http://operationstech.about.com/od/glossary/g/Definition-Of-Mobile-Computing.htm

Bujak, K. R., Radu, I., Catrambone, R., MacIntyre, B., Zheng, R., & Golubski, G. (2013). A psychological perspective on augmented reality in the mathematics classroom. *Computers & Education*, *68*, 536–544. doi:10.1016/j.compedu.2013.02.017

Burdea, G., & Coiffet, P. (2003). *Virtual reality technology*. Wiley. Retrieved April 16, 2016, from https://books.google.ca/books?id=0xWgPZbcz4AC

Cai, H. (2013). *Using augmented reality as motivators for youth environmental education: An American Harts's tongue fern conservation project* (Unpublished Master Thesis). State University of New York College of Environmental Science and Forestry.

Cai, S., Chiang, F. K., & Wang, X. (2013). Using the augmented reality 3D technique for a convex imaging experiment in a physics course. *International Journal of Engineering Education*, *29*(4), 856–865.

Cai, S., Wang, X., & Chiang, F. K. (2014). A case study of augmented reality simulation system application in a chemistry course. *Computers in Human Behavior*, *37*, 31–40. doi:10.1016/j.chb.2014.04.018

Çakal, M.A. & Eymirli, E.B. (2012). *Artırılmış Gerçeklik Teknolojisi*. Kuzey Anadolu Kalkınma Ajansı, TRA1.

Callister, L. C. (2012). Issues in global nursing education. *American Journal of Maternal Child Nursing*, *37*(6), 403. Retrieved June 16, 2016, from http://journals.lww.com/mcnjournal/Citation/2012/11000/Issues_in_Global_NursingEducation.13.aspx.

Cameron, C. (2010). *How augmented reality helps doctors save lives – read write*. Retrieved August 9, 2016, from http://readwrite.com/2010/06/02/how_augmented_reality_helps_doctors_save_lives/

Carlson, K. J., & Gagnon, D. J. (2016). Augmented Reality Integrated Simulation Education in Health Care. *Clinical Simulation in Nursing*, *12*(4), 123–127. doi:10.1016/j.ecns.2015.12.005

Carmigniani, J., Furht, B., Anisetti, M., Ceravolo, P., Damiani, E., & Ivkovic, M. (2011). Augmented reality technologies, systems and applications. *Multimedia Tools and Applications*, *51*(1), 341–377. doi:10.1007/s11042-010-0660-6

Carpenter, G. A. (1989). Neural network models for pattern recognition and associative memory. *Neural Networks*, *2*(4), 243–257. doi:10.1016/0893-6080(89)90035-X

Carpio Cañada, J., Mateo Sanguino, T. J., Merelo Guervós, J. J., & Rivas Santos, V. M. (2015). Open classroom: Enhancing student achievement on artificial intelligence through an international online competition. *Journal of Computer Assisted Learning*, *31*(1), 14–31. doi:10.1111/jcal.12075

Cascales, A., Laguna, I., Pérez-López, D., Perona, P., & Contero, M. (2013). An experience on natural sciences augmented reality contents for preschoolers. In R. Shumaker (Ed.), Virtual, Augmented and Mixed Reality. Systems and Applications (LNCS Vol. 8022, pp. 103-112). Berlin: Springer-Verlag. doi:10.1007/978-3-642-39420-1_12

Castillo, R. I. B., Sánchez, V. G. C., & Villegas, O. O. V. (2015). A pilot study on the use of mobile augmented reality for ınteractive experimentation in quadratic equations. *Mathematical Problems in Engineering*, *2015*, 1–13. doi:10.1155/2015/946034

Catanzariti, P. (2016, March 7). *What the meta 2 means for augmented reality developers*. Academic Press.

Catchoom. (2016). *Augmented Reality SDK*. Retrieved May 27, 2016 from http://catchoom.com/product/craftar/augmented-reality-sdk/

Çetinkaya, H. H., & Akçay, M. (2013). *Eğitim ortamlarında arttırılmış gerçeklik uygulamaları*. Retrieved from http://ab.org.tr/ab13/bildiri/114.pdf

Chang, G., Morreale, P., & Medicherla, P. (2010). Applications of augmented reality systems in education. In D. Gibson & B. Dodge (Eds.), *Proceedings of Society for Information Technology & Teacher Education International Conference 2010*, (pp. 1380- 1385). Chesapeake, VA: AACE.

Chang, G., Morreale, P., & Medicherla, P. (2010). Applications of augmented reality systems in education. In *Society for Information Technology & Teacher Education International Conference 2010* (pp. 1380-1385). Chesapeake, VA: AACE.

Chang, G., Morreale, P., & Medicherla, P. (2010). Applications of augmented reality systems in education. *Proceedings of Society for Information Technology & Teacher Education International Conference 2010*, 1380-1385.

Chang, G., Morreale, P., & Medicherla, P. (2010). Applications of augmented reality systems in education. *Proceedings of Society for Information Technology & Teacher Education International Conference.*

Chang, H.-Y., Wu, H.-K., & Hsu, Y.-S. (2013). Integrating a mobile augmented reality activity to contextualize student learning of a socioscientific issue. *British Journal of Educational Technology*, *44*(3), E95–E99. doi:10.1111/j.1467-8535.2012.01379.x

Chang, K. E., Chang, C. T., Hou, H. T., Sung, Y. T., Chao, H. L., & Lee, C. M. (2014). Development and behavioral pattern analysis of a mobile guide system with augmented reality for painting appreciation instruction in an art museum. *Computers & Education*, *71*, 185–197. doi:10.1016/j.compedu.2013.09.022

Chang, V., & Guetl, C. (2007). E-learning ecosystems (ELES). In *Proceedings of the Inaugural Conference on Digital Ecosystems and Technologies* (pp. 420- 425). IEEE. doi:10.1109/DEST.2007.372010

Chang, Y. L., Hou, H. T., Pan, C. Y., Sung, Y. T., & Chang, K. E. (2015). Apply an augmented reality in a mobile guidance to increase sense of place for heritage places. *Journal of Educational Technology & Society*, *18*(2), 166–178.

Charland, A., & Leroux, B. (2011). Mobile application development: Web vs. native. *Communications of the ACM*, *54*(5), 49–53. doi:10.1145/1941487.1941504

Chee, Y. S. (2016). *Games-to-teach or games-to-learn: Unlocking the power of digital game-based learning through performance*. Dordrecht: Springer. doi:10.1007/978-981-287-518-1

Chen, C. M., & Chung, C. J. (2007). Personalized Mobile English Vocabulary Learning System Based on Item Response Theory and Learning Memory Cycle. *Computers & Education*, *51*(2), 624–645. doi:10.1016/j.compedu.2007.06.011

Chen, C. M., & Tsai, Y. N. (2012). Interactive augmented reality system for enhancing library instruction in elementary schools. *Computers & Education*, *59*(2), 638–652. doi:10.1016/j.compedu.2012.03.001

Chen, C. P., & Wang, C. H. (2015). Employing augmented-reality-embedded instruction to disperse the imparities of individual differences in earth science learning. *Journal of Science Education and Technology*, *24*(6), 835–847. doi:10.1007/s10956-015-9567-3

Chen, P., Liu, X., Cheng, W., & Huang, R. (2017). A review of using Augmented Reality in Education from 2011 to 2016. In *Innovations in Smart Learning*. Singapore: Springer. doi:10.1007/978-981-10-2419-1_2

Chen, Y., Wang, Y., & Kinshuk Chen, N. S. (2014). Is FLIP enough? Or should we use the FLIPPED model instead? *Computers & Education*, *79*, 16–27. doi:10.1016/j.compedu.2014.07.004

Cheon, S. H., & Reeve, J. (2015). A Classroom-Based Intervention to Help Teachers Decrease Students Amotivation. *Contemporary Educational Psychology*, *40*, 99–111. doi:10.1016/j.cedpsych.2014.06.004

Chiang, T. C., Yang, S. H., & Hwang, G. (2014). An augmented reality-based mobile learning system to improve students' learning achievements and motivations in natural science inquiry activities. *Journal of Educational Technology & Society*, *17*(4), 352–365.

Chiang, T. H., Yang, S. J., & Hwang, G. J. (2014). Students online interactive patterns in augmented reality-based inquiry activities. *Computers & Education*, *78*, 97–108. doi:10.1016/j.compedu.2014.05.006

Chinnery, G. (2006). Emerging Technologies - Going to the MALL: Mobile Assisted Language Learning. *Language Learning & Technology*, *10*(1), 9–16.

Chitturi, R., Raghunathan, R., & Mahajan, V. (2008). Delight by Design: The Role of Hedonic Versus Utilitarian Benefits. *Journal of Marketing*, *72*(3), 48–63. doi:10.1509/jmkg.72.3.48

Clark, T. (2014). *The Components of a Digital Age Learning Ecosystem*. Retrieved October 29, 2016, from https://byotnetwork.com/2014/07/06/the-components-of-a-digital-age-learning-ecosystem/

Clarkson, J. (2014, February). *Exploring situated learning in augmented reality apps for adults with autism spectrum disorder.* Paper presented at Graduate Student Research Conference, Oshawa, ON.

Clay, C. A. (2011). Exploring the use of mobile technologies for the acquisition of clinical skills. *Nurse Education Today*, *31*(6), 582–586. doi:10.1016/j.nedt.2010.10.011 PMID:21112132

Cochocki, A., & Unbehauen, R. (1993). *Neural networks for optimization and signal processing*. John Wiley & Sons, Inc.

Cole, M. (2010). Whats culture got to do with it? Educational research as a necessarily interdisciplinary enterprise. *Educational Researcher*, *39*(6), 461–470. doi:10.3102/0013189X10380247

Colpaert, J. (2004). From courseware to coursewear? *Computer Assisted Language Learning*, *17*(3-4), 261–266. doi:10.1080/0958822042000319575

Craig, A. B. (2013). *Understanding augmented reality: Concepts and applications*. Amsterdam: Morgan Kaufmann.

Crompton, H. (2013). Mobile learning: New approach, new theory. In Z. L. Berge & L. Y. Muilenburg (Eds.), *Handbook of Mobile Learning* (pp. 47–57). Florence, KY: Routledge.

Crompton, H. (2015). Using context-aware ubiquitous learning to support students learning about angle. *Journal of Interactive Media in Education*, *1*(13), 1–11. doi:10.5334/jime.aq

Crompton, H., Grant, M. R., & Shraim, K. Y. H. (in press). Technologies to Enhance and Extend Children's Understanding of Geometry: A Configurative Thematic Synthesis of the Literature. *Journal of Educational Technology & Society.*

Cruz-Neira, C., Sandin, D. J., & DeFanti, T. A. (1993). Surround-screen projection-based virtual reality: the design and implementation of the CAVE. In M. C. Whitton (Ed.), *SIGGRAPH '93 Proceedings of the 20th annual conference on Computer graphics and interactive techniques* (pp. 135–142). New York: ACM. doi:10.1145/166117.166134

Çubukcu, Z. (2012). Teachers' evaluation of student-centered learning environments. *Education*, *133*(1), 49–66.

Cuéllar, M. P., & Pegalajar, M. C. (2014). Design and implementation of intelligent systems with LEGO Mindstorms for undergraduate computer engineers. *Computer Applications in Engineering Education*, *22*(1), 153–166. doi:10.1002/cae.20541

Dabbagh, N., Benson, A. D., Denham, A., Joseph, R., Al-Freih, M., Zgheib, G., & Guo, Z. et al. (2016). *Learning Technologies and Globalization: Pedagogical Frameworks and Applications*. Springer International Publishing. doi:10.1007/978-3-319-22963-8

Dalgarno, B., & Lee, M. J. (2010). What are the learning affordances of 3-D virtual environments? *British Journal of Educational Technology*, *41*(1), 10–32. doi:10.1111/j.1467-8535.2009.01038.x

Damasio, A. R. (2010). *L'Autre Moi-même – Les Nouvelles Cartes du Cerveau, de la Conscience et des Emotions*. Paris, F: Odile Jacob.

Daniel, J. (2012). Making sense of MOOCs: Musings in a maze of myth, paradox and possibility. *Journal of Interactive Media in Education*, *2012*(3), 18. doi:10.5334/2012-18

Davenport, C. (2004). *Analysis of PDAs in Nursing: Benefits and barriers*. Retrieved from http://www.pdacortex.com/Analysis_PDAs_Nursing.htm

Davis, F. (1992). Fashion, culture, and identity. Chicago: University of Chicago Press.

Davis, F. D. (1989). Perceived Usefulness, Perceived Ease of Use, and User Acceptance of Information Technology. *Management Information Systems Quarterly*, *13*(3), 319–340. doi:10.2307/249008

De Bono, E. (1995). *Six thinking hats*. Boston: Little Brown and Company.

de Sa, M., & Carriço, L. (2010). Designing and evaluating mobile interaction: Challenges and trends. *Foundations and Trends in Human-Computer Interaction*, *4*(3), 175–243. doi:10.1561/1100000025

Dearnley, C., Haigh, J., & Fairhall, J. (2008). Using mobile technologies for assessment and learning in practice settings: A case study. *Nurse Education in Practice*, *8*(3), 197–204. doi:10.1016/j.nepr.2007.07.003 PMID:17765015

Dede, C. (Speaker). (2008). *Immersive interfaces for learning: Opportunities and perils* [motion picture]. Available from The President and Fellows of Harvard College: http://cyber.law.harvard.edu/interactive/events/luncheon/2008/12/dede

Delello, J. A. (2014). Insights from pre-service teachers using science-based augmented reality. *Journal of Computers in Education*, *1*(4), 295–311. doi:10.1007/s40692-014-0021-y

Demirel, O., Seferoglu, S., & Yagci, E. (2003). Developing Instruction Technologies and Material. Öğretim Teknolojileri ve Materyal Geliştirme. Ankara: PegemA. (in Turkish)

Demski, J. (2013). Jumping off the page. *T.H.E. Journal*, *40*(4), 19–22.

Depari, A., de Dominicis, C. M., Flammini, A., Sisinni, E., Fasanotti, L., & Gritti, P. (2015). Using smartglasses for utility-meter reading.*2015 IEEE Sensors Applications Symposium (SAS)*, 1–6. doi:10.1109/SAS.2015.7133649

Deterding, S., Dixon, D., Khaled, R., & Nacke, L. (2011). From game design elements to gamefulness. In A. Lungmayr (Ed.), *Proceedings of the 15th International Academic MindTrek Conference Envisioning Future Media Environments* (pp. 9–15). New York, NY: ACM. doi:10.1145/2181037.2181040

Dewey, J. (1966). Democracy and Education: An Introduction to the Philosophy of Education. New York: Free Press.

Di Serio, Á., Ibáñez, M. B., & Kloos, C. D. (2013). Impact of an augmented reality system on students motivation for a visual art course. *Computers & Education*, *68*, 586–596. doi:10.1016/j.compedu.2012.03.002

Diaz, C., Hincapié, M., & Moreno, G. (2015). How the Type of Content in Educative Augmented Reality Application Affects the Learning Experience. *Procedia Computer Science*, *75*, 205–212.

Dickens, M. R. (2016). *Exploration of the use of mobile technology in an ADN program* (Doctoral dissertation). Available from ProQuest Dissertations and Theses database. (UMI No. 10017631)

Doane, D. (2001). *Taking flight: The rapid growth of ethical consumerism*. London: New Economics Foundation.

Donald, M. (2001). *A mind so rare: The evolution of human consciousness*. New York: Norton.

Dörnyei, Z., MacIntyre, P., & Henry, A. (Eds.). (2015). Motivational dynamics in language learning. Bristol: Multilingual Matters.

Doswell, J. T., Blake, M. B., & Butcher-Green, J. (2006). Mobile augmented reality system architecture for ubiquitous e-learning. In *2006 Fourth IEEE International Workshop on Wireless, Mobile and Ubiquitous Technology in Education (WMTE'06)* (pp. 121-123). IEEE. doi:10.1109/WMTE.2006.261358

Doswell, J. T. (2006). Context-aware mobile augmented reality architecture for lifelong learning. In *Sixth IEEE International Conference on Advanced Learning Technologies (ICALT'06)* (pp. 372-374). IEEE. doi:10.1109/ICALT.2006.1652448

Ducut, E. & Fontelo, P. (2008). Mobile Devices in Health Education: Current Use and Practice. *Journal of Computing in Higher Education, 20*(2), 59–68. DOI: 008-9003-210.1007/s12528

Due, B. L. (2015). The social construction of a Glasshole: Google Glass and multiactivity in social interaction. *PsychNology, 13*(2-3), 149–178.

Dunleavy, M., & Dede, C. (2014). Augmented reality teaching and learning. In J. Spector, M. Merrill, J. Elen, & M. Bishop (Eds.), *The handbook of research for educational communications and technology* (pp. 735–745). New York: Springer. doi:10.1007/978-1-4614-3185-5_59

Dunleavy, M., Dede, C., & Mitchell, R. (2009). Affordances and limitations of immersive participatory augmented reality simulations for teaching and learning. *Journal of Science Education and Technology, 18*(1), 7–22. doi:10.1007/s10956-008-9119-1

Dünser, A., & Hornecker, E. (2007). An observational study of children interacting with an augmented story book. In *Proceedings of 2nd International Conference of E-Learning and Games (Edutainment 2007)*. CUHK. doi:10.1007/978-3-540-73011-8_31

Duval, R. (2006). A Cognitive Analysis of Problems of Comprehension in a Learning of Mathematics. *Educational Studies in Mathematics, 61*(1), 103–131. doi:10.1007/s10649-006-0400-z

East, M. L., & Havard, B. C. (2015). Mental Health Mobile Apps: From Infusion to Diffusion in the Mental Health Social System. *JMIR Mental Health, 2*(1), e10. doi:10.2196/mental.3954 PMID:26543907

Ebner, M., Schön, M., Taraghi, B., & Steyrer, M. (2014). Teachers little helper: Multi-math-coach. *IADIS International Journal on WWW/Internet, 11*(3), 1-12.

Ebner, M. (2015). *Mobile Learning and Mathematics. Foundations, Design, and Case Studies*. New York: Routledge.

Ebner, M., & Holzinger, A. (2007). Successful implementation of user-centered game based learning in higher education: An example from civil engineering. *Computers & Education, 49*(3), 873–890. doi:10.1016/j.compedu.2005.11.026

Edshelf. (2016). *AR Liver Viewer*. Retrieved August 2016, from edshelf: https://edshelf.com/tool/ar-liver-viewer/

Egitim Teknoloji. (2016). *Anatomy 4D App*. Retrieved from http://egitimteknoloji.net/

Ehrlich, E. M., Eisenach, J. A., & Leighton, W. A. (2010). The Impact of Regulation on Innovation and Choice in Wireless Communications. *Review of Network Economics, 9*(1). doi:10.2202/1446-9022.1194

Eisenkraft, A. (2003). Expanding the 5E model. *Science Teacher (Normal, Ill.), 70*(6), 56–59.

El Sayed, N. A. M., Zayed, H. H., & Sharawy, M. I. (2011). ARSC: Augmented reality student card. An augmented reality solution for the education field. *Computers & Education, 56*(4), 1045–1061. doi:10.1016/j.compedu.2010.10.019

El-Hussein, M. O. M., & Cronje, J. C. (2010). Defining Mobile Learning in the Higher Education Landscape. *Journal of Educational Technology & Society*, *13*(3), 12–21.

Elias, T. (2011). Universal Instructional Design Principles for Mobile Learning. *International Review of Research in Open and Distance Learning*, *12*(2), 143–156. doi:10.19173/irrodl.v12i2.965

Ellis, R. (2003). *Task-Based Language Learning and Teaching*. Oxford.

Embi, P. J. (2001). Information at hand: Using handheld computers in medicine. *Cleveland Clinical Journal of Medicine, 68*(10), 840–842. Retrieved August 10, 2016, from https://www.researchgate.net/profile/Peter_Embi/publication/11752751_Information_at_hand_Using_handheld_computers_in_medicine/links/0deec528f91773f605000000.pdf

Engelke, T., Becker, M., Wuest, H., Keil, J., & Kuijper, A. (2013). MobileAR browser–a generic architecture for rapid AR-multi-level development. *Expert Systems with Applications*, *40*(7), 2704–2714. doi:10.1016/j.eswa.2012.11.003

Entertainment Software Association. (2005). *Essential Facts about the Computer and Video Game Industry*. Retrieved May 4, 2016, from http://www.tntg.org/documents/gamefacts.pdf

Entertainment Software Association. (2015). *Essential Facts about the Computer and Video Game Industry.* Retrieved May 4, 2016, from http://www.theesa.com/about-esa/esa-annual-report/

Enyedy, N., Danish, J. A., Delacruz, G., & Kumar, M. (2012). Learning physics through play in an augmented reality environment. *International Journal of Computer-Supported Collaborative Learning*, *7*(3), 347–378. doi:10.1007/s11412-012-9150-3

Erbaş, Ç., & Demirer, V. (2014). Eğitimde artırılmış gerçeklik uygulamaları: Google Glass örneği. *Journal of Instructional Technologies & Teacher Education*, *3*(2), 8–16.

ESA. (2015). *Essential facts about the computer and video game ındustry*. Retrieved November 16, 2015, from http://www.theesa.com/wp-content/uploads/2015/04/ESA-Essential-Facts-2015.pdf

Estapa, A., & Nadolny, L. (2015). The effect of an augmented reality enhanced mathematic lesson on student achievement and motivation. *Journal of STEM Education: Innovation & Research*, *16*(3), 40–48.

Estapa, A., & Nadolny, L. (2015). The Effect of an Augmented Reality Enhanced Mathematics. Lesson on Student Achievement and Motivation. *Journal of STEM Education*, *16*(3), 40–47.

Eursch, A. (2007). *Increased safety for manual tasks in the field of nuclear science using the technology of augmented reality.* Paper presented at 2007 IEEE Nuclear Science Symposium Conference. doi:10.1109/NSSMIC.2007.4436557

Evans, C. B., & Mixon, D. K. (2015). Students Knowledge of Post-op Pain Management after Participation in Simulation. *Pain Management Nursing*, *16*(6), 930–937. doi:10.1016/j.pmn.2015.07.006 PMID:26697818

Falconi, R. F. (2010). *Usability and game design: Improving the MITAR Game Editor* (Master of Eng. Thesis). Massachusetts Institute of Technology, Cambridge, MA.

Feiner, S., Höllerer, T., Gagas, E., Hallaway, D., Terauchi, T., Güven, S., & MacIntyre, B. (2016). *MARS - Mobile Augmented Reality Systems.* Columbia University. Retrieved from http://graphics.cs.columbia.edu/projects/mars

Ficheman, I. K., & de Deus Lopes, R. (2008). Digital learning ecosystems: authoring, collaboration, immersion and mobility. In *Proceedings of the 7th International Conference on Interaction Design and Children* (pp. 9-12). ACM. doi:10.1145/1463689.1463705

Figure, X. (2016). *Pokémon Go*. Retrieved from http://figurex.net/

Finn, B. (2015). Measuring motivation in low-stakes assessment. *ETS Research Report Series*, *2*(2), 1–17. doi:10.1002/ets2.12067

Fishbein, M., & Ajzen, I. (1975). *Belief, attitude, intention, and behavior: An introduction to theory and research. Addison-Wesley series in social psychology*. Reading, MA: Addison-Wesley Pub. Co.

FitzGerald, E., Adams, A., Ferguson, R., Gaved, M., Mor, Y., & Thomas, R. (2012). Augmented Reality and Mobile Learning: The State of the Art. In *Proceedings of the 11th International Conference on Mobile and Contextual Learning* (pp. 62-69). Retrieved October 29, 2016, from http://ceur-ws.org/Vol-955/

FitzGerald, E., Ferguson, R., Adams, A., Gaved, M., Mor, Y., & Thomas, R. (2013). Augmented reality and mobile learning: The state of the art. *International Journal of Mobile and Blended Learning*, *5*(4), 43–58. doi:10.4018/ijmbl.2013100103

Fjeld, M., & Voegtli, B. M. (2002). *Augmented chemistry: An interactive educational workbench.* Paper presented at International Symposium on Mixed and Augmented Reality, 2002. ISMAR 2002, Darmstadt, Germany. doi:10.1109/ISMAR.2002.1115100

Flasiński, M. (2016). *Introduction to artificial intelligence*. Springer. doi:10.1007/978-3-319-40022-8

Fleck, S., & Simon, G. (2013). *An augmented reality environment for astronomy learning in elementary grades: An exploratory study.* Paper presented at 25ème conférence francophone sur l'Interaction Homme-Machine, IHM'13, Bordeaux, France. doi:10.1145/2534903.2534907

Fleck, S., Hachet, M., & Bastien, J. M. (2015). *Marker-based augmented reality: Instructional-design to improve children interactions with astronomical concepts.* Paper presented at 14th International Conference on Interaction Design and Children, New York, NY. doi:10.1145/2771839.2771842

Freestone, O. M., & MacGoldrick, P. J. (2008). Motivations of the ethical consumer. *Journal of business ethics. JOBE*, *79*(4), 445–467.

Freina, L., & Ott, M. (2015). *A literature review on immersive virtual reality in education: State of the art and perspectives.* Paper presented at the conference of eLearning and Software for Education (eLSE), Bucharest, Romania.

Freitas, R., & Campos, P. (2008). SMART: a System of augmented reality for teaching 2nd grade students.*Proceedings of the 22nd British Computer Society Conference on Human-Computer Interaction (HCI 2008)* (pp. 27-30). Liverpool, UK: John Moores University.

Frenk, J., Chen, L., Bhutta, Z. A., Cohen, J., Crisp, N., Evans, T., & Zurayk, H. et al. (2010). Health professionals for a new: Transforming education to strengthen health systems in an interdependent world. *Lancet*, *376*(9756), 1923–1958. doi:10.1016/S0140-6736(10)61854-5 PMID:21112623

Friberg, M. (2015, September 14). *Augmented reality in the enterprise.* Retrieved from http://blog.metavision.com/augmented-reality-in-the-enterprise/

Friedman, T. (1995). *Making sense of software: Computer games and interactive textuality*. CyberSociety. *Sage (Atlanta, Ga.)*, 73–89.

Furner, J., & Gonzalez-DeHass, A. (2011). How do Students' Mastery and Performance Goals Relate to Math Anxiety? *Eurasia Journal of Mathematics, Science & Technology Education, 7*(4), 227-242. Retrieved from http://www.ejmste.com/

Fyfe, C. (2005). Artificial neural networks. In *Do Smart Adaptive Systems Exist?* (pp. 57–79). Springer Berlin Heidelberg. doi:10.1007/3-540-32374-0_4

Gaebel, M. (2014). *MOOCs: Massive open online courses*. EUA.

Gardner, M., & Elliott, J. (2014). The immersive education laboratory: Understanding affordances, structuring experiences, and creating constructivist, collaborative processes, in mixed-reality smart environments. *EAI Endorsed Transactions on Future Intelligent Educational Environments, 14*(1).

Gardner, H. (2006). *Five minds for the future*. Boston: Harvard Business School Press.

Garrett, B. M., & Jackson, C. (2006). A mobile clinical e-portfolio for nursing and medical students, using wireless personal digital assistants (PDAs). *Nurse Education Today, 26*(8), 647–654. doi:10.1016/j.nedt.2006.07.020 PMID:17011674

Garrett, B. M., Jackson, C., & Wilson, B. (2015). Augmented reality m-learning to enhance nursing skills acquisition in the clinical skills laboratory. *Interactive Technology and Smart Education, 12*(4), 298–314. doi:10.1108/ITSE-05-2015-0013

Gartner Press Release. (2016). *Worldwide Device Shipments to Grow 19 Percent in 2016, While End-User Spending to Decline for the First Time*. Retrieved 30 May, 2016, from http://www.gartner.com/newsroom/id/3187134

Gaudiosi, J. (2016). *7 Ways AR and VR Will Change Tech in 2016*. Retrieved December 18, 2016, from http://fortune.com/2016/01/04/augmented-reality-vr-change-tech/

Gee, J. P. (2005). What would a state of the art instructional video game look like? *Innovate: Journal of Online Education, 1*(6), 1.

Geißler, N., Schönwald, L., Strauß, G., Meixensberger, J., Korb, W., & Bartz, D. (2013). Bewertung visueller Signalgestaltung in der Chirurgie mittels Eye-Tracking / Assessment of Visual Signals in Surgery. *i-com, 12*(2).

Gelibolu, F. (2013). Eğitsel dijital oyunlar: Kuram, tasarım ve uygulama. Pegem Akademi.

George, L. E., Davidson, L. J., Serapiglia, C. P., Barla, S., & Thotakura, A. (2010). Technology in Nursing Education: A Study of PDA Use by Students. *Journal of Professional Nursing, 26*(6), 371–376. doi:10.1016/j.profnurs.2010.08.001 PMID:21078507

Georgiev, T., Georgieva, E., & Smrikarov, A. (2004). M-learning-a New Stage of E-Learning. In *International Conference on Computer Systems and Technologies-CompSysTech* (*Vol. 4*, No. 28, pp. 1-4). Academic Press.

Ghisetti, C., & Pontoni, F. (2015). Investigating policy and R&D effects on environmental innovation: A meta-analysis. *Ecological Economics, 118*, 57–66. doi:10.1016/j.ecolecon.2015.07.009

Ghouaiel, N., Cieutat, J. M., & Jessel, J. P. (2014). *Adaptive augmented reality: Plasticity of augmentations*. Paper presented at Virtual Reality International Conference, Laval, France.

Gibson, J. J. (1979). *The ecological approach to visual perception*. Boston: Houghton Mifflin.

Ginsberg, M. (2012). *Essentials of artificial intelligence*. Newnes.

Google Developers. (2016). *Cardboard SDK for Unit*. Retrieved December 18, 2016, from https://developers.google.com/cardboard/unity

Greeno, J. G. (1994). Gibsons affordances. *Psychological Review, 101*(2), 336–342. doi:10.1037/0033-295X.101.2.336 PMID:8022965

Grimus, M., & Ebner, M. (2014). Learning with Mobile Devices Perceptions of Students and Teachers at Lower Secondary Schools in Austria. In *Proceedings of World Conference on Educational Multimedia, Hypermedia and Telecommunications 2014* (pp. 1600-1609). Chesapeake, VA: AACE.

Griziotti, G. (2011). *Murdoch and Berlusconi: the Fall of Two Media Empires and the Network Multitudes*. Retrieved April 13 2005, from http://www.opendemocracy.net/giorgio-griziotti/murdoch-and-berlusconi-fall-of-two-media-empires-and-network-multitudes

Gu, J., Li, N., & Duh, H. B.-L. (2011). A Remote Mobile Collaborative AR System for Learning in Physics. *IEEE Virtual Reality Conference 2011,* 257-258. Retrieved April 25, 2016, from http://ieeexplore.ieee.org/stamp/stamp.jsp?tp=&arnumber=5759496

Guazzaroni, G. (2013). Emotional mapping of the archaeologist game. Advanced Human-Computer Interaction: Computers in Human Behavior, 29(2), 335-344. doi:10.1016/j.chb.2012.06.008

Guazzaroni, G. (2015). Realtà aumentata un'opportunità di apprendimento. In L. Salvucci (Ed.), Strumenti per la Didattica della Matematica. Ricerche, esperienze, buone pratiche (pp. 145-152). Milano, Italy: Franco Angeli.

Guazzaroni, G., & Compagno, M. (2013). AR moulded-objects performing Giuseppe Verdi's 200th birthday. *Archeomatica*, *4*, 38–41.

Guazzaroni, G., & Leo, T. (2011). Emotional mapping of a place of interest using mobile devices for learning. In I. Arnedillo Sánchez, & P. Isaías (Eds.), *Proceedings of IADIS International Conference on Mobile Learning* (pp. 277-281). IADIS.

Gündoğan, M. B. (2014). E" is for "Ecology": An Ecosystem Approach to Sustainable E-Learning. In T. V. Yuzer & G. Eby (Eds.), *Handbook of Research on Emerging Priorities and Trends in Distance Education: Communication, Pedagogy, and Technology* (pp. 103–119). Hershey, PA: Information Science Reference. doi:10.4018/978-1-4666-5162-3.ch008

Gunes, F. (2012). Turkish instruction from tests to activities. (In Turkish) Testlerden etkinliklere Türkçe öğretimi. *Journal of Language and Literature Education*, *1*(1), 31–42.

Guo, S. H. M., Chong, P. P., & Chang, H. K. (2007). Mobile learning in nursing practical training: An applicability analysis. *International Journal of Mobile Learning and Organisation*, *1*(4), 342–354. doi:10.1504/IJMLO.2007.016175

Guru. (2015). *Top 10 Mobile Phones Operating Systems*. Retrieved May 27, 2016, from http://www.shoutmeloud.com/top-mobile-os-overview.html

Gu, Z., Wei, J., & Xu, F. (2015). An Empirical Study on Factors Influencing Consumers Initial Trust in Wearable Commerce. *Journal of Computer Information Systems*, *56*(1), 79–85. doi:10.1080/08874417.2015.11645804

Haag, J. (2013). Using augmented reality for contextual mobile learning. *Learning Solutions Magazine*. Retrieved January 7, 2016, from https://www.learningsolutionsmag.com/articles/1310/using-augmented-reality-for-contextual-mobile-learning

Haag, J. (2013). *Using Augmented Reality for Contextual Mobile Learning*. Retrieved October 29, 2016, from http://www.learningsolutionsmag.com/articles/1310/using-augmented-reality-for-contextual-mobile-learning

Habok, A., & Nagy, J. (2016). In-service teachers perceptions of project-based learning. *SpringerPlus*, *5*(1), 83. doi:10.1186/s40064-016-1725-4 PMID:26844030

Hakkarainen, M., Woodward, C., & Rainio, K. (2010). Mobile Augmented Reality for Building and Construction.*Mobile World Conference Barcelona*, Barcelona, Spain.

Halverson, E. R., & Sheridan, K. (2014). The Maker Movement in Education. *Harvard Educational Review*, *84*(4), 495–504. doi:10.17763/haer.84.4.34j1g68140382063

Hamari, J., Shernoff, D. J., Rowe, E., Coller, B., Asbell-Clarke, J., & Edwards, T. (2016). Challenging games help students learn: An empirical study on engagement, flow and immersion in game-based learning. *Computers in Human Behavior*, *54*, 170–179. doi:10.1016/j.chb.2015.07.045

Hanesova, D. (2015). Development of CLIL teacher education in UMB (Introduction of European project). In Learning Together: To be a better CLIL Teacher. Banskå Bystrica.

Han, J., Jo, M., Hyun, E., & So, H. J. (2015). Examining young childrens perception toward augmented reality-infused dramatic play. *Educational Technology Research and Development*, *63*(3), 455–474. doi:10.1007/s11423-015-9374-9

Hansen, C., Wieferich, J., Ritter, F., Rieder, C., & Peitgen, H. O. (2010). Illustrative visualization of 3D planning models for augmented reality in liver surgery. *International Journal of Computer Assisted Radiology and Surgery*, *5*(2), 133–141. doi:10.1007/s11548-009-0365-3 PMID:20033519

Harley, J. M., Poitras, E. G., Jarrell, A., Duffy, M. C., & Lajoie, S. P. (2016). Comparing virtual and location-based augmented reality mobile learning: Emotions and learning outcomes. *Educational Technology Research and Development*, *64*(3), 359–388. doi:10.1007/s11423-015-9420-7

Harrison, L. T. (2003). *Introduction to 3D Game Engine Design Using DirectX 9 and C*. New York: Apress. doi:10.1007/978-1-4302-0821-1

Hartmann, G., Stead, G., & DeGani, A. (2011). *Cross-platform mobile development*. Cambridge: Mobile Learning Environment.

Hassoun, M. H. (1995). *Fundamentals of Artificial Neural Networks*. Cambridge, MA: MIT Press.

Hassoun, M. H. (1995). *Fundamentals of artificial neural networks*. MIT Press.

Haughey, D. (n.d.) *Delphi Technique: a Step-By-Step Guide*. Retrieved October 29, 2016, from https://www.projectsmart.co.uk/delphi-technique-a-step-by-step-guide.php

Hegarty, M., & Kriz, S. (2008). Effects of knowledge and spatial ability on learning from animation. *Learning with animation: Research implications for design*, 3-29.

Hein, D. W. E., & Rauschnabel, P. A. (2016). Augmented Reality Smart Glasses and Knowledge Management: A Conceptual Framework for Enterprise Social Networks. In A. Roßmann, M. Besch, & G. Stei (Eds.), *Enterprise Social Networks*. Wiesbaden: Springer. doi:10.1007/978-3-658-12652-0_5

Hennig-Thurau, T., Malthouse, E. C., Friege, C., Gensler, S., Lobschat, L., Rangaswamy, A., & Skiera, B. (2010). The Impact of New Media on Customer Relationships. *Journal of Service Research*, *13*(3), 311–330. doi:10.1177/1094670510375460

Henrysson, A. (2007). *Bringing Augmented Reality to Mobile Phones Anders*. Department of Science and Technology Linköpings Universitet.

Hernandez, J. (2013). *LearnAR. Herramienta de aprendizaje con Realidad Aumentada*. Retrieved from http://creaconlaura.blogspot.com.tr/2013_12_08_archive.html

Hernandez, B., Jimenez, J., & José Martin, M. (2010). Business management software in high-tech firms: The case of the IT services sector. *Journal of Business and Industrial Marketing*, *25*(2), 132–146. doi:10.1108/08858621011017750

HITLab. (2016). *ARToolKit*. Retrieved from https://www.hitl.washington.edu/artoolkit/

Hollerer, T. H. (2004). *User interfaces for mobile augmented reality systems* (Doctoral dissertation). Columbia University.

Höllerer, T. H., & Feiner, S. K. (2004). Mobile Augmented Reality. In Telegeoinformatics: Location-Based Computing and Services. Taylor & Francis Books Ltd.

Höllerer, T., Feiner, S., Terauchi, T., Rashid, G., & Hallaway, D. (1999). Exploring MARS: Developing Indoor and Outdoor User Interfaces to a Mobile Augmented Reality System. *Computers & Graphics*, *23*(6), 779–785. doi:10.1016/S0097-8493(99)00103-X

Honeybourne, C., Sutton, S., & Ward, L. (2006). Knowledge in the Palm of your hands: PDAs in the clinical setting. *Health Information and Libraries Journal*, *23*(1), 51–59. doi:10.1111/j.1471-1842.2006.00621.x PMID:16466499

Hosokawa, S. (1984). The walkman effect. *Popular Music*, *4*, 165. doi:10.1017/S0261143000006218

Hsiao, H. S., Chang, C. S., Lin, C. Y., & Wang, Y. Z. (2016). Weather observers: A manipulative augmented reality system for weather simulations at home, in the classroom, and at a museum. *Interactive Learning Environments*, *24*(1), 205–223. doi:10.1080/10494820.2013.834829

Hsiao, K. F., Chen, N. S., & Huang, S. Y. (2012). Learning while exercising for science education in augmented reality among adolescents. *Interactive Learning Environments*, *20*(4), 331–349. doi:10.1080/10494820.2010.486682

Hsiao, K. F., & Rashvand, H. F. (2011). Body Language and Augmented Reality Learning Environment.*International Conference on Multimedia and Ubiquitous Engineering*. doi:10.1109/MUE.2011.51

Huang, Z., Hui, P., Peylo, C., & Chatzopoulos, D. (2013). *Mobile augmented reality survey: A bottom-up approach*. arXiv preprint arXiv:1309.4413

Huang, K., Liu, T., Graf, S., & Lin, Y. (2008). Embedding Mobile Technology to Outdoor Natural Science Learning Based on The 7e Learning Cycle. In J. Luca, & E. Weippl (Eds.), *Proceedings of World Conference on Educational Multimedia, Hypermedia and Telecommunications* (pp. 2082-2086). Chesapeake, VA: Academic Press.

Huang, T. C., Chen, C. C., & Chou, Y. W. (2016). Animating eco-education: To see, feel, and discover in an augmented reality-based experiential learning environment. *Computers & Education*, *96*, 72–82. doi:10.1016/j.compedu.2016.02.008

Huizenga, J., Admiraal, W., Akkerman, S., & Dam, G. T. (2009). Mobile game-based learning in secondary education: Engagement, motivation and learning in a mobile city game. *Journal of Computer Assisted Learning*, *25*(4), 332–344. doi:10.1111/j.1365-2729.2009.00316.x

Hung, P. H., Lin, Y. F., & Hwang, G. J. (2010). Formative assessment design for PDA integrated ecology observation. *Journal of Educational Technology & Society*, *13*(3), 33–42.

Hutchins, E. (2010). Cognitive Ecology. *Topics in Cognitive Science*, *2*(4), 705–715. doi:10.1111/j.1756-8765.2010.01089.x PMID:25164051

Hwang, S. Y., & Kim, M. J. (2006). A comparison of problem-based learning and lecture-based learning in an adult health nursing course. *Nurse Education Today*, *26*(4), 315–321. doi:10.1016/j.nedt.2005.11.002 PMID:16364510

Ibáñez, M. B., Di Serio, Á., Villarán-Molina, D., & Delgado-Kloos, C. (2014). Experimenting with electromagnetism using augmented reality: Impact on flow student experience and educational effectiveness. *Computers & Education*, *71*, 1–13. doi:10.1016/j.compedu.2013.09.004

Ibáñez, M. B., Di-Serio, Á., Villarán-Molina, D., & Delgado-Kloos, C. (2016). Support for Augmented Reality Simulation Systems: The Effects of Scaffolding on Learning Outcomes and Behavior Patterns. *IEEE Transactions on Learning Technologies*, *9*(1), 46–56. doi:10.1109/TLT.2015.2445761

İbili, E. (2013). *Development, implementation and assessment of the effect augmented reality on geometry teaching materials for geometry classes* (Unpublished doctoral dissertation). Gazi University, Ankara, Turkey.

İbili, E., & Şahin, S. (2013). Artırılmış gerçeklik ile interaktif 3d geometri kitabı yazılımın tasarımı ve geliştirilmesi: ARGE3D. *Afyon Kocatepe Üniversitesi Fen ve Mühendislik Bilimleri Dergisi, 13*, 1–8.

İbili, E., & Şahin, S. (2015). Artırılmış gerçeklik ile interaktif 3D geometri kitabı yazılımın tasarımı ve geliştirilmesi: ARGE3D. *AKÜ Fen ve Mühendislik Bilimleri Dergisi, 13*(1), 1–8.

İbili, E., & Şahin, S. (2015). Investigation of the effects on computer attitudes and computer self-efficacy to use of augmented reality in geometry teaching. *Necatibey Faculty of Education Electronic Journal of Science and Mathematics Education, 9*(1), 332–350.

Indraprastha, A. S., & Shinozaki, M. (2008). Constructing virtual urban environment using game technology. In *Proceedings of the Architecturein Computro, Integrating Methods and Techniques* (pp. 538–541). Antwerpen, Belgium: 26th eCAADe.

International Council of Nursing. (2006). *The Global Nursıng Review Initiative. The Global Nursing Shortage: Priority Areas for Intervention.* Retrieved August 30, 2016, from http://www.icn.ch/images/stories/documents/publications/GNRI/The_Global_Nursing_Shortage-Priority_Areas_for_Intervention.pdf

Ionitescu, S., & Radu, C. (2015). Perspectives on implementing interactive elearning tools using augmented reality in education. *Elearning and Software for Education,* (1), 104-111. doi: 10.12753/2066-026X-15-015

Ivanova, M., & Ivanov, G. (2011). Enhancement of learning and teaching in computer graphics through marker augmented reality technology. *International Journal on New Computer Architectures and Their Applications, 1*(1), 176–184.

Jackson, A. (2013). *Mobile learning: The shortest lived technology fad ever?* Retrieved October 29, 2016, from http://www.trainingzone.co.uk/develop/business/mobile-learning-the-shortest-lived-technology-fad-ever

Jacobsen, E., & Dulsrud, A. (2007). Will Consumers Save The World?: The Framing of Political Consumerism. *Journal of Agricultural & Environmental Ethics, 20*(5), 469–482. doi:10.1007/s10806-007-9043-z

Jaffe, A. B., & Palmer, K. (1997). Environmental Regulation and Innovation: A Panel Data Study. *The Review of Economics and Statistics, 79*(4), 610–619. doi:10.1162/003465397557196

Jain, G. P., Gurupur, V. P., Schroeder, J. L., & Faulkenberry, E. D. (2014). Artificial intelligence-based student learning evaluation: A concept map-based approach for analyzing a students understanding of a topic. *IEEE Transactions on Learning Technologies, 7*(3), 267–279. doi:10.1109/TLT.2014.2330297

Jan, U., Noll, C., Behrends, M., & Albrecht, U. (2012). mARble - augmented reality in medical education. *Biomedizinische Technik. Biomedical Engineering, 18*, 67–7. doi:10.1515/bmt-2012-4252

Jax, K. (1998). Holocoen and Ecosystem: On the Origin and Historical Consequences of Two Concepts. *Journal of the History of Biology, 3*(1), 113–142. doi:10.1023/A:1004261607170

Jeon, J., Kim, S., & Lee, S. (2010). Considerations of Generic Framework for AR on the Web. *Proceedings of W3C Workshop: Augmented Reality on the Web.*

Joan, R. (2015). Enhancing education through mobile augmented reality. *Journal of Educational Technology, 11*(4), 8–14.

Johnson, L., Adams Becker, S., Estrada, V., & Freeman, A. (2015). NMC Horizon Report: 2015 Library Edition. Austin, TX: The New Media Consortium. Retrieved from http://www.nmc.org/publication/nmc-horizon-report-2015-library-edition/

Johnson, L., Levine, A., Smith, R., & Stone, S. (2010). The 2010 Horizon Report. *New Media Consortium.* Retrieved from http://www.nmc.org/pdf/2010-Horizon-Report.pdf

Johnson, L., Levine, A., Smith, R., & Stone, S. (2010). *Simple augmented reality. The 2010 Horizon Report* (pp. 21–24). Austin, TX: The New Media Consortium.

Jovanovic, D., & Jovanovic, S. (2015). An adaptive e-learning system for Java programming course, based on Dokeos LE. *Computer Applications in Engineering Education*, *23*(3), 337–343. doi:10.1002/cae.21603

Juan, C., Beatrice, F., & Cano, J. (2008). *An Augmented Reality System for learning the interior of the human body.* Retrieved August 29, 2016, from http://users.dsic.upv.es/~mcarmen/docs/HumanBody.pdf

Juan, M. C. F. D., Alem, L. A. P., & Cano, J. (2011). ARGreenet and Basic-Greenet: Two mobile games for learning how to recycle.*Proceedings of the 19th International Conference on Computer Graphics, Visualization and Computer Vision*, 25–32.

Jugaru, G. (2016). *5 Top Augmented Reality Apps for Education.* Retrieved from http://www.hongkiat.com/blog/augmented-reality-apps-for-education/

Kalantzis, M., & Cope, B. (2005). *Learning by Design, Common Ground.* Melbourne, Australia: Common Grounds.

Kaleci, D., Kıran, H., & Dinçer, S. (2012). Açık kaynak kodlu 3D oyun motorları. *Akademik Bilişim 2012 Konferansı Bildirileri, Uşak Üniversitesi, Uşak.* Retrieved November 16, 2015, from http://ab.org.tr/ab12/bildiri/139.pdf

Kamboj, V., & Gupta, H. (2012). Mobile Operating Systems. *IJEIR*, *1*(2), 169–174.

Kapp, K. M. (2012). *The gamification of learning and instruction: Game-based methods and strategies for training and education. In Pfeiffer essential resources for training and HR professionals.* San Francisco, CA: Pfeiffer.

Kapp, K. M., Blair, L., & Mesch, R. (2014). *The gamification of learning and instruction fieldbook: Ideas into practice.* San Francisco, CA: Wiley.

Kaufmann, H., & Meyer, B. (2008). *Simulating educational physical experiments in augmented reality.* Interactive Media Systems Group, Institute of Software Technology and Interactive Systems, Vienna University of Technology. Retrieved, 10 June, 2016 from http://publik.tuwien.ac.at/files/PubDat_170658.pdf

Kaufmann, H. (2003). *Collaborative augmented reality in education.* Institute of Software Technology and Interactive Systems, Vienna University of Technology.

Kaufmann, H. (2003). *Collaborative augmented reality in education. Institute of Software Technology and Interactive Systems.* Vienna University of Technology.

Kaufmann, H., & Schmalstieg, D. (2003). Mathematics and geometry education with collaborative augmented reality. *Computers & Graphics*, *27*(3), 339–345. doi:10.1016/S0097-8493(03)00028-1

Kaufmann, H., Schmalstieg, D., & Wagner, M. (2000). Construct3D: A virtual reality application for mathematics and geometry education. *Education and Information Technologies*, *5*(4), 263–276. doi:10.1023/A:1012049406877

Kaya, M., & Koçyigit, A. (2014). *Mobil uygulamalarda vekil tabanlı kod taşıma yönteminin farklı seviyelerdeki bulut bilişim altyapılarının kullanılması durumundaki başarılarının karşılaştırılması.* Retrieved 20 June, 2016 from http://ceur-ws.org/Vol-1221/20_Bildiri.pdf

Kearney, M., Schuck, S., Burden, K., & Aubusson, P. (2012). Viewing Mobile Learning From a Pedagogical Perspective. *Research in Learning Technology*, *20*(0). doi:10.3402/rlt.v20i0.14406

Ke, F., & Hsu, Y. C. (2015). Mobile augmented-reality artifact creation as a component of mobile computer-supported collaborative learning. *The Internet and Higher Education*, *26*, 33–41. doi:10.1016/j.iheduc.2015.04.003

Kenny, R. F., Van Neste-Kenny, J. M. C., Burton, P., & Qayyum, A. (2012). Using Self-Efficacy to Assess the Readiness of Nursing Educators and Students for Mobile Learning. *IRRODL, 13*(3), 277-296. Retrieved June 25, 2016, from http://www.irrodl.org/index.php/irrodl/article/view/1221/2261

Kerawalla, L., Luckin, R., Seljeflot, S., & Woolard, A. (2006). Making it real: Exploring the potential of augmented reality for teaching primary school science. *Virtual Reality (Waltham Cross), 10*(3-4), 163–174. doi:10.1007/s10055-006-0036-4

Keshavarz, B., & Hecht, H. (2014). Pleasant music as a countermeasure against visually induced motion sickness. *Applied Ergonomics, 45*(3), 521–527. doi:10.1016/j.apergo.2013.07.009 PMID:23957932

Kidd, S., & Crompton, H. (2016). Augmented learning with augmented reality. In J. Lu, D. Churchill, & B. Fox (Eds.), *Mobile learning design theories and applications* (pp. 97–108). New York: Springer. doi:10.1007/978-981-10-0027-0_6

Kim, K. J., Shin, D.-H., & Park, E. (2015). Can Coolness Predict Technology Adoption?: Effects of Perceived Coolness on User Acceptance of Smartphones with Curved Screens. *Cyberpsychology, Behavior, and Social Networking, 18*(9), 528–533. doi:10.1089/cyber.2014.0675 PMID:26348813

Kincheloe, J. L., Slattery, P., & Steinberg, S. R. (2000). *Contextualizing Teaching Introduction to Education and Educational Foundations.* Longman.

Kiourexidou, M., Natsis, K., Bamidis, P., Antonopoulos, N., Papathanasiou, E., Sgantzos, M., & Veglis, A. (2015). Augmented Reality for the Study of Human Heart Anatomy. *International Journal of Electronics Communication and Computer Engineering, 6*(6), 658-663. Retrieved August 19, 2016, from http://ijecce.com/administrator/components/com_jresearch/files/publications/IJECCE_3629_Final.pdf

Kirkley, S., & Kirkley, J. (2005). Creating next generation blended learning environments using mixed reality, video games and simulations. *TechTrends, 49*(3), 42–53. doi:10.1007/BF02763646

Kirriemuir, J. (2002). Video gaming, education and digital learning technologies. *D-Lib Magazine, 8*(2). doi:10.1045/february2002-kirriemuir

Kiyokawa, K., Billinghurst, M., Hayes, S., & Gupta, A., Sannohe & Y., Kato, H. (2002). Communication Behaviors of Co-Located Users in Collaborative AR Interfaces. In *Proceedings of the IEEE and ACM International Symposium on Mixed and Augmented Reality* (pp. 139-148). Los Alamitos, CA: IEEE Press. doi:10.1109/ISMAR.2002.1115083

Kiziltan, N., & Ersanli, C. Y. (2007). The contributions of theme-based CBI to Turkish young learners' language development in English. *Journal of Language and Linguistic Studies, 3*(1).

Klerkx, J., Verbert, K., & Duval, E. (2014). Enhancing learning with visualization techniques. In J. M. Spector, M. D. Merrill, J. Elen, & M. J. Bishop (Eds.), *Handbook of research on educational communication and technology* (pp. 791–807). New York: Springer. doi:10.1007/978-1-4614-3185-5_64

Klopfer, E., & Sheldon, J. (2010). Augmenting your own reality: Student authoring of science-based augmented reality games. *New Directions for Youth Development, 28*(128), 85–94. doi:10.1002/yd.378 PMID:21240956

Klopfer, E., & Squire, K. (2008). Environmental detectives: The development of an augmented reality platform for environmental simulations. *Educational Technology Research and Development, 56*(2), 203–228. doi:10.1007/s11423-007-9037-6

Klopfer, E., & Yoon, S. (2004). Developing games and simulations for today and tomorrow's tech savvy youth. *TechTrends, 49*(3), 41–49.

Kochan, S. G. (2011). Programming in Objective-C (3rd. ed.). Addison-Wesley Professional.

Koepsell, D. (2010). On Genies and Bottles: Scientists Moral Responsibility and Dangerous Technology R&D. *Science and Engineering Ethics*, *16*(1), 119–133. doi:10.1007/s11948-009-9158-x PMID:19644770

Korkmaz, Ö. (2013). İlk ve orta öğretimde öğretimsel amaçlı teknoloji kullanımı. In K. Çağıltay & Y. Göktaş (Ed.), Içinde Öğretim teknolojilerinin temelleri: Teoriler, araştırmalar, eğilimler (pp. 431–446). Ankara: Pegem Akademi.

Kose, U. (2009). *An Advanced M-Learning System for Teaching Artificial Intelligence Techniques, International Conference of Education, Research and Innovation 2009* (ICERI 2009). Madrid, Spain: Academic Press.

Kose, U. (2015). Augmented Reality Based E-Learning Applications. In M. Khosrow-Pour (Ed.), *Encyclopedia of Information Science and Technology* (3rd ed.; pp. 7507–7518). Hershey, PA: IGI Global; doi:10.4018/978-1-4666-5888-2.ch739

Kose, U. (2015). Present state of swarm intelligence and future directions. In *Encyclopedia of Information Science and Technology* (pp. 239–252). IRMA International. doi:10.4018/978-1-4666-5888-2.ch023

Kose, U., Koc, D., & Yucesoy, S. A. (2013). An augmented reality based mobile software to support learning experiences in computer science courses. *Procedia Computer Science*, *25*, 370–374. doi:10.1016/j.procs.2013.11.045

Küçük, S., Kapakin, S., & Göktaş, Y. (2015). Medical faculty students views on anatomy learning via mobile augmented reality technology. *Journal of Higher Education and Science*, *5*(3), 316–323. doi:10.5961/jhes.2015.133

Küçük, S., Kapakin, S., & Göktaş, Y. (2016). Learning anatomy via mobile augmented reality: Effects on achievement and cognitive load. *Anatomical Sciences Education*. doi:10.1002/ase.1603 PMID:26950521

Küçük, S., Yılmaz, R. M., & Göktaş, Y. (2014). Augmented reality for learning english: Achievement, attitude and cognitive load levels of students. *Education and Science*, *39*(176), 393–404.

Kuiper, R. (2008). Use of Personal Digital Assistants to Support Clinical Reasoning in Undergraduate Baccalaureate Nursing Students. CIN. *Computers, Informatics, Nursing*, *26*(2), 90–98. doi:10.1097/01.NCN.0000304776.40531.bc PMID:18317260

Kukulska-Hulme, A. (2005). *Mobile learning: A Handbook for Educators and Trainers*. Oxon, UK: Routledge.

Kulik, J. (2003). *Effects of using instructional technology in elementary and secondary schools: What controlled evaluation studies say*. Arlington, VA: SRI International. Retrieved 14 May, 2016 from http://downloads.kennisnet.nl/mediawijzer/Onderzoeken/Kulik_ITinK-12_Main_Report.pdf

Kurshan, B. (2016). *The future of artificial intelligence in education*. Retrieved December 8, 2016, from http://www.forbes.com/sites/barbarakurshan/2016/03/10/the-future-of-artificial-intelligence-in-education/#329b33a81e64

Kuusisto, F. (2015). Vr head-mounted displays. *XRDS: Crossroads. The ACM Magazine for Students*, *22*(1), 65. doi:10.1145/2837754

Laanpere, M., Pata, K., Normak, P., & Põldoja, H. (2014). Pedagogy-driven Design of Digital Learning Ecosystems. *Computer Science and Information Systems*, *11*(1), 419–442. doi:10.2298/CSIS121204015L

Laird, J. (2005). *History of computer games*. Retrieved October 1, 2015, from http://www.emunix.emich.edu/~evett/GameProgramming/History.pdf

Lass, W. (2015, July 13). *The Future of Augmented Reality: Limitations, Possibilities and Hopes*. Retrieved September 02, 2016, from Emergingedtech: http://www.emergingedtech.com/2015/07/future-of-augmented-reality-limitations-possibilities-hopes/

Lave, J., & Wenger, E. (1991). *Situated Learning: Legitimate Peripheral Participation*. Cambridge, UK: Cambridge University Press. doi:10.1017/CBO9780511815355

Lawler, R. (2013). *Get Ready For Even More Google Glasshole Sightings*. Retrieved from http://social.techcrunch.com/2013/01/28/glassholes/

Lazoudis, A., & Agogi, E. (2011). The "Science Center To Go" project. EDEN - 2011 Open Classroom Conference, 7-13.

Learninginspired. (2016). *My Top 5 Educational Augmented Reality Apps*. Retrieved September 1, 2016, from Learninginspired: http://www.learninginspired.com/my-top-five-educational-augmented-reality-apps

Lechner, M., & Tripp, M. (2010). ARML-an Augmented Reality Standard. *MobileARSummit*. Retrieved from http://www.perey.com/MobileARSummit/Mobilizy-ARML.pdf

Lee, K. (2012). The future of learning and training in augmented reality. *InSight: A Journal of Scholarly Teaching, 7*, 31-42.

Lee, K. (2012). The Future of Learning and Training in Augmented Reality. *InSight: A Journal of Scholarly Teaching, 7*, 31-42.

Lee, M., Green, R., & Billinghurst, M. (2008). 3D natural hand interaction for AR applications. *Image and Vision Computing New Zealand,23rd IEEE International Conference*, 1-6.

Lee, T., & Höllerer, T. (2007). Handy AR: Markerless inspection of augmented reality objects using fingertip tracking. *Wearable Computers,11th IEEE International Symposium*, 83-90.

Lee, K. (2012). Augmented reality in education and training. *TechTrends, 56*(2), 13–21. doi:10.1007/s11528-012-0559-3

Lee, M. R., Bojanova, I., & Suder, T. (2015). The New Wearable Computing Frontier. *IT Professional, 5*(September-October), 16–19. doi:10.1109/MITP.2015.84

Lehtinen, S. (2002). Visualization and teaching with state-of-the-art 3Dgame technologies. In *Proceedings of the Connecting the Real and the Virtual-design E-ducation*. (pp. 538–541). Warsaw, Poland: 20th eCAADe.

Leone, S., & Guazzaroni, G. (2010). Pedagogical sustainability of interoperable formal and informal learning environments. In F. Lazarinis, S. Green, & E. Pearson (Eds.), *Developing and Utilizing E-Learning Applications*. doi:10.4018/978-1-61692-791-2.ch001

Leo, T., Manganello, F., & Chen, N. S. (2010). From the Learning Work to the Learning Adventure. In *Proceedings of EDEN 2010*. Academic Press.

Lester, J. C., Converse, S. A., Kahler, S. E., Barlow, S. T., Stone, B. A., & Bhogal, R. S. (1997). The persona effect: affective impact of animated pedagogical agents.*Proceedings of conference on human factors in computing systems*, 359-366. doi:10.1145/258549.258797

Leue, M. C., Jung, T., & Dieck, D. T. (2015). Google Glass Augmented Reality: Generic Learning Outcomes for Art Galleries. In I. Tussyadiah & A. Inversini (Eds.), *Information and Communication Technologies in Tourism* (pp. 463–476). Vienna: Springer Computer Science. doi:10.1007/978-3-319-14343-9_34

Libow Martinez, S., & Stager, G. (2013). *Invent to learn: Making, tinkering, and engineering in the classroom*. Torrance, CA: Constructing Modern Knowledge Press.

Liestol, G. (2011). Learning through situated simulations: Exploring mobile augmented reality. *Research Bulletin (Sun Chiwawitthaya Thang Thale Phuket), 1*, 2011.

Lin, H. C. K., Chen, M. C., & Chang, C. K. (2015). Assessing the effectiveness of learning solid geometry by using an augmented reality-assisted learning system. *Interactive Learning Environments, 23*(6), 799–810. doi:10.1080/10494820.2013.817435

Lin, T. J., Duh, H. B. L., Li, N., Wang, H. Y., & Tsai, C. C. (2013). An investigation of learners collaborative knowledge construction performances and behavior patterns in an augmented reality simulation system. *Computers & Education, 68*, 314–321. doi:10.1016/j.compedu.2013.05.011

Liyanagunawardena, T. R., Adams, A. A., & Williams, S. A. (2013). MOOCs: A systematic study of the published literature 20082012. *The International Review of Research in Open and Distributed Learning, 14*(3), 202–227. doi:10.19173/irrodl.v14i3.1455

Lohse, G. L. (1997). Consumer Eye Movement Patterns on Yellow Pages Advertising. *Journal of Advertising, 26*(1), 61–73. doi:10.1080/00913367.1997.10673518

Loyd, J. W. (2012). *Mobile virtual patients for nurse training* (Master dissertation). Available from ProQuest Dissertations and Theses database. (UMI No. 1531108)

Mace, W. M. (2005). James J. Gibson's Ecological Approach: Perceiving What Exists. *Ethics & the Environment, 10*(2), 195-216.

Mahadzir, N. N., & Phung, L. F. (2013). The Use of Augmented Reality Pop-Up Book to Increase Motivation in English Language Learning For National Primary School. *IOSR Journal of Research & Method in Education, 1*(1), 26-38.

Makumburage, T. A. (2015). *State of the Art in Mobile Augmented Reality Development Tools/APIs.* Retrieved May 30, 2016 from http://www.slideshare.net/maku13s1/stateoftheart-in-mobile-augmented-reality-development-toolsapis

Mann, S., Nolan, J., & Wellman, B. (2003). Sousveillance: Inventing and Using Wearable Computing Devices for Data Collection in Surveillance Environments. *Surveillance & Society, 1*(3), 331–355.

Mano, M. R., & Ciletti, M. D. (2013). *Digital Design with an Introduction to Verilog HDL.* Pearson.

Manzelli, P. (2013). *Futuri orizzonti dell'arte quantistica e della realtà aumentata.* Paper presented at the meeting of Arte quantistica, scienza e realtà aumentata: sinergie d'innovazione, Pietrasanta, Italy.

Markham, T., Larmer, J., & Ravitz, J. L. (2003). *Project based learning handbook: a guide to standards-focused project based learning for middle and high school teachers* (2nd ed.). Novato, CA: Buck Institute for Education.

Marsh, D. (2002). *CLIL/EMILE - The European Dimension: Actions, Trends and Foresight Potential. DG Education & Culture*. European Commission.

Martín, S., Díaz, G., Cáceres, M., Gago, D., & Gibert, M. (2012). A Mobile Augmented Reality Gymkhana For Improving Technological Skills And History Learning: Outcomes And Some Determining Factors. In T. Bastiaens & G. Marks (Eds.), *Proceedings of E-Learn: World Conference on E-Learning in Corporate, Government, Healthcare, and Higher Education 2012* (pp. 260-265). Chesapeake, VA: Association for the Advancement of Computing in Education (AACE).

Martinez, H., Skournetou, D., Hyppölä, J., Laukkanen, S. and Heikkilä, A. (2014). Drivers and Bottlenecks in the Adoption of Augmented Reality Applications. Journal of Multimedia Theory and Application 2(1), 2014, pp.27-44.

Martín-Gutiérrez, J., Fabiani, P., Benesova, W., Meneses, M. D., & Mora, C. E. (2015). Augmented reality to promote collaborative and autonomous learning in higher education. *Computers in Human Behavior, 51*, 752–761. doi:10.1016/j.chb.2014.11.093

Martin, S., Diaz, G., Sancristobal, E., Gil, R., Castro, M., & Peire, J. (2011). New technology trends in education: Seven years of forecasts and convergence. *Computers & Education, 57*(3), 1893–1906. doi:10.1016/j.compedu.2011.04.003

Matcha, W., & Rambli, D. R. A. (2013). Exploratory study on collaborative interaction through the use of augmented reality in science learning. *Procedia Computer Science, 25*, 144–153. doi:10.1016/j.procs.2013.11.018

Mayer, R. E. (2002). Multimedia learning. *Psychology of Learning and Motivation*, *41*, 85–139. doi:10.1016/S0079-7421(02)80005-6

Mayer, R. E. (2005). Principles for Reducing Extraneous Processing in Multimedia Learning: Coherence, Signaling, Redundancy, Spatial Contiguity, and Temporal Contiguity Principles. In R. Mayer (Ed.), *The Cambridge Handbook of Multimedia Learning* (pp. 183–200). Cambridge, UK: Cambridge University Press. doi:10.1017/CBO9780511816819.013

Mayer, R. E. (2008). Applying the science of learning: Evidence-based principles for the design of multimedia instruction. *The American Psychologist*, *63*(8), 760–769. doi:10.1037/0003-066X.63.8.760 PMID:19014238

McAlpine, H., Lockerbie, L. Ramsay, D, & Beaman, S. (2002). Evaluating a Web based Graduate Level Nursing Ethics Course: Thumbs Up or Thumbs Down? *The Journal of Continuing Education in Nursing, 33*(1), 12-18. DOI: 20020101-0410.3928/0022-0124

McCarthy, J. (2007). *What is Artificial Intelligence?* Online: http://www-formal.stanford.edu/jmc/whatisai/whatisai.html

McGrath, R. (2016). *The Pace of Technology Adoption Is Speeding Up*. Retrieved December 18, 2016, from https://hbr.org/2013/11/the-pace-of-technology-adoption-is-speeding-up

McKinney, A. A., & Karen, P. (2009). Podcasts and videostreaming: Useful tools to facilitate learning of pathophysiology in undergraduate nurse education? *Nurse Education in Practice*, *9*(6), 372–376. doi:10.1016/j.nepr.2008.11.003 PMID:19124275

Medicherla, P. S., Chang, G., & Morreale, P. (2010). *Visualization for ıncreased understanding and learning using augmented reality*. Paper presented at 11th ACM SIGMM International Conference on Multimedia Information Retrieval (MIR '10), Philadelphia, PA. doi:10.1145/1743384.1743462

Medley, C. F., & Horne, C. (2005). Using Simulation Technology for Undergraduate Nursing Education. *The Journal of Nursing Education*, *44*(1), 31–34. PMID:15673172

Mehrotra, K., Mohan, C. K., & Ranka, S. (1997). *Elements of artificial neural networks*. MIT Press.

Meta. (2014, September 18). *Meta pioneers: Simx* [Video file]. Retrieved from https://www.metavision.com/meet-meta

Microsoft HoloLens. (2015, July 8). *Microsoft Hololens: Partner spotlight with Case Western Reserve University* [Video file]. Retrieved from https://www.microsoft.com/microsoft-hololens/en-us/commercial#fullbleed-volvo

Milgram, P., & Kishino, F. (1994). A taxonomy of mixed reality visual displays. *IEICE Transactions on Information and Systems*, *77*(12), 1321–1329.

Miller, J., Shaw-Kokot, J. R., Arnold, M. S., Boggin, T., Crowell, K. E., Allegri, F., & Berrier, S. B. et al. (2005). A study of personal digital assistants to enhance undergraduate clinical nursing education. *The Journal of Nursing Education*, *44*, 119–126. PMID:15673170

Miller, W. T., Werbos, P. J., & Sutton, R. S. (1995). *Neural networks for control*. MIT Press. doi:10.1007/978-3-642-57760-4

Mobil Artırılmış Gerçeklik. (2016). Retrieved from http://www.matbilisim.com/index.php/mobil-artirilmis-gerceklik-mobile-augmented-reality/, 14.09.2016).

Möhring, M., Lessig, C., & Bimber, O. (2004). Video See-Through AR on Consumer Cell-Phones.*IEEE International Symposium on Mixed and Augmented Reality (ISMAR)*. doi:10.1109/ISMAR.2004.63

Moreno-Armella, L. (2014). *Educación Matemática: del signo al pixel*. Bucaramanga, Colombia: Universidad Industrial de Santander.

Moreno-Armella, L., & Hegedus, S. J. (2013). From Static to Dynamic Mathematics: Historical and Representational Perspectives. In S. J. Hegedus & J. Roschelle (Eds.), *The SimCalc Vision and Contributions: Democratizing access to important mathematics* (pp. 27–46). Springer-Verlag; doi:10.1007/978-94-007-5696-0_3

Moreno, R. (2006). Does the modality principle hold for different media? A Test of the method-affects-learning hypothesis. *Journal of Computer Assisted Learning*, *22*(3), 149–158. doi:10.1111/j.1365-2729.2006.00170.x

Moshtaghi, O., Kelley, K. S., Armstrong, W. B., Ghavami, Y., Gu, J., & Djalilian, H. R. (2015). Using google glass to solve communication and surgical education challenges in the operating room. *The Laryngoscope*, *125*(10), 2295–2297. doi:10.1002/lary.25249 PMID:25775933

Motiwalla, L. F. (2007). Mobile learning: A framework and evaluation. *Computers & Education*, *49*(3), 581–596. doi:10.1016/j.compedu.2005.10.011

Muensterer, O. J., Lacher, M., Zoeller, C., Bronstein, M., & Kübler, J. (2014). Google Glass in pediatric surgery: An exploratory study. *International journal of surgery (London, England)*, *12*(4), 281–289. doi:10.1016/j.ijsu.2014.02.003 PMID:24534776

Munnerley, D., Bacon, M., Wilson, A., Steele, J., Hedberg, J., & Fitzgerald, R. (2012). Confronting an augmented reality. *Research in Learning Technology, ALT-C 2012 Conference Proceedings*, 39-48.

Nabiyev, V. V. (2005). *Artificial Intelligence (Problems – Methods – Algorithm)* [Yapay Zekâ (Problemler – Yöntemler – Algoritma)]. Ankara, Turkey: Seçkin Press.

Naeyc & Fred Rogers Center. (2012, January). *Technology and Interactive Media as Tools in Early Childhood Programs Serving Children from Birth through Age 8*. Retrieved September 20, 2016, from Naeyc.org: http://www.naeyc.org/files/naeyc/file/positions/PS_technology_WEB2.pdf

Nesbit, J. C., Adesope, O. O., Liu, Q., & Ma, W. (2014). How Effective are Intelligent Tutoring Systems in Computer Science Education? In *Advanced Learning Technologies (ICALT), 2014 IEEE 14th Int. Conference on* (pp. 99-103). IEEE.

Nesloney, T. (2013, April 11). *Augmented Reality Brings New Dimensions to Learning*. Retrieved August 2016, from Edutopia: http://www.edutopia.org/blog/augmented-reality-new-dimensions-learning-drew-minock

New Media Consortium. (2005). *The Horizon Report. National Learning Infrastructure Initiative*. Stanford, CA: McGraw Hill.

Newcombe, N. S. (2013). Seeing Relationships: Using Spatial Thinking to Teach Science, Mathematics, and Social Studies. *American Educator, 37*(1), 26–31 and 40. Retrieved from http://silccenter.org/publications_pdfs/Newcombe final 3-4-13.pdf

Newholm, T., & Shaw, D. (2007). Studying the ethical consumer: A review of research. *Journal of Consumer Behaviour*, *6*(5), 253–270. doi:10.1002/cb.225

Nilsson, N. J. (2014). *Principles of artificial intelligence*. Morgan Kaufmann.

Nincarean, D., Alia, M. B., Halima, N. D., & Rahman, M. H. (2013). Mobile Augmented Reality: the potential for education. *13th International Educational Technology Conference*, 103, 657 – 664.

Norman, D. A. (1993). *Things that make us smart: Defending human attributes in the age of the machine*. Reading, MA: Addison-Wesley Pub. Co.

Norman, D. A. (1999). Affordance, conventions, and design. *Interaction*, *6*(3), 38–43. doi:10.1145/301153.301168

Norris, C., & Soloway, E. (2014). *Personalized Learning, Flipped Classrooms, Video Watching: Last Gasps of the Old Education*. Retrieved from http://thejournal.com/articles/2014/03/31/broken-educational-model.aspx#6V6xhizZguxtx5hW.99

November, A., & Mull, B. (2012). *Flipped Learning: A Response To Five Common Criticisms*. Retrieved from http://novemberlearning.com/educational-resources-for-educators/teaching-and-learning-articles/flipped-learning-a-response-to-five-common-criticisms-article/

Núñez, M., Quirós, R., Núñez, I., Carda, J. B., & Camahort, E. (2008, July). *Collaborative augmented reality for inorganic chemistry education*. Paper presented at 5th WSEAS / IASME International Conference on Engineerıng Education (EE'08), Heraklion, Greece.

Nye, B. D. (2015). Intelligent tutoring systems by and for the developing world: A review of trends and approaches for educational technology in a global context. *International Journal of Artificial Intelligence in Education*, *25*(2), 177–203. doi:10.1007/s40593-014-0028-6

O'Brien, H. L., & Toms, E. G. (2005). *Engagement as process in computer mediated environments*. Paper presented at ASIS&T, Charlotte, NC.

O'Driscoll, J. (2004). The discourse of English as T-Rex. In *Language and the Future of Europe: Ideologies Policies and Practices*. University of Southampton.

Odum, E. P. (1971). *Fundamentals of Ecology*. New York: Saunders.

Oh, S., & Woo, W. (2008). *ARGarden: Augmented edutainment system with a learning companion* (Vol. 5080). Transactions on Edutainment I Lecture Notes in Computer Science.

Olssona, T., Kärkkäinena, T., Lagerstama, E., & Ventä-Olkkonen, L. (2012). User evaluation of mobile augmented reality scenarios. *Journal of Ambient Intelligence and Smart Environments*, *4*, 29–47. doi:10.3233/AIS-2011-0127

OpenEducationWeek.Org. (2016). *What is Open Education?* Online: https://www.openeducationweek.org/page/what-is-open-education

OpenSource.Com. (2016). *What is Open Education?* Online: https://opensource.com/resources/what-open-education

Ortega, L. D. M., Barachino, P. R., Rodriguez, J., Lourdes, M., Gonzalez, H., Ramon, J., & Salvador, O. T. et al. (2011). Using m-learning on nursing courses to improve learning, CIN. *Computers, Informatics, Nursing*, *29*(5), 311–317. doi:10.1097/NCN.0b013e3181fcbddb PMID:21084973

Ostrow, L., & DiMaria-Ghalili, R. A. (2005). Distance education for graduate nursing: One state school's experience. *The Journal of Nursing Education*, *44*, 5–10. PMID:15673168

Overmars, M. (2012). *A brief history of computer games*. Retrieved September 9, 2015, from http://www.cs.uu.nl/docs/vakken/b2go/literature/history_of_games.pdf

Özarslan, Y. (2011). Öğrenen içerik etkileşiminin genişletilmiş gerçeklik ile zenginleştirilmesi. *International Computer & Instructional Technologies Symposium (ICITS 2011)*. Fırat Üniversitesi.

Özarslan, Y. (2013). *The effect of augmented reality enhanced learning materials on learners' achievement and learners' satisfaction* (Unpublished Doctoral Dissertation). Anadolu University, Eskişehir, Turkey.

Özdamar Keskin, N., & Kılınç, H. (2015). A comparison of development platforms for mobile learning applications and case studies. *Journal of Open Education Practices and Research*, *1*(3), 68–90.

Özdamar Keskin, N., & Kuzu, A. (2015). Development and testing of a m-learning system for the professional development of academics through design-based action research. *The International Review of Research in Open and Distributed Learning*, *16*(1), 193–220. doi:10.19173/irrodl.v16i1.1613

Paavilainen, J. (2004). *Mobile games: Creating business with N-GAGE Nokia*. New Riders.

Padzensky, R. (2014). *06: Market Outlook*. Retrieved May 27, 2016 from http://augmera.com/?p=214

Palmieri, M., Singh, I., & Cicchetti, A. (2012). Comparison of cross-platform mobile development tools. *Intelligence in Next Generation Networks (ICIN),16th IEEE International Conference*, 179-186.

Papagiannakis, G., Singh, G., & Magnenat-Thalmann, N. (2008). A survey of mobile and wireless technologies for augmented reality systems. *Computer Animation and Virtual Worlds*, *19*(1), 3–22. doi:10.1002/cav.221

Papert, S. (1986). *Constructionism: A New Opportunity for Elementary Science Education*. Academic Press.

Park, C. W., MacInnis, D. J., Priester, J., Eisingerich, A. B., & Iacobucci, D. (2010). Brand attachment and brand attitude strength: Conceptual and empirical differentiation of two critical brand equity drivers. *Journal of Marketing*, *74*(6), 1–17. doi:10.1509/jmkg.74.6.1

Pasaréti, O., Hajdú, H., & Matuszka, T. Jámbor,i A., Molnár, I. & Turcsányi-Szabó, M. (2011). *Augmented reality in education*. Paper presented at INFODIDACT Informatika Szakmódszertani konferencia cd-kiadvány.

Peña-Ayala, A. (2013). Intelligent and Adaptive Educational-Learning Systems. *Smart Innovation, Systems and Technologies*, 17.

Perey, C., Engelke, T., & Reed, C. (2011). Current status of standards for augmented reality. In Recent Trends of Mobile Collaborative Augmented Reality Systems, (pp. 21-38). Springer. doi:10.1007/978-1-4419-9845-3_2

Perez-Lopez, D., & Contero, M. (2013). Delivering educational multimedia contents through an augmented reality application: A case study on its impact on knowledge acquisition and retention. *Turkish Online Journal of Educational Technology, 12*(4), 19–28.

Perry, B. (2015). Gamifying French Language Learning: A Case Study Examining a Quest-based, Augmented Reality Mobile Learning-tool. *Procedia: Social and Behavioral Sciences*, *174*, 2308–2315. doi:10.1016/j.sbspro.2015.01.892

Petronio, S. S. (2002). *Boundaries of privacy: Dialectics of disclosure. SUNY series in communication studies*. Albany, NY: State University of New York Press.

Pieters, R., Rosbergen, E., & Wedel, M. (1999). Visual Attention to Repeated Print Advertising: A Test of Scanpath Theory. *JMR, Journal of Marketing Research*, *36*(4), 424. doi:10.2307/3151998

PoparI. S. T. http://popartoys.com

Porter, M., & Heppelmann, J. E. (2015, October). How smart, connected products are transforming companies. *Harvard Business Review*.

Preece, J. (2000). *Online communities: Designing usability and supporting socialbilty*. John Wiley & Sons, Inc.

Premadasa, K., & Bhatia, K. (2013). Real life applications in mathematics: What do students prefer? *International Journal for the Scholarship of Teaching and Learning*, *7*(2), 20. doi:10.20429/ijsotl.2013.070220

Prensky, M. (2001). *Fun, play and games: What makes games engaging. From digital game-based learning*. Retrieved September 9, 2015, from http://www.marcprensky.com/writing/Prensky%20-%20Digital%20Game-Based%20Learning-Ch5.pdf

Prensky, M. (2007). Digital game-based learning. St. Paul, MN: Paragon House.

PriceWaterhouseCoopers. (2015). *The wearable future*. Author.

PTC Inc. (2016). *Vuforia Target Manager,* Retrieved August 18, 2016 from https://library.vuforia.com/articles/Training/Getting-Started-with-the-Vuforia-Target-Manager

Quinn, C. (2001). Get ready for m-learning. *Training & Development*, *20*(2), 20–21.

Quint, F., & Loch, F. (2015). Using Smart Glasses to Document Maintenance Processes. In A. Weisbecker, M. Burmester, & A. Schmidt (Eds.), *Mensch und Computer 2015. Workshopband* (pp. 203–208). Stuttgart: Oldenbourg Wissenschaftsverlag. doi:10.1515/9783110443905-030

Radkowski, R. (2016). *AR for Physics Education. The Augmented Reality Physics Experiments Toolbox*. Retrieved from http://arlab.me.iastate.edu/ar-physics.html

Radu, I. (2014). Augmented reality in education: A meta-review and cross-media analysis. *Personal and Ubiquitous Computing*, *18*(6), 1533–1543. doi:10.1007/s00779-013-0747-y

Ramakrishnamurthy, R. (2016). *How to Develop Augmented Reality Based Mobile Apps?* Retrieved May 27, 2016 from http://blog.contus.com/augmented-reality-apps-development/

Rambli, D. R. A., Matcha, W., & Sulaiman, S. (2013). Fun learning with AR alphabet book for preschool children. *Procedia Computer Science*, *25*, 211–219. doi:10.1016/j.procs.2013.11.026

Ramey, K. (2012). *Technological Advancements and Their Effects on Humanity*. Retrieved October 29, 2016, from http://www.useoftechnology.com/technological-advancements-effects-humanity/

Raskar, R., Welch, G., & Fuchs, H. (1998). Spatially Augmented Reality. *First International Workshop on Augmented Reality*. Retrieved June 12, 2016, from http://citeseerx.ist.psu.edu/viewdoc/download?doi=10.1.1.17.4260&rep=rep1&type=df

Rathburn, M. K. (2015). Building connections through contextualized learning in an undergraduate course on scientific and mathematical literacy. *International Journal for the Scholarship of Teaching and Learning*, *9*(1). Retrieved from http://digitalcommons.georgiasouthern.edu/ij-sotl/vol9/iss1/11

Ratneshwar, S., & Mick, D. G. (2005). *Inside consumption: Consumer motives, goals, and desires*. London: Routledge.

Rauschnabel, P. A., Brem, A., & Ro, Y. K. (2015). *Augmented Reality Smart Glasses. Definition, Conceptual Insights, and Managerial Importance*. unpublished working paper.

Rauschnabel, P. A., Brem, A., & Ivens, B. S. (2015). Who will buy smart glasses? *Computers in Human Behavior*, *49*, 635–647. doi:10.1016/j.chb.2015.03.003

Rauschnabel, P. A., & Ro, Y. K. (2016). Augmented reality smart glasses: An investigation of technology acceptance drivers. *International Journal of Technology Marketing*, *11*(2), 123–148. doi:10.1504/IJTMKT.2016.075690

Reeves, T. C., Herrington, J., & Oliver, R. (2004). A development research agenda for online collaborative learning. *Educational Technology Research and Development*, *52*(4), 53–65. doi:10.1007/BF02504718

Regis. (2013). *Artırılmış Gerçeklik Nedir?* Retrieved from http://www.regisapp.com/augmented-reality-nedir

Reinelt, P., Hadish, S., & Ernst, C.-P. H. (2016). How Design Influences Headphone Usage. In C.-P. H. Ernst (Ed.), *Progress in IS. The Drivers of Wearable Device Usage* (pp. 59–68). Cham: Springer International Publishing. doi:10.1007/978-3-319-30376-5_6

Reyna, J. (2011) Digital Teaching and Learning Ecosystem (DTLE): A Theoretical Approach for Online Learning Environments. In *Proceedings ASCILITE 2011* (pp. 1083 – 1088). Hobart: The University of Tasmania. Retrieved October 29, 2016, from http://www.ascilite.org/conferences/hobart11/downloads/ProceedingsV3.pdf

Reynes, J., & Yerli, C. (2007).The CRYENGINE technology applied to architecture. In *Imagina 2007*. Monaco: the 25th European Digital Content Creation.

Rheingold, H. (1992). *Virtual Reality*. Simon & Schuster. Retrieved August 20, 2016, from Morton Heiling: http://www.mortonheilig.com/SensoramaPatent.pdf

Richards, D., & Taylor, M. (2015). A Comparison of learning gains when using a 2D simulation tool versus a 3D virtual world: An experiment to find the right representation involving the Marginal Value Theorem. *Computers & Education, 86*, 157–171. doi:10.1016/j.compedu.2015.03.009

Riesman, D., Glazer, N., Denney, R., & Gitlin, T. (2001). The lonely crowd: A study of the changing American character (Abridged and rev. ed.). Yale Nota bene. New Haven, CT: Yale University Press.

Ripley, B. D. (1994). Neural networks and related methods for classification. *Journal of the Royal Statistical Society. Series B. Methodological*, 409–456.

Ritchie, D. (2007). *Yes, a video game contributed to Unix Development.* Retrieved September 9, 2015, from http://sites.fas.harvard.edu/~lib215/reference/history/spacetravel.html

Rosenbaum, E., Klopfer, E., & Perry, J. (2006). On location learning: Authentic applied science with networked augmented realities. *Journal of Science Education and Technology, 16*(1), 31–45. doi:10.1007/s10956-006-9036-0

Roukounaki, K. (2014, Mar 16). *Top 5 Tools for Augmented Reality in Mobile Apps* [Weblog]. Retrieved August 18, 2016 from https://www.developereconomics.com/top-5-tools-for-augmented-reality-in-mobile-apps/

Rouse, M. (2016, February). *Augmented Reality (AR).* Retrieved August 2016, from Whatis.com: http://whatis.techtarget.com/definition/augmented-reality-AR

Rouse, R. (2005). *Game Design Theory & Practice* (2nd ed.). Wordware Publishing.

Rubashkina, Y., Galeotti, M., & Verdolini, E. (2015). Environmental regulation and competitiveness: Empirical evidence on the Porter Hypothesis from European manufacturing sectors. *Energy Policy, 83*, 288–300. doi:10.1016/j.enpol.2015.02.014

Russell, S. J., Norvig, P., Canny, J. F., Malik, J. M., & Edwards, D. D. (2003). *Artificial intelligence: a modern approach* (Vol. 2). Upper Saddle River, NJ: Prentice hall.

Şahin, M. C. (2009). Yeni bin yılın öğrencilerinin özellikleri. *Anadolu University Journal Of Social Sciences, 9*(2), 155–172.

Salinas, P., González-Mendívil, E., Quintero, E., Ríos, H., & Morales, S. (2013). 2013 International Conference on Virtual and Augmented Reality in Education. The Development of a Didactic Prototype for the Learning of Mathematics through Augmented Reality. *Procedia Computer Science, 25*, 62–70.

Salinas, P., Quintero, E., & González-Mendívil, E. (2014). How can Augmented Reality Favor the Learning of Calculus? *Proceedings of FECS 2014. World Academy of Science*. CSREA Press.

Salinas, P., Quintero, E., Ramirez, P. G., & Mendívil, E. G. (2015). Fostering spatial visualization through augmented reality in calculus learning. In *ASEE Annual Conference and Exposition, Conference Proceedings* (Vol. 122nd ASEE Annual Conference and Exposition: Making Value for Society). Retrieved from https://www.scopus.com/inward/record.uri?eid=2-s2.0-84941996184&partnerID=40&md5=6ab8d46079c7bf40467d03c0ba4bd417

Santos, M. E. C., Chen, A., Taketomi, T., Yamamoto, G., Miyazaki, J., & Kato, H. (2014). Augmented reality learning experiences: Survey of prototype design and evaluation. IEEE Transactions on Learning Technologies, 7(1), 38-56.

Santos, M. E. C., Chen, A., Taketomi, T., Yamamoto, G., Miyazaki, J., & Kato, H. (2014). Augmented reality learning experiences: Survey of prototype design and evaluation. *Learning Technologies. IEEE Transactions on*, *7*(1), 38–56.

Saran, M., Cagiltay, K., & Seferoglu, G. (2008). Use of Mobile Phones in Language Learning: Developing Effective Instructional Materials. In *Proceedings of the Fifth IEEE International Conference on Wireless, Mobile, and Ubiquitous Technology in Education*. IEEE. doi:10.1109/WMUTE.2008.49

Schank, R. C. (1995). What We Learn When We Learn by Doing. *Technical Report No. 60*. Northwestern University, Institute for Learning Sciences.

Schawabe, G., & Goth, C. (2005). Mobile Learning with a Mobile Game: Design and Motivational Effects. *Journal of Computer Assisted Learning*, *21*(3), 204–216. doi:10.1111/j.1365-2729.2005.00128.x

Schmalstieg, D., & Wagner, D. (2007). *Experiences with handheld augmented reality*. Paper presented at 6th IEEE and ACM International Symposium on Mixed and Augmented Reality.

Schnetter, V. A., Lacy, D., Jones, M. M., Bakrim, K., Allen, P. E., & O'Neal, C. (2014). Course development for web-based nursing education programs. *Nurse Education in Practice*, 1–6. doi:10.1016/j.nepr.2014.06.007 PMID:25154660

Schön, S., Ebner, M., & Kumar, S. (2014). *The Maker Movement. Implications of new digital gadgets, fabrication tools and spaces for creative learning and teaching*. Retrieved May 6, 2016, from http://www.openeducationeuropa.eu/de/article/The-Maker-Movement.-Implications-of-new-digital-gadgets,-fabrication-tools-and-spaces-for-creative-learning-and-teaching

Schreppegrell, M., Achugar, M., & Oteiza, T. (2004). The grammar of history: Enhancing content based instruction through a functional focus on language. *TESOL Quarterly*, *38*(1), 67–93. doi:10.2307/3588259

Serrano, N., Hernantes, J., & Gallardo, G. (2013). Mobile web apps. *Software, IEEE*, *30*(5), 22–27. doi:10.1109/MS.2013.111

Shamsudin, S., & Nesi, H. (2006). Computer-mediated communication in English for specific purposes: A case study with computer science students at Universiti Teknologi Malaysia. *Computer Assisted Language Learning*, *19*(4-5), 317–339. doi:10.1080/09588220601043164

Sharples, M. (2002). Disruptive devices: Mobile technology for conversational learning. *International Journal of Continuing Engineering Education and Lifelong Learning*, *12*(5-6), 504–520. doi:10.1504/IJCEELL.2002.002148

Sharples, M., & Pea, R. (2014). Mobile Learning. In K. Sawyer (Ed.), *The Cambridge Handbook of the Learning Sciences* (2nd ed.; pp. 501–521). New York, NY: Cambriudge University Press. doi:10.1017/CBO9781139519526.030

Shaw, D., Grehan, E., Shiu, E., Hassan, L., & Thomson, J. (2005). An exploration of values in ethical consumer decision making. *Journal of Consumer Behaviour*, *4*(3), 185–200. doi:10.1002/cb.3

Shaw, D., Newholm, T., & Dickinson, R. (2006). Consumption as voting: An exploration of consumer empowerment. *European Journal of Marketing*, *40*(9/10), 1049–1067. doi:10.1108/03090560610681005

Shelton, B. E., & Hedley, N. R. (2002). *Using augmented reality for teaching earth-sun relationships to undergraduate geography students*. Paper presented at The First IEEE International Augmented Reality Toolkit Workshop, Darmstadt, Germany. doi:10.1109/ART.2002.1106948

Shelton, B. E., & Stevens, R. (2004). *Using coordination classes to interpret conceptual change in astronomical thinking*. Paper presented at 6th International Conference for the Learning Sciences, Santa Monica, CA.

Sherry, J. L., & Lucas, K. (2001). *Video game uses and gratifications as predictors of use and game preference.* Retrieved September 22, 2015, from http://icagames.comm.msu.edu/vgu%26g.pdf

Shirazi, A., & Behzadan, A. (2014). Design and assessment of a mobile augmented reality-based information delivery tool for construction and civil engineering curriculum. *Journal of Professional Issues in Engineering Education and Practice, 04014012.* doi:10.1061/(ASCE)EI.1943-5541.0000229

Siemens, G. (2013). Massive open online courses: Innovation in education. *Open Educational Resources: Innovation, Research and Practice, 5.*

Silberman, M. L., & Biech, E. (2015). *Active Training: A Handbook of Techniques, Designs, Case Examples, and Tips.* John Wiley & Sons. doi:10.1002/9781119154778

Siltanen, S. (2012). *Theory And Applications Of Marker-Based Augmented Reality.* Retrieved from http://www.vtt.fi/inf/pdf/science/2012/S3.pdf

Sime, J. D. (1999). Crowd facilities, management and communications in disasters. *Facilities, 17*(9/10), 313–324. doi:10.1108/02632779910278728

Singh, D., Tripathi, G., & Jara, A. J. (2014). *A survey of internet-of-things: Future vision, architecture, challenges and services.* Paper presented at the IEEE World forum on Internet of things (WF-IoT), Seoul, South Korea.

Sırakaya, M. (2015). *Effects of augmented reality applications on students' achievement, misconceptions and course engagement* (Unpublished doctoral dissertation). Gazi University, Ankara, Turkey.

Sırakaya, M., & Seferoğlu, S. (2016). Öğrenme ortamlarında yeni bir araç: Bir eğitlence uygulaması olarak artırılmış gerçeklik. In A. İşman, H. F. Odabaşı, & B. Akkoyunlu (Eds.), *Eğitim teknolojileri okumaları 2016* (pp. 417–438). Ankara: Salmat Basım Yayıncılık.

Smith, M. (2013). Developing Mobile Apps. *Journal of Management & Engineering Integration, 6*(2), 79.

Smutni, P. (2012). Mobile development tools and cross-platform solutions.*Carpathian Control Conference (ICCC), 13th IEEE International Conference*, 653-656.

Solomon, C., & Papert, S. (1971). *Twenty Things to Do With a Computer.* Retrieved May 6, 2016, from http://www.stager.org/articles/twentythings.pdf

Soloway, E., Norris, C., Blumenfeld, P., Fishman, B., Krajcik, J., & Marx, R. (2001). Log on education: Handheld devices are ready-at-hand. *Communications of the ACM, 44*(6), 15–20. doi:10.1145/376134.376140

Sommerauer, P., & Müller, O. (2014). Augmented reality in informal learning environments: A field experiment in a mathematics exhibition. *Computers & Education, 79*, 59–68. doi:10.1016/j.compedu.2014.07.013

Somyürek, S. (2014). Gaining the attention of generation Z in learning process: Augmented reality. *Eğitim Teknolojisi Kuram ve Uygulama, 4*(1), 63–80.

Spagnolli, A., Guardigli, E., Orso, V., Varotto, A., & Gamberini, L. (2014). Measuring User Acceptance of Wearable Symbiotic Devices: Validation Study Across Application Scenarios. In G. Jacucci, L. Gamberini, J. Freeman, & A. Spagnolli (Eds.), Lecture Notes in Computer Science: Vol. 8820. *Symbiotic Interaction* (pp. 87–98). Cham: Springer International Publishing. doi:10.1007/978-3-319-13500-7_7

SparcOpen.Org. (2016). *Open Education.* Online: http://sparcopen.org/open-education/

Specht, M., Ternier, S., & Greller, W. (2011). Dimensions of Mobile Augmented Reality for Learning: A First Inventory. *Journal of the Research Center for Educational Technology, 7*(1), 117–127.

Specht, M., Ternier, S., & Greller, W. (2011). Mobile augmented reality for learning: A case study. *Journal of the Research Center for Educational Technology*, *7*(1), 117–127.

Specht, M., Ternire, S., & Greller, W. (2011). Dimensions of mobile augmented reality for learning: A first inventory. *Journal of the Research Center for Educational Technology*, *7*(1), 117–127.

Spencer, J. (2012, July 14). *11 reasons teachers aren't using technology*. Retrieved from http://www.spencerauthor.com/2012/07/11-reasons-teachers-arent-using.html/

Spitzer, M., & Ebner, M. (2015). Collaborative Learning Through Drawing on iPads. In *Proceedings of World Conference on Educational Multimedia, Hypermedia and Telecommunications 2015* (pp. 633-642).Chesapeake, VA: AACE.

Spitzer, M., & Ebner, M. (2016). Use cases and architecture of an information system to integrate smart glasses in educational environments. In *EdMedia:World Conference on Educational Media and Technology,* (1), 57-64.

Squire, K. D., & Jan, M. (2007). Mad city mystery: Developing scientific argumentation skills with a place-based augmented reality game on handheld computers. *Journal of Science Education and Technology*, *16*(1), 5–29. doi:10.1007/s10956-006-9037-z

Squire, K., & Klopfer, E. (2007). Augmented reality simulations on handheld computers. *Journal of the Learning Sciences*, *16*(3), 371–413. doi:10.1080/10508400701413435

Sridharan, B., Deng, H., & Corbitt, B. (2010). Critical success factors in e-learning ecosystems: A qualitative study. *Journal of Systems and Information Technology*, *12*(4), 263–288. doi:10.1108/13287261011095798

Stake, R. E. (2010). *Qualitative research: Studying how things work*. Guilford Publications.

Stanković, S. (2015). Virtual Reality and Virtual Environments in 10 Lectures. In A. C. Bovik (Ed.), *Synthesis Lectures on Image, Video, and Multimedia Processing* (Vol. 8, pp. 1–197). Morgan & Claypool Publishers. doi:10.2200/S00671ED1V01Y201509IVM019

Statista. (2016). *Global market share held by the leading smartphone operating systems in sales to end users from 1st quarter 2009 to 1st quarter 2016*. Retrieved May 27, 2016 from http://www.statista.com/statistics/266136/global-market-share-held-by-smartphone-operating-systems/

Statt, N. (2015, January 24). Microsoft's Hololens explained: How it works and why it's different. *CNET*. Retrieved from http://www.cnet.com/news/microsoft-hololens-explained-how-it-works-and-why-its-different/

Steuer, J. (1992). Defining virtual reality: Dimensions determining telepresence. *Journal of Communication*, *42*(4), 73–93. doi:10.1111/j.1460-2466.1992.tb00812.x

Stockwell, G., & Hubbard, P. (2013). *Some emerging principles for mobile-assisted language learning*. The International Research Foundation for English Language Education. Retrieved from http://www.tirfonline.org/wp-content/uploads/2013/11/TIRF_MALL_Papers_StockwellHubbard.pdf

Strong, A. I. (2016). *Applications of Artificial Intelligence & Associated Technologies. Science*. Academic Press.

Sumadio, D. D., & Rambli, D. R. A. (2010). Preliminary evaluation on user acceptance of the augmented reality use for education.*Proceedings of Second International Conference on Computer Engineering and Applications*, 461–465. doi:10.1109/ICCEA.2010.239

Sutherland, I. E. (1968). A head-mounted three dimensional display. In AFIPS (Ed.), *Proceedings of the December 9-11, 1968, fall joint computer conference, part I* (pp. 757–764). doi:10.1145/1476589.1476686

Sutherland, I. I. (1968). A head-mounted three dimensional display. *Proceedings of AFIPS*, *68*, 757–764. Retrieved from http://design.osu.edu/carlson/history/PDFs/p757-sutherland.pdf

Tansley, A. G. (1935). The Use and Abuse of Vegetational Concepts and Terms. *Ecology*, *16*(3), 284–307. doi:10.2307/1930070

Tarng, W., & Ou, K.-L. (2012). *A study of campus butterfly ecology learning system based on augmented reality and mobile learning.* Paper presented at 2012 IEEE Seventh International Conference on Wireless, Mobile and Ubiquitous Technology in Education. doi:10.1109/WMUTE.2012.17

Taşkıran, A., Koral, E., & Bozkurt, A. (2015). *Artırılmış gerçeklik uygulamasının yabancı dil eğitiminde kullanılması.* Paper presented at Akademik Bilişim 15, Eskişehir, Turkey.

Taşkıran, A., Koral, E., & Bozkurt, A. (2015). *Artırılmış Gerçeklik Uygulamasının Yabancı Dil Öğretiminde Kullanılması. Akademik Bilişim 2015*. Eskişehir: Anadolu Üniversitesi.

Technopedia. (2016). *Mobile device*. Retrieved 20 May, 2016 from https://www.techopedia.com/definition/23586/mobile-device

The Economist Intelligence Unit. (2008). *The Future of Higher Education: How Technology Will Shape Learning.* Retrieved October 29, 2016, from https://www.nmc.org/pdf/Future-of-Higher-Ed-(NMC).pdf

Thomas, R. G., John, N. W., & Delieu, J. M. (2010). Augmented Reality for Anatomical Education. *Journal of Visual Communication in Medicine*, *33*(1), 6–15. doi:10.3109/17453050903557359 PMID:20297908

Thompson, G. (2014). The Maker Movement Conquers the Classroom. *THE Journal*. Retrieved May 6, 2016, from https://thejournal.com/Articles/2014/04/30/The-MakerMovement-Conquers-the-Classroom.aspx?Page=2

Thompson, B. (2005). The transforming effect of handheld computers on nursing practice. *Nursing Administration Quarterly*, *29*(4), 308–324. doi:10.1097/00006216-200510000-00004 PMID:16260994

Tian, K., Endo, M., Urata, M., Mouri, K., & Yasuda, T. (2014). Multi-viewpoint smartphone AR-based learning system for astronomical observation. *International Journal of Computer Th eory and Engineering*, *6*(5), 396–400.

Tomi, A. B., & Rambli, D. R. A. (2013). An interactive mobile augmented reality magical playbook: Learning number with the thirsty crow. *Procedia Computer Science*, *25*, 123–130. doi:10.1016/j.procs.2013.11.015

Tomiuc, A. (2014). Navigating Culture. Enhancing Visitor Museum Experience through Mobile Technologies. From Smartphone to Google Glass. *Journal of Media Research*, *20*(3), 33–46.

Tooey, M. J., & Mayo, A. (2004). *Handheld technologies in a clinical setting: State of the technology and resources.* Critical Care Nurse. Retrieved May 25, 2016, from http://web.b.ebscohost.com/ehost/pdfviewer/pdfviewer?vid=3&sid=ae11ccab-4ef24341-9045-3f2e4d66eb80%40sessionmgr102&hid=116

Top Best Apps for Kids. (2016). *Best Kids Apps – AR Flashcards.* Retrieved from http://topbestappsforkids.com/best-kids-apps-ar-flashcards/, 2016

Tosik-Gün, E. (2014). *Effects of augmented reality applications on students' spatial ability* (Unpublished Master Thesis). Gazi University, Ankara, Turkey.

Tri-Council for Nursing. (2010). *Tri-Council for Nursing Issues New Consensus Policy Statement on the Educational Advancement of Registered Nurses.* Retrieved January 25, 2016, from http://www.aacn.nche.edu/Educationresources/TricouncilEdStatement.pdf

Trifonova, A. (2003). *Mobile Learning – Review of the Literature.* University of Trento Technical Report.

Tufekci, A., Ekinci, H., & Kose, U. (2013). Development of an internet-based exam system for mobile environments and evaluation of its usability. *Mevlana Int. J. of Education*, *3*(4), 57–74. doi:10.13054/mije.13.59.3.4

Tuğtekin, U., & Kaleci, D. (2011). 3D modelleme tekniği kullanılarak bilgisayar oyunu tasarımı. *Akademik Bilişim 2011 Konferansı Bildirileri, İnönü Üniversitesi, Malatya*. Retrieved November 16, 2015, from http://ab.org.tr/ab11/kitap/tugtekin_kaleci_AB11.pdf

TUIK. (2013). *Turkish Statistical Institute*. Retrieved January 2, 2016, from http://www.tuik.gov.tr/PreTablo.do?alt_id=1095

Tulke, J., Nour, M., & Beucke, K. (2008). A Dynamic Framework for Construction Scheduling Based on BIM Using IFC. In *Proceedings of 17th IABSE Congress, Creating and Renewing Urban Structures – Tall Buildings, Bridges and Infrastructure*. Chicago: IABSE.

Uddin, M. S., & Yoon, S. (2002). Peter Eisenman's House X, Scheme G: 3D game engine for portable virtual representation of architecture. In *Proceedings of the Connecting the Real and the Virtual-design E-ducation*. (pp. 526–531). Warsaw, Poland: 20th eCAADe.

Udell, C., & Woodill, G. (2015) *The Mobile Learning Ecosystem*. Retrieved October 29, 2016, from https://www.td.org/Publications/Newsletters/Links/2015/12/The-Mobile-Learning-Ecosystem

Unity Technologies. (2016). *Prefabs*. Publication 5.4-U. Retrieved August 18, 2016, from http://docs.unity3d.com/Manual/Prefabs.html?_ga=1.244525509.1541532845.1453019544

University of Calgary. (2016). *Augmented reality in education*. Retrieved 10 July, 2106 from http://679augmentedreality.wikispaces.com/History+of+AR

Usta, E., & Korkmaz, Ö. (2010). Öğretmen adaylarının bilgisayar yeterlikleri ve teknoloji kullanımına ilişkin algıları ile öğretmenlik mesleğine yönelik tutumları. *Uluslararası İnsan Bilimleri Dergisi*, *7*(1), 1335–1349.

Van der Wal, M. M., de Kraker, J., Kroeze, C., Kirschner, P. A., & Valkering, P. (2016). Can computer models be used for social learning? A serious game in water management. *Environmental Modelling & Software*, *75*, 119–132. doi:10.1016/j.envsoft.2015.10.008

van Krevelen, D. W. F., & Poelman, R. (2010). A survey of augmented reality technologies, applications and limitations. *International Journal of Virtual Reality*, *9*(2), 1.

Van Krevelen, D. W. F., & Poelman, R. (2010). A survey of augmented reality technologies, applications and limitations. *The International Journal of Virtual Reality*, *9*(2), 1–19.

Van Rossum, G. (2007). Python Programming Language.*USENIX Annual Technical Conference*, 41.

Vandebona, U. V. (2015). *State-of-the-Art in Mobile Augmented Reality Development Tools/APIs*. Retrieved May 30, 2016, from http://www.slideshare.net/UpekhaVandebona/literature-survey-report-13440722

Venkatesh, V., & Bala, H. (2008). Technology Acceptance Model 3 and a Research Agenda on Interventions. *Decision Sciences*, *39*(2), 273–315. doi:10.1111/j.1540-5915.2008.00192.x

Venkatesh, V., & Davis, F. D. (2000). A Theoretical Extension of the Technology Acceptance Model: Four Longitudinal Field Studies. *Management Science*, *46*(2), 186–204. doi:10.1287/mnsc.46.2.186.11926

Venkatesh, V., Morris, M. G., Davis, G. B., & Davis, F. D. (2003). User Acceptance Technology. Toward a unified view. *Management Information Systems Quarterly*, *27*(3), 425–478.

Venkatesh, V., Thong, J. Y. L., & Xu, X. (2012). Consumer Acceptance and use of information technology: Extending the unified theory of acceptance and use of technology. *Management Information Systems Quarterly*, *36*(1), 157–178.

Vera, M. D. M. S., Breis, J. T. F., Sánchez, J. L. S., & Espinosa, M. P. P. (2013). Practical experiences for the development of educational systems in the semantic web. *Journal of New Approaches in Educational Research*, *2*(1), 23–31. doi:10.7821/naer.2.1.23-31

Vilkoniene, M. (2009). Influence of augmented reality technology upon pupils' knowledge about human digestive system: The results of the experiment. *Online Submission*, *6*(1), 36–43.

Virtual-Realms. (2016). *New Gallery 2016*. Retrieved from http://www.virtual-realms.com/new-gallery/2016

Vitell, S. J. (2003). Consumer Ethics Research: Review, Synthesis and Suggestions for the Future. *Journal of Business Ethics*, *43*(1/2), 33–47. doi:10.1023/A:1022907014295

Vitotechnology. (2013, December). *User Manual Star Walk*. Retrieved August 2016, from Vitotechnology: http://vitotechnology.com/manual_/pdf/StarWalk_manual_en.pdf

Vlahakis, V., Karigiannis, J., Tsotros, M., Gounaris, M., Almeida, L., Stricker, D., & Ioannidis, N. et al. (2001). Archeoguide: First results of an augmented reality, mobile computing system in cultural heritage sites.*Proceedings of the 2001 Conference on Virtual Reality, Archeology, and Cultural Heritage*. New York: ACM. doi:10.1145/584993.585015

Vygotsky, L. S. (1978). *Mind and society: The development of higher mental processes*. Cambridge, MA: Harvard University Press.

Waclawski, E. (2012). How I Use It: Survey Monkey. *Occupational Medicine*, *62*(6), 477–477. doi:10.1093/occmed/kqs075 PMID:22915570

Wagner, D., Schmalstieg, D., & Billinghurst, M. (2006). Handheld AR for collaborative edutainment. *Advances in Artificial Reality and TeleExistence,* 85–96.

Wagner, D., Schmalstieg, D., Szalavári, Z., Traxler, C., & Stifter, M. (2014). *Mobile augmented reality quest (MARQ) 2005-2007*. Retrieved August 2016, from Handheld Augmented Reality: http://handheldar.icg.tugraz.at/marq.php

Wagner, D., & Schmalstieg, D. (2003). *First Steps Towards Handheld Augmented Reality. IEEE International Symposium on Wearable Computers*, 127-135.

Walia, A., Singhal, N., & Sharma, A. K. (2015). A Novel E-learning Approach to Add More Cognition to Semantic Web. In *Computational Intelligence & Communication Technology (CICT), 2015 IEEE Int. Conference on* (pp. 13-17). IEEE.

Wall, D., Ray, W., Pathak, R. D., & Lin, S. M. (2014). A google glass application to support shoppers with dietary management of diabetes. *Journal of Diabetes Science and Technology*, *8*(6), 1245–1246. doi:10.1177/1932296814543288 PMID:25015954

Wang, X. (2012). Augmented reality: A new way of augmented learning. *E Learn Magazine*. Retrieved 10 April, 2016 from http://elearnmag.acm.org/featured.cfm?aid=2380717

Wang, H. Y., Duh, H. B. L., Li, N., Lin, T. J., & Tsai, C. C. (2014). An investigation of university students collaborative inquiry learning behaviors in an augmented reality simulation and a traditional simulation. *Journal of Science Education and Technology*, *23*(5), 682–691. doi:10.1007/s10956-014-9494-8

Wang, X., Love, P. E. D., Kim, M. J., Park, C.-S., Sing, C.-P., & Hou, L. (2013). A conceptual framework for integrating building information modeling with augmented reality. *Automation in Construction*, *34*, 37–44. doi:10.1016/j.autcon.2012.10.012

Wang, Y., Cui, S., Yang, Y., & Lian, J. A. (2009). Virtual reality mathematic learning module for engineering students. *Technology Interface Journal*, *10*(1), 1–10.

Warschauer, M. (2011). *Eventually tablets will facilitate more personalized and interactive learning*. Retrieved 10 May, 2016 from http://edutechdebate.org/tablet-computers-in-education/eventually-tablets-will-facilitate-more-personalized-and-interactive-learning/

Wassom, B., & Bishop, A. (2015). *Augmented reality law, privacy, and ethics: Law, society, and emerging AR technologies*. Amsterdam: Syngress.

Watanuki, K. (2010). A Mixed Realitybased Emotional Interactions and Communications for Manufacturing Skills Training. In S. Fukuda (Ed.), Emotional Engineering: Service Development (p. 58). Springer Science & Business Media.

Wegerif, R. (2013). *Dialogic: Education for the Internet Age*. London: Routledge.

Weinbaum, G. S. (2007, October 5). *Project Gutenberg's Pygmalion's Spectacles*. Retrieved Agust 2016, from Gutenberg: http://www.gutenberg.org/files/22893/22893-h/22893-h.htm

Wei, X., Weng, D., Liu, Y., & Wang, Y. (2015). Teaching based on augmented reality for a technical creative design course. *Computers & Education*, *81*, 221–234. doi:10.1016/j.compedu.2014.10.017

Weiz, D., Anand, G., & Ernst, C.-P. H. (2016). The Influence of Subjective Norm on the Usage of Smartglasses. In C.-P. H. Ernst (Ed.), *Progress in IS. The Drivers of Wearable Device Usage* (pp. 1–11). Cham: Springer International Publishing. doi:10.1007/978-3-319-30376-5_1

Wenglinsky, H. (2006). Technology and achievement: The bottom line. *Learning in the Digital Age*, *63*(4), 29–32.

Widder, B. (2016). *Best Augmented Reality Apps*. Retrieved October 29, 2016, from http://www.digitaltrends.com/android/best-augmented-reality-apps/

Wikipedia. (2015). *History of video games*. Retrieved September 29, 2015, from https://en.wikipedia.org/wiki/History_of_video_games

Wikipedia. (2016a). *Augmented reality*. Retrieved December 18, 2016, from https://en.wikipedia.org/wiki/Augmented_reality

Wikipedia. (2016a, August). *Augmented Reality*. Retrieved from Wikipedia: https://en.wikipedia.org/wiki/Augmented_reality

Wikipedia. (2016b). *Virtual reality*. Retrieved December 18, 2016, from https://en.wikipedia.org/wiki/Virtual_reality

Wikipedia. (2016b). *Word Lens*. Retrieved August 20, 2016, from Wikipedia: https://en.wikipedia.org/wiki/Word_Lens

Wikipedia. (2016c, September). *Pokémon Go*. Retrieved September 23, 2016, from Wikipedia: https://en.wikipedia.org/wiki/Pok%C3%A9mon_Go

Wilde, M., Hategan, M., Wozniak, J. M., Clifford, B., Katz, D. S., & Foster, I. (2011). Swift: A language for distributed parallel scripting. *Parallel Computing*, *37*(9), 633–652. doi:10.1016/j.parco.2011.05.005

Williamson, O. E. (1985). The economic institutions of capitalism: Firms, markets, relational contracting. New York: Free Press.

Willis, J. (1996). *A Framework for Task-Based Language Teaching*. New York: Longman.

Winn, W. (1993). A conceptual basis for educational applications of virtual reality. In Human Interface Technology Laboratory. Washington Technology Center, University of Washington.

Winn, W. D., Windschitl, M., Fruland, R., & Lee, Y. L. (2002). When does immersion in a virtual environment help students construct understanding? *Proceedings of the International Conference of the Learning Science*. Mahwah, NJ: Erlbaum. Retrieved April 25, 2016, from http://www.hitl.washington.edu/people/tfurness/courses/inde543/READINGS-03/WINN/winnpaper1.pdf

Winters, N. (2013). Mobile learning in the majority world: A critique of the GSMA position. In S. Price, C. Jewitt, & B. Brown (Eds.), *SAGE Handbook of Digital Technology Research* (pp. 402–412). London: SAGE Publications Ltd. doi:10.4135/9781446282229.n27

Witkowski, T. H., & Reddy, S. (2010). Antecedents of ethical consumption activities in Germany and the United States. *Australasian Marketing Journal*, *18*(1), 8–14. doi:10.1016/j.ausmj.2009.10.011

Wojciechowski, R., & Cellary, W. (2013). Evaluation of learners attitude toward learning in ARIES augmented reality environments. *Computers & Education*, *68*, 570–585. doi:10.1016/j.compedu.2013.02.014

Wu, H. K., Lee, S. W. Y., Chang, H. Y., & Liang, J. C. (2013). Current status, opportunities and challenges of augmented reality in education. *Computers & Education*, *62*, 41–49. doi:10.1016/j.compedu.2012.10.024

Wu, P.-H., Hwang, G.-J., Su, L.-H., & Huang, Y.-M. (2012). A Context-Aware Mobile Learning System for Supporting Cognitive Apprenticeships in Nursing Skills Training. *Journal of Educational Technology & Society*, *15*(1), 223–236.

Yalcin, B., Inal, C., Kalayci, I., & Tarihi, M. G. (2008). GPS ölçü süresinin nokta konum doğruluğu ile ilişkisi. *J. Fac. Eng. Arch. Selcuk Univ*, *23*(1).

Yang, T., & Choi, Y. M. (2015). Study on the Design Characteristics of Head Mounted Displays (HMD) for Use in Guided Repair and Maintenance. In R. Shumaker & S. Lackey (Eds.), *Lecture Notes in Computer Science. Virtual, augmented and mixed reality. 7th international conference, vamr* (Vol. 9179, 535–543). Springer.

Yan, W., Culp, C., & Graf, R. (2011). Integrating BIM and gaming for real-time interactive architectural visualization. *Automation in Construction*, *20*(4), 446–458. doi:10.1016/j.autcon.2010.11.013

Yegnanarayana, B. (2009). *Artificial neural networks*. PHI Learning Pvt. Ltd.

Yen, J. C., Tsai, C. H., & Wang, J. Y. (2012). *The effects of augmented reality on students' moon phases concept learning and their conceptual changes of misconception*. Paper presented at 2012 International Conference on Business and Information, Sapporo, Japan.

Yen, J. C., Tsai, C. H., & Wu, M. (2013). Augmented reality in the higher education: Students science concept learning and academic achievement in astronomy. *Procedia: Social and Behavioral Sciences*, *103*, 165–173. doi:10.1016/j.sbspro.2013.10.322

Yeo, T. C., Ungi, T., Thainual, U.-P., & Lasso, A. (2011). The Effect of Augmented Reality Training on Percutaneous Needle Placement in Spinal Facet Joint Injections. *IEEE Transactions on Bio-Medical Engineering*, *58*(7), 2031–2037. doi:10.1109/TBME.2011.2132131 PMID:21435970

Yılmaz, R. M. (2014). *Effects of three dimensional storytelling developed with augmented reality technology on narrative skill and creativity* (Unpublished doctoral dissertation). Atatürk University, Erzurum, Turkey.

Yiwen, G., Li, H., & Luo, Y. (2015). An empirical study of wearable technology acceptance in healthcare. *Industrial Management & Data Systems*, *115*(9), 1704–1723. doi:10.1108/IMDS-03-2015-0087

Yoon, S. A., Elinich, K., Wang, J., Steinmeier, C., & Tucker, S. (2012). Using augmented reality and knowledge-building scaffolds to improve learning in a science museum. *International Journal of Computer-Supported Collaborative Learning*, *7*(4), 519–541. doi:10.1007/s11412-012-9156-x

Yuan, L., Powell, S., & Cetis, J. (2013). *MOOCs and open education: Implications for higher education*. JISC Cetis Publications.

Yuen, S., Yaoyuneyong, G., & Johnson, E. (2011). Augmented reality: An overview and five directions for AR in education. *Journal of Educational Technology Development and Exchange*, *4*(1), 119–140.

Yusoff, Z., & Dahlan, H. M. (2013). *Mobile based learning: An integrated framework to support learning engagement through Augmented Reality environment*. Paper presented at 2013 International Conference on Research and Innovation in Information Systems (ICRIIS). doi:10.1109/ICRIIS.2013.6716718

Zarzuela, M. M., Pernas, F. J. D. P., Martínez, L. B., Ortega, D. G., & Rodríguez, M. A. (2013). Mobile serious game using augmented reality for supporting childrens learning about animals. *Procedia Computer Science*, *25*, 375–381. doi:10.1016/j.procs.2013.11.046

Zatarain-Cabada, R., Barrón-Estrada, M. L., Alor-Hernández, G., & Reyes-García, C. A. (2014). Emotion Recognition in Intelligent Tutoring Systems for Android-Based Mobile Devices. In *Mexican International Conference on Artificial Intelligence* (pp. 494-504). Berlin: Springer International Publishing. doi:10.1007/978-3-319-13647-9_44

Zhang, J., Sung, Y. T., Hou, H. T., & Chang, K. E. (2014). The development and evaluation of an augmented reality-based armillary sphere for astronomical observation instruction. *Computers & Education*, *73*, 178–188. doi:10.1016/j.compedu.2014.01.003

Zheng, S. (2015). Research on Mobile Learning Based on Augmented Reality. *Open Journal of Social Sciences*, *3*(12), 179–182. doi:10.4236/jss.2015.312019

Zhong, L., Xue, Q., Liu, H., & Yuan, J. (2010) Research on 3GMobileLearning Based on Cloud Service. *International Conference on E-Product E-Service and E-Entertainment (ICEEE)*, 1-4. Doi:10.1109/ICEEE.2010.5660179

Zhou, F., Duh, H. B., & Billinghurst, M. (2008). Trends in augmented reality tracking, interaction and display: A review of ten years of ISMAR. *2008 7th IEEEACM International Symposium on Mixed and Augmented Reality*, *2*(4), 193–202. DOI doi:10.1109/ISMAR.2008.4637362

Zhou, F., Duh, H. B. L., & Billinghurst, M. (2008). Trends in augmented reality tracking, interaction and display: A review of ten years of ISMAR. In *IEEE and ACM ISMAR 2008* (pp. 193–202). Cambridge, UK: IEEE and ACM.

Zhou, Z., Cheok, A. D., & Pan, J. (2004). 3D story cube: An interactive tangible user interface for storytelling with 3D graphics and audio. *Personal and Ubiquitous Computing*, *8*(5), 374–376. doi:10.1007/s00779-004-0300-0

Zhu, E., Hadadgar, A., Masiello, I., & Zary, N. (2014). *Augmented reality in healthcare education: An integrative review*. Doi:10.7717/peerj.469

ZooKazam Magical Animals. http://zookazam.com

About the Contributors

Gulsun Kurubacak is a professor in Distance Education at the College of Open Education of Anadolu University. Dr. Kurubacak undertook graduate studies at Anadolu University, Turkey (MA. Educational Technology) and the University of Cincinnati, USA (Ed.D. Curriculum & Instruction), and also has worked a post-doctoral fellow at the College of Education at New Mexico State University, USA (2001-2002). Dr. Kurubacak earned her B.S. degree in Computer Engineering from the College of Informatics Technologies and Engineering of Hoca Ahmet Yesevi International Turk-Kazakhstani University in the year 2012-2013. Also, she is currently a graduate student in the Department of Computer and Instructional Technologies of Anadolu University. Dr. Kurubacak has thirty years of experience in focusing on the egalitarian and ecological aspects of open and distance education; finding new answers, viewpoints and explanations to online communication problems through critical pedagogy; and improving learner critical and creative thinking skills through project-based online learning, universal design principles and new communication technologies. She continues to manage and provide pedagogical support for distance learning.

Hakan Altınpulluk is a Research Assistant in Distance Education at the College of Open Education of Anadolu University, Turkey. He undertook undergraduate studies in the field of Computer Education and Instructional Technologies (CEIT) between the years of 2005-2009 at Anadolu University. Also, he is currently a doctoral student in the Department of Distance Education at Anadolu University since 2010. Hakan Altınpulluk continues to work in the field of Open and Distance Education, Augmented Reality, Virtual Reality, Mobile Learning, Mobile Health, Massive Open Online Courses, Learning Management Systems, Open Educational Resources, Personal Learning Environments, and E-Learning Systems.

* * *

Derya Bozdoğan is with TED University English Language Education program working as an Assistant Professor, her research areas are ICT in education, Content and Language Integrated Learning (CLIL) and Teaching English to Young Learners.

Helen Crompton is an Assistant Professor of Instructional Technology at Old Dominion University Virginia. She is a highly experienced researcher, educator, author and presenter in the field of educational technology. She draws from 23 years in education, with 16 years as a full time K-10 classroom teacher, and she has a PhD in educational technology and mathematics education from the University of North Carolina at Chapel Hill. Her research focuses on the affordances of mobile learning in mathematics. Dr. Crompton works as a consultant for two United Nations Agencies (United Nations, Educational, Scientific, and Cultural Organization: UNESCO and International Telecommunication Union: ITU) to research, author and edit publications summarizing research on mobile learning. Dr. Crompton is also a faculty member for the International Society for Technology in Education (ISTE), teaching the ISTE Standards academy, consulting, and recently designing ISTE's self –paced Mobile Learning Academy and Verizon's Mobile Learning Academy.

Martin Ebner is currently head of the Department for Educational Technology at Graz University of Technology and therefore responsible for all university wide e-learning activities. He holds an Assoc. Prof. on media informatics and works also at the Institute for Information System Computer Media as senior researcher. His research focuses strongly on e-learning, mobile learning, learning analytics, social media and Open Educational Resources. Martin has given a number of lectures in this area as well as workshops and keynotes at international conferences. For publications as well as further research activities, please visit his website: http://martinebner.at.

Gülay Ekren has been working as a lecturer at the Department of Computer Technologies of Vocational School of Ayancık, Sinop University since 2009. She has a BA degree from the Computer and Instructional Technologies Department of Ege University, and MA degree from the Management Information Systems Department of Gazi University. She also has an online MA degree from the Distance Education Department of Anadolu University. She is now a Phd student in Department of Management and Information Systems in Sakarya University. Her research interests are open and distance learning, distance education management, mobile learning, quality assurance and accreditation in open and distance learning, management and information systems.

Bülent Gürsel Emiroğlu is working as an Instructor and Researcher in Computer Engineering Departments of high-ranked universities of Turkey since 2002. His research areas include educational technology, technologies for disabled students, e-learning, Internet / Web / Mobile technologies, Computer Networks, Human Computer Interaction and Usability. He has many cited articles printed in well known academic journals and presented many papers in national and international conferences.

M. Banu Gundogan is an instructor in Computer Education and Instructional Technology Department of Middle East Technical University. As an Industrial Design graduate, she started her career working on hands on science exhibits and managed Turkey's first science centre, Feza Gürsey Science Centre project in Ankara. Working on the development, implementation and management of instructional design projects regarding interactive science education, she has developed 'bilimce' (science language) project and managed science school - science camp projects. While transferring 'bilimce' activities to the web, she started studying distance education and is continuing her studies focusing on the ecology and sustainability issues regarding open and distance learning.

Daniel W. E. Hein is a research assistant and PhD candidate at the University of Bamberg at the chair of Prof. Dr. Björn Ivens. After receiving his bachelor's from the "Westfälische Wilhems-Universität" (Munster), he turned to the Otto-Friedrich-University for consecutive studies. His research focus lies on the management of customer relations and the impact of business digitization thereon, as well as how digitization affects further aspects of Marketing. Research Center for Business Models in the Digital World, a joint research unit established by Fraunhofer IIS and the University of Bamberg run by the Fraunhofer Institute for Integrated Circuits (IIS) in cooperation with the University of Bamberg. He is involved in teaching courses for both graduate and undergraduate students.

Björn S. Ivens is Chair Professor of Marketing at Otto-Friedrich-University Bamberg (Germany). Prior to this he was Full Professor of Management at Université de Lausanne (Switzerland). His research interests include branding, digitization of business models, marketing organization, and pricing. His research has been published in journals such as Industrial Marketing Management, Journal of Business Research, Psychology & Marketing, Marketing Theory, European Journal of Marketing, Journal of Marketing Management and the International Journal of Electronic Commerce. He has taught in graduate and executive programs at universities and business schools in Germany, Austria, Switzerland, France, Italy, and the UK. He consults with and conducts research projects for several large European corporations. He is a founding member of the European Foundation for Key Account Management as well as the Research Center for Digital Business Models in the Digital World, a joint research unit established by Fraunhofer IIS and the University of Bamberg.

Jennah L. Jodoin was a student of Marketing in the College of Business at University of Michigan-Dearborn (USA) at the time this publication was being written. She completed her Bachelor of Business Administration (BBA) in December 2016. Before that, Jennah maintained a research assistant role for iLabs – University of Michigan-Dearborn's center for innovation research. It is there that she learned to analyze and present research findings. This is her first publication and has spurred her interest in New Technologies.

Devkan Kaleci, PhD, is an assistant professor at the University of İnönü. His research interests are: virtual reality, augmented reality, game based learning and distance education. https://www.researchgate.net/profile/Devkan_Kaleci.

Buket Kasap teaches English at Ankara University. She is currenlyt interested in mobile applications of CLIL and ESP.

Utku Kose received the B.S. degree in 2008 from computer education of Gazi University, Turkey as a faculty valedictorian. He received M.S. degree in 2010 from Afyon Kocatepe University, Turkey and D.S. / Ph. D. in 2016 from Selcuk University, Turkey in the field of computer engineering. Between 2009 and 2011, he has worked as a Research Assistant in Afyon Kocatepe University. Following, he has also worked as a Lecturer and Vocational School - Vice Director in Afyon Kocatepe University between 2011 and 2012. Currently, he is a Lecturer in Usak University, Turkey and also the Director of Computer Sciences Application and Research Center of the same university. His research interest includes artificial intelligence, the chaos theory, distance education, e-learning, computer education, and computer science.

Adile Aşkım Kurt is an associate professor in the Department of Computer Education and Instructional Technology at Anadolu University, Turkey. She has an M.A. in Statistics with emphasis on factor analysis, and a Ph.D. in Computer Education & Instructional Technology with emphasis on comparison of web based education to face to face education. Dr. Kurt has tried to create and disseminate knowledge about the use of emerging communication technologies. She has authored or co-authored more than 25 international journal articles, 14 book chapters and over 50 national and international conference presentations. She has been the principal investigator of three research projects and participated in three national projects. She lists her primary research interests as ICT integration, digital literacy, globalization, technology use in special education and data analysis.

Lisabeth Leighton is a Master of Education Secondary Education and Teaching student at Old Dominion University and a Data Technician/Disciplinary Secretary at Virginia Beach City Public Schools. Ms. Leighton has a keen research interest in the use of technology in education, in particular how augmented reality tools can be used to extend and enhance learning. She has experience in K-12 schools from working as a teacher-permanent substitute and as a teacher assistant.

Nilgun Ozdamar Keskin received her Ph.D. in Anadolu University Educational Sciences Institution, Department of Computer Education and Instructional Technologies in 2011. She served as a visiting researcher at University of Central Florida (2009-2010), and as a research assistant at Open Education Faculty, Anadolu University (2005-2012). She still continues to serve as an assistant of professor at Anadolu University. She is currently working as an executive coordinator of Photography and Cameraman Associate Degree Program, an E-portfolio System Coordinator of Public Service Practice Course for Preschool Teacher Training Program and an assistant of Erasmus Coordinator for Open Education Faculty. Also, she is managing an international project on mobile learning granted by British Council and several national projects. She got a best paper award at mLearn Conference 2011 in China. Her research interests include instructional design, open and distance education, online learning, mobile learning, and seamless learning design.

Muzaffer Ozdemir received his bachelor's degree in Electrical and Electronics Engineering from the Firat University in Turkey in year 1997. He got his master's degree from Institute of Science and Technology at Çanakkale Onsekiz Mart University in 2008. He received his Ph.D. degree from Çanakkale Onsekiz Mart University at the same Institute. He has been working as an assistant professor in Computer Education and Instructional Technology Department of Çanakkale Onsekiz Mart University for seven years. He took an active role in the establishment of the distance education center of Çanakkale Onsekiz Mart University. Currently, he working as assistant manager of this center. He has been teaching both undergraduate courses such as "Instructional Technologies and Material Design", "Web Design", "Multimedia design", "Graphics and Animation in Education" and graduate courses such as "Virtual and Augmented Reality in Education: Theory and Practice" and "E-Learning: From Theory to Practice."

Philipp A. Rauschnabel is an Assistant Professor of Marketing in the College of Business at University of Michigan-Dearborn (USA). He received his PhD in Marketing from University of Bamberg and holds a Master of Science Degree in Marketing and Channel Management (University of Goettingen). Prof. Rauschnabel's research interests include Branding and New Technologies. He published the first academic consumer acceptance studies on Augmented Reality Smart Glasses. He also consults regularly with, and presents research findings at, various companies and organizations on these topics. Blog: http://www.philipprauschnabel.com.

Patricia Salinas is Professor and Researcher in the Mathematics Department at Tecnológico de Monterrey, Campus Monterrey. Her background includes a Bachelor in Mathematics, a Master Degree in Science with specialization in Mathematics Education, a Master Degree in Education with specialization in Mathematics, and a PhD. in Mathematics Education. She is coauthor of several books where a new didactic for mathematics has been broadcasting since 2003. Her research agenda includes the integration of digital technology for the learning of Mathematics. She is working in the transformation of curriculum through the incorporation of visual and gestural elements in the cognitive process related with mathematical thinking. This is possible through the integration of emergent digital technologies in the interaction with mathematical knowledge. Her goal is the creation of ecological environments in formal education for the appropriation of a mathematical culture.

Michael Spitzer received his MSc in Information and Computer Engineering from Graz University of Technology in 2015. As his master's thesis he implemented a collaborative sketch tool (Teamsketch) for iPads to train collaborative work with primary school pupils. Since then he focused his work on technology enhanced learning (TEL). In 2016 he started the PhD program at Graz University of Technology as a researcher in the field technology enhanced learning with augmented reality (AR).

Christof Sternig received his BSc in Computer Science from Graz University of Technology (TU) in 2011. In his master's degree he focused on multimedia information systems and computer graphics and experimented with mobile technology and virtual reality. From 2011 to 2015 he also worked at the Institute for Physical and Theoretical Chemistry at the TU Graz as a developer for Fortran programs during his master studies in Computer Science. Currently he is writing his master thesis about virtual reality learning apps to finish his MSc degree in Computer Science.

Irfan Sural is an Assistant Professor in Department of Computer Education & Instructional Technology at Osmangazi University, Eskisehir-Turkey and serves as deputy director of the university's Distance Education Center. He received both B.S and M.S. degree in Electrical and Electronics Engineering Department and a PhD degree in Distance Education Program of Social Sciences Institute of Anadolu University. He has a strong practical as well as theoretical background in the use of technology for e-learning. His research interests mainly focus on personalized learning systems, building distance education infrastructures, mobile learning, augmented reality, artificial intelligence in distance education, learning management systems and open-source software projects in education. He has contributed to many national and international projects and presented papers on the integration of technology in education.

Tansel Tepe, PhD Student, is a research assistant at the University of Kilis 7 Aralık. His research interests are: Educational 3D games, 3D virtual worlds, virtual reality and augmented reality. https://www.researchgate.net/profile/Tansel_Tepe.

Nilgün Tosun has completed her bachelor, master and PhD programs in Trakya University, Computer Engineering Department. In the year 2006-2009, she served as Assistant Director of Trakya University Distance Education Center where she is one of the founders. She served as Assistant Director of Tunca Vocational High School (Distance Education) and Head of Technical Programs Department of Tunca VHS, in the year 2007-2008. And she also served as Head of Computer Education and Instructional Department in the year 2008-2012. She is currently working at the same department as Assist. Prof. Dr. She has been working on distance education, technology integration in education, social networks and security in social networks and flipped learning issues.

Belgin Boz Yuksekdag has been working at Anadolu University College of Open Education Testing & Research Center as a lecturer doctor. She graduated from the School of Nursing of Hacettepe University in 1990. She earned her MS degree from the Graduate School of Istanbul University (Health Sciences Department), Internal Medicine Nursing in 1995, and also the MA degree from the Graduate School of Anadolu University (Distance Education Department), in 2008. She earned her PhD degree from the Graduate School of Marmara University (Health Sciences Department), Psychiatry Nursing, in 2013.

Index

H

I

K

L

M

N

O

P

R

S

T

U

V

W

U

V

W

www.ingramcontent.com/pod-product-compliance
Lightning Source LLC
LaVergne TN
LVHW062141200926
848484LV00007B/1841
* 9 7 8 1 5 2 2 5 2 1 1 0 5 *